Writing Software Documentation

THE ALLYN AND BACON SERIES IN TECHNICAL COMMUNICATION

Series Editor: Sam Dragga, Texas Tech University

Thomas T. Barker
*Writing Software Documentation:
A Task-Oriented Approach, Second Edition*

Carol M. Barnum
Usability Testing and Research

Deborah S. Bosley
*Global Contexts: Case Studies in International
Technical Communication*

Melody Bowdon and J. Blake Scott
*Service-Learning in Technical and
Professional Communication*

Paul Dombrowski
Ethics in Technical Communication

David K. Farkas and Jean B. Farkas
Principles of Web Design

Laura J. Gurak
Oral Presentations for Technical Communication

Sandra W. Harner and Tom G. Zimmerman
Technical Marketing Communications

Richard Johnson-Sheehan
Writing Proposals: Rhetoric for Managing Change

Dan Jones
Technical Writing Style

Charles Kostelnick and David D. Roberts
*Designing Visual Language: Strategies for
Professional Communicators*

Carolyn Rude
Technical Editing, Third Edition

Gerald J. Savage and Dale L. Sullivan
*Writing a Professional Life: Stories of
Technical Communicators On and Off the Job*

Writing Software Documentation

A Task-Oriented Approach

Second Edition

Thomas T. Barker
Texas Tech University

New York   San Francisco   Boston
London   Toronto   Sydney   Tokyo   Singapore   Madrid
Mexico City   Munich   Paris   Cape Town   Hong Kong   Montreal

Senior Vice President/Publisher: Joseph Opiela
Marketing Manager: Christopher Bennem
Production Manager: Denise Phillip
Project Coordination, Text Design, and Electronic Page Makeup: WestWords, Inc.
Cover Design Manager: John Callahan
Cover Designer: Teresa Ward
Manufacturing Buyer: Roy Pickering
Printer and Binder: Hamilton Printing
Cover Printer: Coral Graphics

For permission to use copyrighted material, grateful acknowledgment is made to the copyright holders on
pp. 459–460, which are hereby made part of this copyright page.

Many of the designations used by manufacturers and sellers to distinguish their products are claimed as trademarks.
Where those designations appear in this book, Longman Publishers was aware of the trademark claim, and the designations have been printed with an initial capital. Designations within quotation marks represent hypothetical products.

Library of Congress Cataloging-in-Publication Data

Barker, Thomas T.
    Writing software documentation : a task-oriented approach / Thomas T. Barker.—2nd ed.
        p. cm.—(Allyn and Bacon series in technical communication)
    Includes bibliographical references and index.
    ISBN 0–321–10328–9
    1. Software documentation. I. Title. II. Series.

QA76.76.D63 B37 2002
005.1'5—dc21

                                                        2002016296

Please visit our website at http://www.ablongman.com

ISBN 0–321–10328–9

1 2 3 4 5 6 7 8 9 10—HT—05 04 03 02

For Emily and John

CONTENTS

CHAPTER 1   **Understanding Task Orientation**   1

CHAPTER 6   **Planning and Writing Your Documents**   173

CHAPTER 7   **Getting Useful Reviews**   **217**

How to Read This Chapter   217

Example   217

Guidelines   218

Discussion   229

CHAPTER 8   **Conducting Usability Tests**   **240**

How to Read This Chapter   240

Guidelines   240

CHAPTER 9   Editing and Fine Tuning   266

CHAPTER 13 **Using Graphics Effectively** **405**

CHAPTER 14 **Designing Indexes** **436**

FOREWORD

by the Series Editor

The Allyn & Bacon Series in Technical Communication is designed for the growing number of students enrolled in undergraduate and graduate programs in technical communication. Such programs offer a wide variety of courses beyond the introductory technical writing course—advanced courses for which fully satisfactory and appropriately focused textbooks have often been impossible to locate. This series will also serve the continuing education needs of professional technical communicators, both those who desire to upgrade or update their own communication abilities as well as those who train or supervise writers, editors, and artists within their organization.

The chief characteristic of the books in this series is their consistent effort to integrate theory and practice. The books offer both research-based and experienced-based instruction, describing not only what to do and how to do it but explaining why. The instructors who teach advanced courses and the students who enroll in these courses are looking for more than rigid rules and ad hoc guidelines. They want books that demonstrate theoretical sophistication and a solid foundation in the research of the field as well as pragmatic advice and perceptive applications. Instructors and students will also find these books filled with activities and assignments adaptable to the classroom and to the self-guided learning processes of professional technical communication.

To operate effectively in the field of technical communication, today's students require extensive training in the creation, analysis, and design of information for both domestic and international audiences, for both paper and electronic environments. The books in the Allyn & Bacon Series address those subjects that are most frequently taught at the undergraduate and graduate levels as a direct response to both the educational needs of students and the practical demands of business and industry. Additional books will be developed for the series in order to satisfy or anticipate changes in writing technologies, academic curricula, and the profession of technical communication.

Sam Dragga
Texas Tech University

PREFACE

This textbook covers the subject of software documentation, which comes in many forms—from the familiar print *User's Manual* or *Installation Guide* to the online help program, Wizards, and embedded help. I focus exclusively on creating documents (or *information products*) that help software users learn program features and use them to work productively. Most manuals and help documents that accompany software programs are written by technical writers in the software industry. Students or professionals interested in doing this interesting and challenging work will find in this book a basic foundation in the principles of writing these kinds of documents.

What's New in the Second Edition

This edition improves substantially on the first by the addition of and revision of a number of features:

- *Updated examples.* Where possible examples were selected (many from STC award winners) to reflect current trends in documentation.

- *Discussions of communicating across cultures.* The sections on task orientation and documentation planning contain discussions of techniques for communicating across boundaries of organizational and national cultures.

- *Reorganized table of contents.* The second edition puts the three main forms of software documentation at the front of the book so that students and others new to the field can get a clear idea of the forms before tackling the process of creating them.

- *Increased emphasis on task orientation.* The second edition eliminates an entire chapter on constructing an inventory of the software program (called a "task list") that, while useful, fostered a "system" orientation in some documents.

- *Stronger theoretical basis.* This edition focuses on the three-part hierarchy of activity, action, and operation implicit in an "activity theory" approach to work. This approach focuses the user manual and other help documents on activities and actions rather than on operations; on user tasks rather than interface features.

- *Emphasis on process.* The book is divided into Part I: The Forms, Part II: The Process, and Part III: The Tools so readers can clearly see the components of software documentation as a field.

- *Increased emphasis on utility.* The Guidelines in all chapters sections have been revised to reflect a series of easy-to-follow steps. In Part III: The Tools, the chapters include suggestions for planning, executing, and testing the design features discussed in the chapter.

The Approach in the Book

Two words sum up my approach in this book: *task orientation*. The software manual that encourages productive software use should reflects the user's workplace tasks, not just the software interface. For example, if your manual presents the features of a word processing program organized according to the menu structure of the program, then the user must already know a great deal about the software to find the right command. But if your manual presents the features of the same word processing program organized around the usual tasks office workers perform, then the user already possesses some knowledge about how to apply the software productively. In this book I show you how to apply this simple principle to all the elements of documentation. Using this principle will help you design successful projects.

In this edition I introduce the concept of the *default manual:* the manual that is based on the structure of the program and represents little actual user analysis. The default manual follows the naïve principle that the purpose of software documentation is to record what appears on the screen accurately and completely—applying the term "documentation" in the sense of being a comprehensive "documentary" of the on-screen events, menus, dialog boxes, and so on. The problem with such a manual is that it often presents too much information to the user causing a information overload or *information anxiety*. "Why do I need to know all this?" or "Where do I start?" the frustrated user might ask. In contrast, the *task-oriented manual* is essentially *about the user* in that it applies extensive user analysis to the organization, content, and layout of the manual. The task-oriented manual follows the principle that users bring meaning to the software features in the form of workplace activities and actions. The goal is not the use of the software, but the performance of work. Thus, the subject of the manual becomes the context of software usage, not the software itself. The priority for the writer becomes usability rather than mindless, formalistic accuracy.

What Can This Book Do for You?

The following list covers some of the ways this book can help students and professionals learn to write successful, task-oriented manuals and help.

- *Introduce the basic concepts of task orientation and the forms it takes.* Chapters 1 through 4 illustrate the importance of the user's context in shaping the content and design of manuals and help. They show how principles of task orientation apply to manuals that teach, manuals that guide, and manuals that support software work.

- *Give you a place to start if you don't already have one.* Chapters 5 through 9 follow a complete step-by-step process for analyzing users, learning a software program, and designing task-oriented documents.

- *Teach you the basics about page and screen design.* Chapters 10 through 14 provide you with the right background in designing useful information, laying out pages and screens, and designing various forms of manuals and online help.

- *Provide useful tools for writing manuals and online help.* Each chapter has a checklist of the main points you want to remember in preparing your documents. These checklists can keep you organized and on track.

- *Provide practice in writing.* The chapters all have useful and interesting exercises you can use to build useful skill sets that are valuable in the technical communication job market.
- *Help you understand software user.* All chapters relate in important ways to one guiding idea: Understand your user and you'll have the key to helping that person use software productively in the workplace.

Why Should You Read This Book?

This book meets the needs of two groups of people: those interested in finding practical steps to help them start and complete a project successfully, and those interested in exploring the ideas behind software documentation as a discipline and as a profession.

If you're a project-oriented reader facing a manual or help project and needing assistance, this book can guide you through an entire successful documentation project sequence.

If you're relatively experienced with writing but new to software documentation and you want to broaden your understanding of the ideas behind designing product-support documents for software programs, you will find a discussion of the ideas and research behind the idea of task orientation.

Perhaps your motivation contains a combination of these two situations. Both of these activities can be relevant to the education and training of a professional technical writer in the software industry. To help you meet these goals in reading, the chapters have two tracks, represented by the two main expository sections of each chapter: the reading-to-do track (the Guidelines section) and the reading-to-understand track (the Discussion section).

These two tracks complement one another, meeting the needs of the project-oriented reader, the understanding-oriented reader, and the needs of the reader who is reading for both of these purposes.

The Reading-to-Do Track

The ideas in the Guidelines section of the chapter give the person facing a project immediate, step-by-step advice on how to proceed. Often Guidelines will focus on the steps for achieving some documentation objective, such as getting useful review information after you've written a draft, or finding out if your procedures work right.

Readers who are interested in getting a jump start on a project will also find inspiration by browsing the Discussion section of each chapter, where I've included many useful examples. Similarly, when you're in the midst of a project, you'll find the Checklists at the end of each chapter very useful as reminders of the principles in the chapter.

The Reading-to-Understand Track

The Discussion section of each chapter looks at key issues related to the chapter topic. For instance, in Chapter 2, "Writing to Teach—Tutorials," the Discussion section explores the background in cognitive psychology that informs much tutorial design in the profession today, contrasting the elaborative approach and the minimalist approach.

The reader interested in deepening his or her understanding of the principles behind task-oriented documentation will appreciate the Practice/Problem Solving suggestions at the end of each chapter. Also, where relevant, I have tried to include cross references to material in the Guidelines sections.

What Is Software Documentation?

Here is an important definition:

> Software documentation *is a form of writing for both print and online media that supports the efficient and effective use of software in its intended environment.*

Software documentation, as many researchers have shown and as technical writers and software documenters know from their work in the business, contributes significantly to the value of the software product. In this sense the documentation contributes to the user's efficiency in the workplace and thus has an important role to play in modern business. Think of how often you hear people complain about manuals and online help. To me, this speaks for a need for documentation—more useful and practical documentation than has characterized the software industry in the past.

Over its evolution, software documentation has expanded to take on the challenge of providing useful and practical information products for users. Whereas documentation once aimed to satisfy the support needs of the experienced user, documentation in the 2000s aims also to make software useful. This means not just teaching features but supporting workplace tasks with step-by-step relevancy.

In changing from the goal of supporting experts to guiding and teaching beginning and intermediate users, researchers looked to a number of resource disciplines, including document design, instructional psychology, cognitive psychology, ergonomics and human factors, and traditional rhetoric. These explorations created a great number of design innovations that, coupled with technological advances in page design and functionality, have given us the exciting world of single-sourced documents (online documents with dynamically generated content and adaptive interfaces) and embedded help files (documents that present information at the point of need through features of the software interface).

But of all the innovations that sprang from the rapid rise of computer and software technology during the 1980s and 1990s, task orientation has provided the most dependable and useful tool for manual design. Task orientation, as an organizing principle in manuals and online help and as a goal in their design and writing, informs the approach I take in this book.

Consider another important definition:

> Task orientation *is an approach to software documentation that presents information in chronological order based on the user's workplace sequences.*

Task orientation encourages the successful application of software to workplace objectives. Other terms used for task orientation include how-to, step-by-step, procedures, walk-throughs, and tutorials. This approach to documentation is shown in a variety of print and online forms: tutorials, "getting started" booklets, instruction steps, job performance aids, and online help procedures.

In this book I use the principles of task orientation to show you the benefits of using this strategy in every part of the design of your information product. By following these principles you will leverage the user's interest in performing the job successfully, not just learning a new piece of software. The next section illustrates some of the benefits of the task-orientation design strategy.

Who Can Benefit from Reading This Book?

Those who can benefit from this book include any students or professionals associated with writing for the software industry. The section below describes characteristics of some of these people and points out how the information in *Writing Software Documentation* can help them in their learning and work.

Students Preparing for Careers in the Computer Industry

If you want to succeed as a writer in the computer industry you need to know how to design user documentation from a task-orientation point of view. Consider this scenario: A medical management software company keeps incurring high support costs from a client who still has an old version of its manual. That manual contains screen shots of the menus with explanations, in arcane computer terms, of what each of the menu functions does. Other clients using the same system log only half the support calls of this client, but they have received and used the newer step-by-step version of the manual organized after patterns of activities in the user's workplace. They can find, by skimming the table of contents, tasks that relate to their work and that have practical value in the workplace. They see listed there things they get paid to do, like "Print a Patient-Tracking Report" or "Create a Treatment Analysis Graph." When they turn to these procedures they find steps leading them logically through the task, whether it's printing a document in a special way or converting a document from Microsoft Word to HTML format.

Which manual would you want to have produced? The one that *doubles* the support calls (at an average cost of $75 each) from perplexed users, or the manual that *lowers* the number of support calls or otherwise shows an improvement in user productivity—productivity you can measure convincingly and repeat in other projects?

Engineers, Computer Scientists, Managers, Trainers, Usability Specialists

This book has also something to offer readers from the technical side of the computer industry (IT and software development) who have a great deal of technical expertise in software programming, system design, and hardware training but may not have a full range of documentation resources at their fingertips. For these readers this book can offer a number of benefits:

- *Current examples.* The examples, many of which won awards or represent current page or screen designs, can help the software engineer keep up with current designs.
- *Overview of the standard documentation process.* The reader accustomed to engineering processes will feel right at home following the standard documentation procedure outlined in Chapter 6, "Planning and Writing Your Documents." This procedure has helped me and many others see ways to keep development costs down.

- *Insight into making their products useful.* The ideas of task orientation have a broad application in many areas. Technical employees and designers need to understand the approach used by writers interested in building a bridge between users and technologies.

- *Useful tips and techniques.* The programmer who wants to document his or her new application may appreciate helpful hints on structuring a help file for maximum usability or efficiently using information in the manual and the online help.

The Structure of the *Writing Software Documentation*

Chapter 1: Understanding Task Orientation

Chapter 1, "Understanding Task Orientation," describes the nature of software-mediated work and analyzes ways to design task orientation into manuals and online help. It describes a system for analyzing workplace activities and actions and matching them with software operations (descriptions of features). It offers a set of guidelines to direct the document development process.

Part I: The Forms of Software Documentation

Chapters 2 through 4 introduce the reader to three basic forms of software documentation: tutorials, user's guides, and reference documents. Each of these forms has a different purpose. These purposes are: to teach as preparation for using the software, to guide while using the software, and to support after the user has experience with the software. These purposes require different kinds of user analysis which leads to different strategies for document organization, page layout, language, review, and testing.

Chapter 2, "Writing to Teach—Tutorials," focuses on how to write to help users memorize basic program features in order to guide them from being novices to being experienced users, or from being experienced with a program, to being expert. The chapter shows how to organize the two main types of print tutorials in use today: direct instruction and minimalist. The principles of skill selection and tutorial structure apply to teaching documentation in online and multi-media formats as well as print.

Chapter 3, "Writing to Guide—Procedures," focuses on how to write procedures: step-by-step tasks organized around workplace activities and actions that form the heart of the task-orientated approach. This chapter discusses various formats for presenting procedural information and the elements of a typical task.

Chapter 4, "Writing to Support—Reference," focuses on how to create technical support pages and screens for expert users, using the strategy of the structured reference entry. The chapter also looks at methods of organizing reference information and the psychology behind reference support.

Part II: The Process of Software Documentation

Chapters 5 through 9 present information in the sequence a writer would need while writing a manual or help system. Although the phases of the process overlap considerably and some require more time than others, the process roughly follows that used by writers in the software industry.

Chapter 5, "Analyzing Your Users," shows how to conduct a thorough user analysis, thus forming the basis for the design work on the documents that will follow. Because task orientation implies a thorough knowledge of the user's workplace, the chapter focuses on special techniques for getting the right information.

Chapter 6, "Planning and Writing Your Documents," guides you through the stages of writing a manual or help document and covers how to organize people and resources and design a document with maximum usability.

Chapter 7, "Getting Useful Reviews," covers the process of sending out a draft for review by team members and users. This crucial process helps insure usability and task orientation, but you need to handle it carefully to make it anything but a waste of your time and the client's time.

Chapter 8, "Conducting Usability Tests," discusses types of usability tests you can perform to measure how well your manual supports user tasks. It offers an easy-to-follow ten-step process for planning and conducting valuable usability tests.

Chapter 9, "Editing and Fine Tuning," covers the basics of switching from the writer mode to the editor mode. It looks at the industry standard method of editing and shows how that method can contribute to the overall task orientation of the final product. In fact, much of the polishing of a task-oriented approach occurs during editing.

Part III: The Tools of Software Documentation

Chapters 10 through 14 present information by topics selected for their importance in designing task-oriented manuals and online help. These chapters function as a tool reference for the writer: a place to read about design techniques and ways to apply them to real-world writing problems. The reader moving through the progression of Part I will want to consult these chapters as necessary to fill in the background.

Each of these chapters takes a process approach to the topic. The guidelines are arranged in a step-by-step manner to provide a methodology for planning and executing these important parts of the documentation process.

Chapter 10, "Designing for Task Orientation," presents techniques for structuring documents in a way that allows for ease of use and productivity. It ties in with Chapter 5, "Analyzing Your Users," by showing how each of the eight information areas relevant to the user analysis can get converted into useful and productive document designs.

Chapter 11, "Laying Out Pages and Screens," tackles the basic elements you need to know about pages: layout and words. It focuses on layout (how to arrange text on pages and screens) and text (how to pick the right fonts for the right job). It contains a number of examples of common formats and provides a methodology for designing pages and screens.

Chapter 12, "Getting the Language Right," contains guidelines that show you how to maintain a high degree of task orientation by selecting language related to task work and by structuring sentences and paragraphs for easy comprehension and job performance.

Chapter 13, "Using Graphics Effectively," puts graphics—screens, drawings, diagrams, and icons—into the context of the user's questions about a software product and shows ways to answer those questions using images. It elaborates on five ways to use graphics and gives descriptions of the most popular forms in manuals and help.

Chapter 14, "Designing Indexes," examines one of the most important elements of software documentation: the index, or, if online, the keyword search. This chapter shows how to increase the usability of a manual or online help system through indexes.

How Are the Chapters Organized?

Each chapter contains the following sections:

- *How to Read This Chapter.* The introductory section helps you identify which topics you can use for particular documentation tasks or problems. It also includes specific advice for reading the chapter, whether you're new to software documentation or have some experience.

- *Examples.* The examples section of the chapter presents a page or other element relevant to the chapter topic. The examples serve to set the stage for the subsequent guidelines and discussion.

- *Guidelines.* The guidelines section breaks the work presented in the chapter into a process or methodology. The guidelines often contain many examples and practical tips for putting documentation features to work or preparing to write documentation.

- *Discussion.* The discussion section steps back from the process and looks at the underlying principles of the chapter topic.

- *Checklist.* The checklist section summarizes the chapter's contents in checklist format to aid the reader who's actively working on a documentation project.

- *Glossary.* The chapter glossary collects all the terms of a specific chapter that relate to the topic and warrant definition. (These terms are shown in bold italic in the text.) Terms also are represented in the book's index.

- *Practice/Problem Solving.* The practice/problem solving section of each chapter poses cases for applying the chapter ideas or starting interesting discussions of the chapter topics.

An *Instructor's Manual* is available from the publisher to adopters of this text. Please contact your local Allyn & Bacon–Longman representative.

Acknowledgments

I would like to thank the many people who directly and indirectly contributed to this book.

At Allyn & Bacon: Joe Opiela for his professional managing of the review process, Julie Hallett for her professional manuscript preparation, Teresa Ward for her help in preparing the *Instructor's Manual,* and Stacy Dorgan for her help in preparing the companion website.

I would also like to thank the reviewers for this edition: Daryl Grider, West Virginia State College; Sandi Harner, Cedarville University; Dan Jones, University of Central Florida; and Kim Lambdin, Metropolitan State College of Denver.

At WestWords, Inc.: Pat McCutcheon for handling the production process and Deborah Jelinek for her careful and professional copy editing.

At Texas Tech University: Sam Dragga for arranging my schedule so I would have time to work on the project and to my colleagues for their suggestions and support.

In addition to these specific people, I also want to thank those who have provided inspiration in software documentation. Barbara Mirel and Joann Hackos are two whose work is a continuing source of ideas and innovation. I also am eternally grateful to my students for their faith in the task-oriented approach and their ability to help me see how to improve the book and its approach.

CHAPTER 1

Understanding Task Orientation

This chapter looks first at the principles that shape the effort of writing manuals and help files. We then turn to the application of those principles to the dominant forms of documentation: tutorials, users guides, and reference guides. We also look briefly at how these principles apply to the process of documentation writing.

This chapter helps software documentation writers achieve two goals: encourage users to learn the program (proficiency) and encourage users to apply the program to problems in the workplace (efficiency). This chapter defines *task orientation* and gives two examples. It describes and explains nine characteristics of manuals that provide workplace solutions for users. It explains the five characteristics of the *default user manual* and explains five characteristics of the *task-oriented* user manual.

How to Read This Chapter

If you're unfamiliar with software documentation, study the Examples, then read the Discussion section, then the Guidelines.

If you have some experience in software documentation, read the Guidelines, then compare your work to Figures 1.1 and 1.2 in the Examples section. Then ponder the Discussion.

Examples

A number of things determine the success of software documentation; put another way, you can easily find a number of ways to mess up a documentation project. This book examines as many dos and don'ts as possible in software documentation, but focuses on one overriding principle: Make the software usable. A manual that does this adapts the software to the user's job, rather than making the user adapt to the software. What kind of manual encourages adaptation to the user's job? We can begin our exploration with an example.

This example comes from a tutorial manual for a program called PV-WAVE P&C. This program enables scientists and engineers to manipulate research data and view it in charts and graphs. These highly technical users may have used other programs or methods to manipulate their data, and may not easily see how this program can make a difference to them. To accommodate such users, the writers went to considerable lengths to make the integration of the program easier.[1] A number of features, indicated in Figure 1.1 encourage such use. Don't get the wrong idea—that this manual doesn't offer procedural information. If you read this example carefully you can see that the writers have represented the steps of a procedure, but

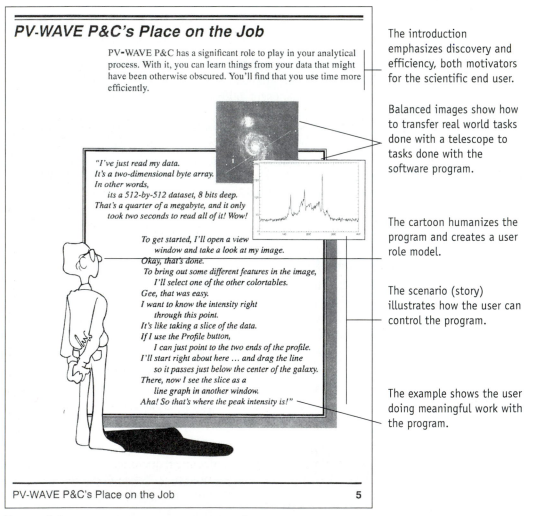

PV-WAVE P&C's Place on the Job

PV-WAVE P&C has a significant role to play in your analytical process. With it, you can learn things from your data that might have been otherwise obscured. You'll find that you use time more efficiently.

The introduction emphasizes discovery and efficiency, both motivators for the scientific end user.

Balanced images show how to transfer real world tasks done with a telescope to tasks done with the software program.

"I've just read my data.
It's a two-dimensional byte array.
In other words,
   its a 512-by-512 dataset, 8 bits deep.
That's a quarter of a megabyte, and it only
   took two seconds to read all of it! Wow!

To get started, I'll open a view
   window and take a look at my image.
Okay, that's done.
To bring out some different features in the image,
   I'll select one of the other colortables.
Gee, that was easy.
I want to know the intensity right
   through this point.
It's like taking a slice of the data.
If I use the Profile button,
   I can just point to the two ends of the profile.
I'll start right about here … and drag the line
   so it passes just below the center of the galaxy.
There, now I see the slice as a
   line graph in another window.
Aha! So that's where the peak intensity is!"

The cartoon humanizes the program and creates a user role model.

The scenario (story) illustrates how the user can control the program.

The example shows the user doing meaningful work with the program.

PV-WAVE P&C's Place on the Job   5

FIGURE 1.1   Getting Started with PV-Wave
This software encourages user control through a scenario that suggests efficient application of the software to work.

they have done so in a way that couches that procedure in the context of the user's workplace.

Software users often need both how-so and how-to information while working with a program. In the example in Figure 1.2, the electronic controls over the presentation of information allow the user to choose the level of detail appropriate to the user's problem-solving needs. Users prefer to choose the level of detail because doing so allows them to relate the program to their workplace, instead of dryly cataloging the system features. The electronic controls make it possible for users to get to the correct steps quickly.

Guidelines

All software documentation should do what the page shown in Figures 1.1 and 1.2 do: Explain and show the connections between the user's professional work and the computer program. *Scenarios,* examples, and page layout can all contribute to this

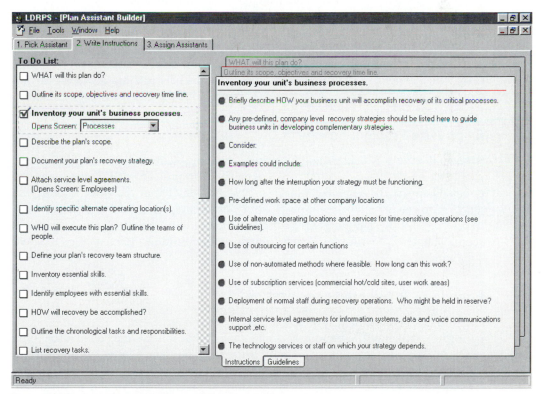

FIGURE 1.2   Getting Assistance While You Work
This help screen encourages software use by highlighting step-by-step information and by providing background information to uninitiated users. It suggests the use of "Guidelines" as an efficient way to get work done.

FIGURE 1.3   Guidelines for a Successful Manual or Help System

1. Emphasize problem-solving.

2. Provide task-oriented organization.

3. Encourage user control of information.

4. Orient pages semantically.

5. Facilitate both routine and complex tasks.

6. Design for users.

7. Facilitate communication tasks.

8. Encourage user communities.

9. Support cognitive processing.

explanation. A manual that does this can be described as "task oriented," because it helps the user manage and communicate information related to his or her task. This book contains many strategies for encouraging task-oriented, integrated software use. Figure 1.3 presents a summary of the main strategies.

As this book progresses we will explore these and other techniques that can help you create a manual or help system that focuses on helping users solve complex tasks, one that helps your users clearly see the relation between this new program and their workplace. The following paragraphs briefly explain some of these techniques.

1 Emphasize Problem Solving

A manual or help system should help users solve problems in the workplace. Some problems might include: "How can I organize this project?" "Where should our company invest development time?" or "Where can I find the peak intensity of data?" You can help the user through introductory paragraphs that preview not only the steps to follow, but the goals and objectives of their software work. As you will see in Chapter 2, "Writing to Teach," you have opportunities to encourage creative solutions.

2 Provide Task-Oriented Organization

Organize a manual or help system in a way that matches the kinds of tasks a user will perform. For example, a word processing manual that follows the "open a file, type in words, save the file, exit the program" sequence would seem more logical than one organized alphabetically, for example, or according to the menus of the program. A task-oriented arrangement begins in the table of contents (of your manual) or the introductory screen (of your help system). As the following chapters will show, a task orientation should pervade the design of your manual or help system so that even the seemingly mechanical forms of reference have the right touch to make them very functional in the workplace.

3 Encourage User Control of Information

"User control of information" means the feeling, among software users, that *they* decide what the program does for them. To encourage this, the manual should show users how to make key decisions, supply key information, or determine key program outputs. Examples include: specifying what a database program will search for and identifying which data the program will process (as in Figure 1.1, which shows the scientist selecting just the right "slice" of data to use). Users need to feel in control of the program. Cross-references in manuals and hypertext links in online systems can help maintain the user's sense of control over the documentation because these document design elements allow users to choose where they go for additional information, or where to proceed after they have finished a section. Chapter 3, "Writing to Guide—Procedures," explores the importance of emphasizing the actions the user takes.

4 Orient Pages Semantically

Semantic orientation in page design means you arrange the elements of the page meaningfully, according to elements of the job the user needs to perform. Figure 1.1, showing the non-computerized image of the galaxy and the computerized chart of the peak intensity, illustrates a semantic orientation. The juxtaposition of the two items mirrors the user's work progress from the telescope to the graph. Other examples of semantic organization include putting important elements first and making important elements larger to help users apply the program to their work.

One of the best way to orient pages semantically employs visuals and graphics to balance text in a complementary way. Chapter 11, "Laying Out Pages and Screens," discusses ways to balance graphics and text and also to maintain the appropriate density of print and legibility of letters. These basic elements can contribute to the usability of information in the workplace.

5 Facilitate Both Routine and Complex Tasks

As Barbara Mirel points out, users of software face both **routine tasks** and **complex tasks.** Routine tasks include repeatable tasks that are easily represented by conventional procedures. She points out, however, that "Complex tasks differ from routine tasks such as data entry of standard accounting calculations because they are not performed the same way every time."[2] Routine tasks are often those represented by menu functions in a program ("Save a file," "Delete a record," and so on) but complex tasks require the user to apply knowledge that isn't easily codified in step-by-step procedures. This knowledge, called tacit or pragmatic knowledge consists of insider knowledge that comes from years of experience. Using a spreadsheet to schedule employees fairly, or using a word processor to identify trends in annual reports: these tasks are highly dependent on the situation surrounding the user and can not be easily represented by manuals that cover only menu functions. However, the more you can help users apply software to complex tasks the more users will value your manual or help system. You learn about users' routine and complex tasks during the user analysis, covered in Chapter 5, "Analyzing Your Users."

6   Design for Users

The concept of user-driven design means that the organization of a manual comes from the user's needs rather than from models or templates of what a user's guide *should* look like or from schemes based on giving some users one kind of information and other users other information. According to Janice Redish, user-driven design should allow users to:

- Find what they need
- Understand what they find
- Use what they understand appropriately[3]

User-driven design requires the kind of extensive user analysis discussed in Chapter 5, "Analyzing Your Users" and in Chapter 8, "Conducting Usability Tests." It means that each manual or help system presents you with new design challenges that you can only meet by involving the user in the document development process.

When you study your users following the seven suggestions in Chapter 5, you will find that each of the topic areas discussed in that chapter ties in to specific techniques, such as using icons to suggest key points or tables to show how features and their uses can add to the task orientation, and thus the value, of your documents.

7   Facilitate Communication Tasks

Users of software programs work in contexts that require them to communicate about their work. These tasks are called *communication tasks* because they depend on the user's workplace demands rather than on a narrow view of program features. Face it, just opening a file is only part of the picture from the user's perspective—the user opens a file so that he or she can communicate information to another person, not for its own sake. Document designers can help users see the *why* behind the program features by analyzing what kinds of information users need and how they communicate, and then identifying those program features—for example, print functions, report functions, or disk output functions—that support communication tasks. Also, communication tasks are facilitated by tasks that transfer data from one application to another.

How do you help users with communication tasks? Learning about the user's communication tasks presents a great opportunity to record the common terminology for procedures and tasks—the "jargon" that you can use in glossaries and for writing steps and explanations. The specifics of using communication-oriented language are discussed in Chapter 12, "Getting the Language Right."

8   Encourage User Communities

Users often need encouragement to rely on other users of the program, their *user community;* task-oriented documentation encourages users to identify and get help from others. Other users of the program, while not exactly experts in the software, can render valuable help because they understand the user's job demands. Companies

FIGURE 1.4   An Example of a User Community

foster user communities with mailing lists, article archives, and contact lists and other information specifically for users of their products. Figure 1.4 shows an example of a user community web page for users of a popular media playing program. For a person writing, say, a tutorial for this program, this site would provide valuable information about the users of the product.

User communities, discussed in detail in Chapter 5, "Analyzing Your Users," can provide a wealth of information about user tasks. Users can also help support your development effort. Chapter 6, "Planning and Writing Your Documents," discusses the idea of including users in the development process through interviews and client reviews. User communities can help provide candidates for this kind of user-involved document development.

9   ## Support Cognitive Processing

Since the advent of computerized work in business and industry, we have learned much about how users process information. We have learned that people use mental models, called *cognitive schema,* that help them learn new information, process the information, and apply the information that comes at them at an alarmingly faster and faster rate. The task-oriented manual uses principles of knowledge representation, parallelism, and analogy to convey software features and applications to workplace tasks. These techniques, described in Chapter 4, "Writing to Support—Reference," allow users to absorb what your manual or help system has to say with as little effort as possible.

For the modern professional in business and industry, the software program lies at the heart of the knowledge management problem because computers both represent the cause and the solution to the problem. Computers generate complex information, and computer software allows the user to store, transfer, and present it. Our understanding of how computers affect work grows through research on the impact of computers on people's work and workplace roles. If we explore software use from the point of view of the user's information environment, we see a strong need for good software documentation.

Discussion

The principles that underlie the guidelines above derive from studies of the use of software to perform productive work. A good manual or help system has many features that make it succeed, but the bottom line is this: *the more a manual can support productive work, the greater the chance of acceptance and satisfaction by a user.* First, consider the overall goals of software documentation and examine some problems inherent in doing software work, then examine some ways a good online or hard-copy manual can help users overcome these problems.

The Principles of Software Documentation

This book addresses an important need in the business and professional workplace today to help people use software efficiently and effectively. According to the Nua Internet Surveys the numbers of persons online in use in the United States since 1995 has risen from 18 million to 166 million.[4] Like it or not, those technical communicators who choose to write software documentation—manuals of all sorts, help files, user's guides, reference cards, job performance aids, FAQs, installation guides, and other information products that accompany software programs—find themselves confronted with no less of a challenge: make users *proficient* with software and *efficient* in their jobs. While such factors as training, individual motivation, peer pressure, the boss's dictate, or fear of falling behind in a career can certainly contribute to people using their software (and thus their computer) efficiently, the computer manual (print or online) remains the single most common form of support. When in doubt and often as a last resort, most users turn to the manual.

Researcher Barbara Mirel, and others, approach the goals of software documentation from a perspective of the user doing "knowledge work" or work requiring that users do more than just learn to use the menus of a computer program. According to Mirel, the job of the manual writer consists of constructing documents that help users "learn to work with their programs in new ways, adapting software capabilities to the specific purposes and goals of their jobs."[5] This kind of software work is called "adaptive computing" and means that users have to apply software to complex tasks not represented by menu selections. Complex tasks require problem solving, analytical skills, and knowledge derived from experience—the kinds of learning that is situated in real workplace activities. In terms of our discussion of the goals of the software manual or help file, supporting "adaptive computing" means going beyond just saying how a program works. You have to show how to apply the program to complex workplace tasks.

Table 1.1 refers to two goals of the software user: the goal of learning a program and the goal of applying the program. The writer has to convey the correct instruction on learning to use the program, helping users navigate menus, learn commands and terminology of a program. But good manuals and help files also have a second goal: telling how to apply the program to complex tasks that you can't just perform by selecting menu functions. Helping the user solve problems—improve his or her performance in the workplace—requires that the manual writer learn the user's workplace goals, problem solving strategies, and other kinds of knowledge that make people effective workers, and then showing them how the software can help them do that.

This chapter offers a strategy for addressing the needs of software users when they turn to the manual. First you will look at a definition of a design strategy for manuals that attempts to address the user's needs productively to encourage efficient use of word processors and database programs. Then you will examine examples of manuals that exhibit some of the features that contribute to efficient software use. To understand how these manuals work, you need to look at changes that have taken place in the modern workplace since computers and computer software arrived. You will see that working with software requires a significant shift in thinking and learning, a shift that requires users to develop new skills and job roles and documentation writers to adjust their approach to writing manuals and online help.

TABLE 1.1   Goals of the Software User and Manual

Goals of the Software User	Goals of the Manual or Help
Learn to use the program	Teach the features of the program
Apply the program to useful work	Tell how to apply the program to complex tasks

A Definition of Task Orientation

This book uses the term *task orientation* to indicate the writer's purpose. The following definition expresses how task orientation helps articulate this purpose.

> **Task orientation:** *A design strategy for software documentation that attempts to increase user knowledge of and application of a program by integrating the software with the user's work environment.*

When confronted with a new piece of software, most users have one question to ask: "How will this program help me in my job?" An informed answer to that question, one that points out exactly the greater job efficiency, or the savings in time, or the greater accuracy of production, can provide just the motivation a new user needs. Those who reject a software program often do so with a parallel observation: "This program did not help me in my job." Often the complaint gets worse: "This program slowed my production time down," or "This program alienates my employees and makes them feel like subordinates to a machine." But when a user sees that learning and using a program can *increase* job efficiency, most will take the time to read the manual and learn the program. As Patricia Anson points out, the full potential of a manual is realized when "technical writers take an approach to developing documentation (online and on paper) that models the natural cognitive processes of users who are seeking to fill knowledge gaps through the right information, presented at the right time, and in the right place to meet task goals."[6] Clearly, the manual that encourages this kind of integration with task goals will also increase job efficiency.

The Theory Behind Task Orientation

While the idea that software should help people do meaningful work may seem obvious, it is nonetheless helpful to explore the theory behind the approach. Exploring the theory may help you understand the principles that can guide your design of manuals and online help, and provide the foundation for techniques you will find in the chapters that follow.

The Default Manual

As you can see, there are two ways to define the user of any software: as a person who needs to learn about menu functions and commands, and as a person who uses software for workplace ends. In the past, manuals and help systems tended to focus on the former definition, assuming that if the user could understand how the program worked, he or she could figure out how to apply it in the workplace. The idea of basing computer manuals on program features was and is very prevalent in the computer industry today. Such an approach creates manuals with sections like "Using the File Menu," or "Creating a Table." While these topics are necessary for using the program, they convey a subtle message to users: "Learn the menus and features of this program and that's all you need to do to be productive." This may or may not be true. While these basic tasks are important, they create an implicit role for the user. That role is one of a person who is technologically deficient or ignorant, a mere

TABLE 1.2   Characteristics of the Default Manual

Characteristic	Example
Covers the features of the program	Using the File Menu... Saving a document...
Implicit role of technological ignorance imposed on the user	"Read this manual before proceeding..."
Ignores the user's workplace use of the program	"Understanding Net Hog Pro"
Assumes one way of learning	"This tutorial will help you understand the function of the Setup menu..."
Overly simplified approach to program operation	Step-by-step organization

receiver of information, and an operator instead of a thinker, someone who needs to accommodate his or her behavior to the program rather than vice versa.

In this book I refer to the user defined by such a manual as the *default user* and the manual written in such a way as the *default manual*. Defining the user as "a person who operates a computer" tends to have a limiting effect on how that person sees their job and work. This limiting effect stems from the tendency in manuals to isolate the user from his or her environment and concentrate on defining program features exclusively.

The Default User

The following are some characteristics of the default user:

- perceives job skills as decreasing in importance
- sees computer use as separate from job goals
- becomes isolated from other employees
- fears remote supervision
- suffers from information overload

Often, people resist using computers and software because of the inherent complexities of abstraction and information overload. A brief overview of these areas can demonstrate the challenges facing software documenters who are determined to help software users be efficient in their jobs.

DECREASED IMPORTANCE OF JOB SKILLS: "MY EXPERIENCE ISN'T ANY GOOD ANY MORE." When we speak of workers' skills losing their importance we often speak of it as *job deskilling,* which means that the computer program can perform many of the tasks a person used to perform so the job requires less skilled people. Consider the example of the maintenance worker in a plant. Before the advent of the computerized inventory control system, keeping track of parts for machines required experience acquired over years of repairing motors. With the computer system in place, parts are reordered automatically once the levels of inventory fall to predetermined levels. The decision to order new parts, for example, appears to have *gone into the computer,* so the company can now afford to hire persons with less job experience, with fewer special skills.

Some managers and professional workers report in research studies that they perceive their jobs as less meaningful than before[7]. But often the case is that because the computer can provide information to the experienced user on re-ordering parts, that user is, in fact, challenged with new problems: "How can I make re-ordering parts even more efficient," or "Perhaps I can concentrate my efforts on deciding among parts vendors now instead of having to keep track of inventory." The skill doesn't go away, but the problem changes for the user, requiring that user skills be applied in new and more challenging ways.

Much resistance to computers derives from the perception by professional persons that the computer, often in company-wide systems, has begun to take over some tasks that employees used to perform, as in the case of deciding which parts to re-order. The default user is the user who "lets the computer do it." Often the default user gets this limited idea from a manual that merely describes features like "reordering parts." A better way to approach this would be to help the user see ways to apply the "reordering parts" feature to the new circumstance of managing parts re-ordering, or making parts re-ordering as efficient as possible. These larger tasks suggest to the user that skills don't decrease in importance, but that they change, grow in complexity, and require new learning instead of thoughtless following of steps.

INCREASINGLY ABSTRACT TASKS: "I JUST CAN'T UNDERSTAND HOW THIS THING WORKS." Part of the reason people have trouble seeing the link between doing by hand and doing with a computer lies in the abstract nature of computer work. Anyone who has tried to learn a computer programming language has experienced the abstractness of the way computers do things. By-hand work (writing your name with a pencil) and computer work (writing your name with a word processor) embody a contrast of the concrete and the abstract. The pencil creates marks simply, when a piece of graphite is dragged across a page leaving a visible trail. The computer uses a highly complex, electronic system of buffers, wires, computer chips, and circuits to leave its visible phosphorescent trail on the screen. How does it happen? The computer does things in a very abstract way. You can't touch it; it's not concrete.

The same feeling of loss of control faces all computer users. Without a feeling of control over their work, workers feel that it loses most of its simplicity. And this apparent loss creates resistance to software and threatens efficient use. Writers of manuals and help systems need to develop techniques—such as decision trees, lists of suggested uses, examples in different disciplines—to re-awaken the computer user to a confident awareness of the computer as a flexible tool.

Whereas increased abstraction relates to how people see their jobs through their tools—computerized or not—work also takes place in a social domain. And that, too, appears threatened by computer-mediated work.

INCREASINGLY ISOLATED FROM OTHER EMPLOYEES: "I'M STUCK IN FRONT OF THIS COMPUTER." Business organizations embody a complex web of social structures that have evolved over history and often have to change because of work done at a computer screen. Social structures—the people we relate to at work, the work communities we inhabit, the coalitions we form—play a major role in our job satisfaction. In some companies, social groups take on names: the front office, the back office, the first floor, and so on. But now, Zuboff asserts that the computer screen has now become the primary focus of a person's interaction with a company, and with others in the company. No more chatting over the cubicle walls, no more friendly errands to run to different parts of the building. One person, a benefits analyst at an insurance company, put it this way: "No talking, no looking, no walking. I have a cork in my mouth, blinders for my eyes, chains on my arms. With the radiation [from the computer screen, supposedly] I have lost my hair. The only way you can make your production goals is to give up your freedom."[8]

People need others to communicate with, to get feedback from, and to get rewards and other incentives that make work enjoyable. They create useful dialogs with others to help share and solve problems. But people using computers risk a diminished importance of their co-workers in their job. Many potential software users, understandably, resist this isolation. They lose their social contact, even if, before, they may not have realized the social aspect of their work. The software documenter, as we will see below, faces a challenge to introduce the isolated user to new possibilities of interaction with co-workers *through* the computer.

REMOTELY SUPERVISED: "MY BOSS HAS AN ELECTRONIC LEASH ON ME." Ironically, the computer-mediated user will feel both increased isolation because he or she seems chained to a computer screen, but can also feel increasingly exposed to the manager or supervisor. For example, before the secretary had a computer to work on the boss had to physically walk to the secretary's desk to check on the status of a typing job. Now the boss can check on the secretary by looking up the file on the network. Before, the manager had to catch you loafing or had to come to your desk and pull files to make sure you kept up with your work. Now the manager can access your files electronically, check on your productivity, even organize your work day for you without ever showing up physically Figure 1.5.

This kind of remote supervision through the computer system can result in a number of detrimental effects. Some computer users may feel that they can't think up new ways of doing things because the computer "has it already figured out." Others may get an ambiguous sense of their actual boss, and may attribute authority to the computer system itself. They may lose their sense of control over their work because of the increased supervision exercised *through* the computer system. Whatever the effect, computer users often resist using software because of the control they perceive it has over their work.

OVERLOADED WITH INFORMATION: "WHY DO I NEED TO KNOW THAT?" Some users resist computer use because they feel overloaded by information. (It's not uncommon for a frequent email user to accumulate hundreds of unread messages while away from the computer on vacation.) Consider the writing student who can't decide which of the suggestions made by a grammatical analysis program to follow up on. Or consider the researcher faced with volumes of descriptive statistics but little idea as to which ones to regard as significant. Similarly, a computer network management program can provide a supervisor with a full screen of information about network use, but such a screen can also intimidate the supervisor who can't tell which statistics mean more than others. Having volumes of information does not always solve problems for

FIGURE 1.5   Remote Supervision Doesn't Work For Everyone
Workers in computerized environments often feel a sense of lacking a place of their own away from the boss and co-workers. Often our best creative work occurs in such private spaces.

© Ed Stein, reprinted by permission of Newspaper Enterprise Association, Inc.

users. In fact, according to author Richard Wurman, having too much information without the ability to understand its significance can cause ***information anxiety.***[9] This anxiety can afflict computer users who find themselves flooded by information without knowing which they should try to understand.

Often the default manual will contain a statement at the beginning of the document that says "read this manual before proceeding." Rarely do users follow these directions because the default manual is a manual that focuses on the system features, menu items, and other information about the software product. But users are only secondarily interested in the software product; their main focus is on their jobs. So the reader often finds him or herself in the situation of trying to use the program without really having "read" all the manual, experimenting with the interface, trying to figure out how to apply the program. The problem occurs when the user gets stuck or lost. Then the user has the feeling that just maybe he or she *should have* read the manual (but the manual doesn't say anything about applying the program anyway). This feeling of having information but not seeing how to apply it leads to information anxiety, a condition that causes many users to restrict their use of software and, ultimately, give up trying to learn and apply it.

The Task-Oriented User

As indicated earlier, software documenters face the challenge of making programs easy to use and applicable to workplace objectives. This book will help you meet that challenge. As a beginning, we can examine the difficulties that face the users of computer software, and see how we can address them through documentation design. The following discussion of the characteristics of the task-oriented user, whose software

TABLE 1.3   The Default User Versus the Task-Oriented User

Default User	Task-Oriented User
Decreased importance of job skills	Challenged by redefined work activities
Increasingly abstract tasks	Conceptually oriented
Isolated from social networks	Aware of user communities
Remotely supervised	Self-managing
Overloaded with information	Supplied with resources

use fits with his or her work environment, can help you write manuals that support efficient and productive software work.

CHALLENGED BY SKILL DEMANDS: "THIS PROGRAM MAKES ME A BETTER MANAGER." While software can perform some of the skills of trained employees, software use requires users to engage in complex tasks, tasks that require a human mind and that call forth sophisticated professional knowledge. A database program, for instance, requires sophisticated workplace skills in organizing and categorizing information, understanding relationships between sets of data, and analyzing numerical trends. A computer can sort and categorize but it cannot handle ideas. In fact, software can expand the kinds of activities that a person can engage in at work, if only the manual or help system points the user in the right direction.

Applying an action/operation framework to computer work can help documenters see the kinds of complex tasks they need to support. Computer activities require two kinds of efforts: actions and operations (Figure 1.6). *Actions* are tasks that grow out of work situations that often require communication and thought. For example, in the activity of using a word processing program, the actions would consist of writing a letter, writing an analysis, contributing content to a quarterly report, taking notes during an interview, and so on. When the accountant thinks of "using the word processor" he or she thinks of these kinds of actions (or complex tasks). *Operations,* on the other hand, consist of program functions; limited efforts, often defined by the menu items of a program. In a word processor, for example, tasks like "opening a file," "checking the spelling," "setting page margins" and so on comprise operations.

Using the framework of an activity such as in Figure 1.6, you can see that to the reader, the activity of "using a word processor" has most meaning at the level of *actions.*

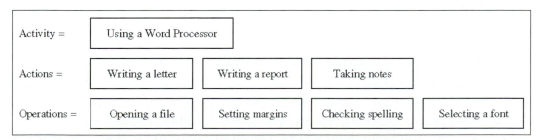

FIGURE 1.6   Activities, Actions, and Operations in Using a Word Processor

Actions like these have meaning to users because they relate to meaningful work, they help the user become a better communicator, they suggest thinking and reflect complexity. They help define the social conventions of the workplace ("what a letter looks like," "what reports should contain," and so on.) The task-oriented manual instructs and appeals to the user "where the action is" or at the action level rather than the operation level. At the operation level tasks are generic, isolated, and applicable to any number of actions.[10] Any given action could employ a combination of operations to complete it. But the point is that nobody uses a word processor simply to "open a file."

The software documenter needs to find ways to reinforce the skill challenges inherent in actions, because these represent efficient computer work. Making the program functions or operations easy to use can help, because it allows the writer to focus on higher-level, advanced actions that challenge the user and that the user is familiar with. In this way, users begin to see computer work as more than mere step-by-step keyboarding. Users need to see their work as significant: to see that what they do with a software program can have an impact on their work, their organization, and others within their organization. Challenging the user often requires teaching computer skills in the context of a person's job so that the user can see the benefit of the software (Table 1.4.)

CONCEPTUALLY ORIENTED: "THIS GIVES ME SOMETHING NEW TO THINK ABOUT." Work with software requires handling abstract concepts such as data types and processing instructions that make computer work difficult and can cause users to reject programs. Technical writers have always faced the challenge of explaining abstract and highly technical information to novice readers. Fortunately, researchers in instructional design have found ways to help explain abstract concepts. The paragraphs below present some of the new approaches available to the software documenter.

Part of the difficulty in using a software program or web application lies in the kinds of knowledge resources it calls on in the user. For example, if I have to learn how to use a billing application for a web site, I need to type in my name, credit card information, billing and shipping information and so on. These require simply following explicit steps and are easy to teach. But that's not all there is to it. To use this program I have to understand about Internet security, my secure browser, encryption technology and so on. I may not understand all about these complex areas, but I know enough about them and respect them enough to know that they are issues in this particular software activity. Researchers call this kind of knowledge *tacit knowledge* because it's not easily represented by steps and it's assumed and understood by me instead of made explicit. They can become explicit, as in if I happen onto a web site

TABLE 1.4   Statements that Reinforce Complex Actions

The LOGDAT file contains oil well information organized according to geographical areas familiar to well analysts.

"Owners of print shops can give you valuable advice that will affect how you create the publication—for example, which printer to compose the publication for. (There's more about this in "Using an Outside Printing Service" on page 197.)"

that seems to me in some way to lack the appropriate security. Then I think "should I follow up with a phone call" or "should I use a different method of payment here."

The manual for the task-oriented user should evoke or suggest structures of tacit knowledge by giving the user control over the program, teaching in ways the user expects, putting software instructions in the context of workplace actions, and aligning software work with workplace goals, and helping the user solve problems.

AWARE OF USER COMMUNITIES: "SHAKESPEARE WOULD USE EMAIL…NOW." The inherent communicative nature of computing provides numerous ways for the documenter to help users overcome the isolation they experience when converting their work to a computer. For one thing, computer users automatically become part of a user community. The term refers loosely to those who use the same program within an organization, but it can also refer to others who use software in their work. Computer support divisions in corporations have discovered an increase in user acceptance of software when they encourage the formation of user groups. The term *user groups* refers to groups that meet, either electronically or in person, to discuss issues with their computing and exchange ideas to increase efficiency and productivity. In an R&D organization, for example, you might find UNIX operating system user's groups or WordPerfect user's groups. Meeting with these groups allows users to increase their social contacts within an organization and overcome the sense of isolation they may feel.

Most employees work in groups and as a result have to coordinate their activities, share work in progress, and store the results. This collaboration often creates situations in which managers must produce group reports or engineers must contribute designs of parts to an overall project. Documentation plays a key role in supporting collaborative work by indicating ways that users can communicate information. In fact, Bødker argues, software usage implies communicative activities in which users interact with other users socially as part of projects.[11] In fact, given the prevalence of

TABLE 1.5   Ways to Help Users Think About Software Work

Actions Evoking Tacit Knowledge	Example
Give users control over software functions	"Using the data visualization feature requires you to select an appropriate data model, as described below."
Select the right training method	"Experimenting with the *Net Hog Plus* will help you see ways to adapt your user profile to different Internet environments."
Suggest workplace goals	"Depending on how you organize your project, you can either create smaller graphics files or larger graphics files. The size of the graphics files depends on the objectives of your project."
Help with problem solving	"Look over the options in the Special Effects menu and select one that helps you convey the theme you want to portray."

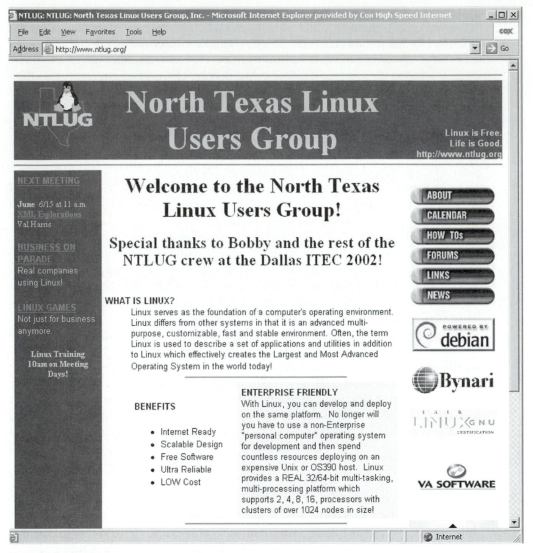

FIGURE 1.7   An Example of a User Community
User groups like this one foster communication among software users and promote sharing of information.

computer networks in companies, many users prefer this method of coordinating activities and sharing information.

INFORMATION RICH: "MY SOFTWARE GIVES ME A BETTER VIEW OF MY TASKS." In modern corporations, information represents the source of power and authority. But for users to realize this fact they need help in managing information means putting it to productive use. Good software documenters should find ways to reinforce workplace skills by showing users the potential use for information that programs generate. For example, a phone utility program that records time spent on the phone can generate valuable information for sales reps about time of contact with clients. But just how to apply this information may elude the novice user. Pointing out how to use phone tracking time in progress reports to supervisors can help salespersons understand the value of this program feature. It is up to manual and help writers to alert users to new information work they face. Not only does software increase the ways employees can manipulate information, it can open them up to new kinds of tasks. Table 1.6 below lists and describes a number of new kinds of tasks facing the task-oriented software user.

The Forms of Software Documentation

So far in this chapter we have looked at the principles that underlie any form of software documentation: principles of instruction in workplace tasks using techniques to help the reader relate software to complex tasks that face users in contemporary business settings. When you apply those principles to the creation of actual manuals and help files you need to consider the types of documents that you will create, and the processes that you will follow to create them.

Let's look at how people use software. First they usually learn how to apply the program to their work, how the features match their workplace goals, and how the

TABLE 1.6   New Types of User Tasks

Type of User Task	What the Person Does	Example
Planning Tasks	Identifying goals and manipulating time and resources in the abstract, to find ways to meet the goal. Articulating future events using various computer programs.	"What's the best way to organize the shop inventory?"
Decision-Making Tasks	Assembling rich alternatives without giving in to one solution for doing something. Clarity of evaluation of alternatives in settling on an action or stance.	"Which supplier provides the most efficient delivery times?"
Problem-Solving Tasks	Identifying elements that block progress in business or organizations and identifying and evaluating ways to accomodate the unexpected.	"How many strawberries can we ship in each box?"
Operating Tasks	Keypunching and inputting of information and using menu items to manipulate the program and the data. Essential work in information processing, involving questions of transfer and storage.	"How do I translate my design into transferable format?"
Knowledge Work Tasks	Identifying information of value to an organization or department, with the intention of accumulating valuable wisdom.	"I would like to open this meeting with figures, showing last quarter's increase in productivity in our department."

actions they perform correlate to the menus and screens. Next they use the program, sometimes on a daily basis, and sometimes intermittently, using features they know well and some they don't know well but can follow with a little guidance from the manual or help system if they get stuck. Finally, as advanced users, they just need to look up information about a program, find examples, and troubleshoot error messages. These users focus on the technical aspects of a program, how it works, what its components consist of and so on.

Forms of documentation follow these functions in the workplace. Initial users of a program, or novice users of software in general, require tutorial documents, the intention of which is to teach basic functions and their application. Intermediate users of a program require procedural documentation, the intention of which is to help them during actual use of the program in their workplace. Advanced users of a program require reference documentation, the intention of which is to further their understanding about how the program operates and how they can manipulate and adapt it to highly specialized uses.

The following three chapters explore in greater detail the three forms of manuals and help that arise from these kinds of usage patterns. For our purposes in this chapter it is useful to see how the three forms relate to the principles of task orientation. Table 1.7 shows an overview of the three forms of documentation.

Tutorial Documentation

Tutorial documentation is documentation that intends to teach the basic functions and features of a program to a user in such a way that the person can begin applying the program to workplace tasks. Examples of this form include getting started guides and online and printed tutorials. Because the intention is to teach, the relationship between the writer and the user resembles that of a teacher and learner. The tutorial document embodies all sorts of instructional design tools to assist in learning: sample scenarios, examples of usage, walk-throughs, demonstrations, rewards for learning, structured "lessons," worksheets, self-evaluation forms, and memory aids. Tutorial documentation focuses heavily on *actions* that the user can take that evoke problem solving and other productive workplace behaviors.

Procedural Documentation

Procedural documentation is documentation that intends to guide the user in the every day use of the program, often when the user needs information at the time of use. Examples of this form include users guides and help files consisting of step-by-step procedures, tips and help embedded in the user interface, context-sensitive help available at the click of a mouse, and wizards that assist users in performing difficult, important, but seldom used tasks. Procedural documentation, because of its intention to guide the reader, implies a more distant relationship between the manual writer and user: one of an informed assistant. This type of document employs many tools to help users in actual use of the program, including step-by-step procedures, suggestions and tips, descriptions of

TABLE 1.7   The Forms of Software Documentation

Tutorial	Procedural	Reference
the user motivation is to learn	the user motivation is perform routine tasks	the user motivation is to obtain information "about" the program
intention to teach the features of the program	intention to guide through step-by-step procedures for using the program	
relationship of teacher and learner	relationship of guide and mentor	relationship of resource and client
defines the task through scenarios, cases, examples; narrative structures	defines the task through chronological, step-by-step structures following the menu choices or fields in a pane or screen	lets the user define the task
focus on basic *actions*	focus on *operations* organized around workplace actions	focus on the program

fields and screens, pop ups and screen regions containing definitions and usage ideas. The focus on *actions* in procedural documentation comes through organization of procedures around key decision-making or problem solving activities of the user. Using electronic interfaces, performance support systems can "read" a user's keystrokes and behaviors and automatically provide just the right information at the right time.

Reference Documentation

Reference documentation is documentation that intends to supply information "about" the program for advanced users. Reference users rarely consult the user's guide or tutorial but need from time to time to look up information about the program. Examples of this form include alphabetical listings of program features, lists of examples, file formats, technical troubleshooting data, data for using an application with related programs, and special program settings. Because reference documentation serves advanced program users, the relationship of write to user is that of an information resource to client. The "task" the user needs to accomplish is not defined by the writer (as is often the case with tutorial documentation) but by the user. The user brings the task to the document as opposed to having tasks outlined by the document. Of all the three main forms of documentation, reference documentation is more purely descriptive of the program itself than of the user or the user's application of the program. It focuses on interface elements more than the other two forms.

The Processes of Software Documentation

The writer who faces the challenge of creating truly useful task-orientated documentation needs to look to the process of writing itself and find ways to learn about users. Can a writer write a document without ever consulting a user. Of course. You can study the program diligently, and comb over the specifications that programmers use to write it. If you're part of a development team, you can ask the programmers themselves how the program operates and compile that information into a manual or help system. Earlier in this chapter I described the kind of manual that you would produce following this process—the default manual. And because the default manual builds its model of the user from the program itself, you end up writing to the default user: a disembodied, context-neutral, logical abstraction. This user results when you default to the excuse that "anybody could use this program." If you write to that user all you're doing is putting the program into textual form; "documenting" it in the most simplistic way by just telling what one sees on the computer screen when the program is running. Task-oriented documentation consists of something else.

Task-oriented documentation consists of manuals and help that reflect actual users in all their variety and human forms. This means that the process you follow does not begin with the program, but with the users themselves. The process of task-oriented documentation requires that you analyze the user in his or her actual work environment, to discover the rich texture of activities within which your program and your manual must fit.

Take for example, Aubrey's exploration of palm pilot software users. Aubrey could have just written down what the features of the software are and how they work, but instead he spends some time with actual users. When he does he discovers that palm pilot software plays an important but small role in users' workplace activities. He learns that Michael's pilot sits on his desk most of the time until Michael takes a trip to the Ukraine to do educational consulting. He learns that Susan uses her palm pilot to keep track of research lab usage on the fly because she's too busy to record it on a laptop or desktop computer and that she needs to know the fastest ways to download information. He learns that nobody really uses certain functions because they are so poorly designed that users avoid them. He finds out that, for security reasons, international construction users need to download additional encryption programs. Aubrey's writing process doesn't start with the software but with *people*.

As you will see in the second section of this book, the process of writing task-oriented software documentation is one of exploration of user needs, and then of constant involvement of the user in the process of writing and testing. This process is called a ***usability process*** (see Figure 1.8) and it means that users and their needs drive the writing. Early in the process user interviews help you learn about actions and activities in the workplace, later, user reviews help you refine narratives and cases that you will use to orient your readers to the right procedures and help topics. Usability tests give you feedback on how well your manual or help files meet your readers objectives of supporting problem solving and integrating with workplace activities. Finally, usability evaluations help you understand how well your techniques—organization, examples, information design, page layout, hypertext structures—worked in actual user settings.

Planning stages

- User interviews to find out what actions users take using the software
- Focus groups to find out user needs and organizational constraints

Development stages

- User reviews to see how well the manual fits with workplace tasks
- User lab tests to gauge the accuracy of manual and help information
- User field tests to gather additional information about workplace uses

Evaluation stages

- User field evaluation to asses the overall value of documents
- User usage reports to help adjust writing and research processes for subsequent manual releases

FIGURE 1.8   **The Usability Process**

Glossary

actions: tasks that require a combination of various menu functions and program features to accomplish, but that grow out of the user's actual work environment. Actions associated with a fitness tracking program would be getting into shape, controlling one's diet, deciding on an exercise program. Actions arise out of the user's workplace or activity context and consist of one or more *operations.*

automating: a process of converting a manufacturing or business task from one done by human action to one done through a machine such as a robot or a computer. Tasks such as calculation, writing, and analysis are automated by computers.

cognitive schema: in cognitive psychology, this term refers to mental models of people, things, organizations, and so forth that people form as a way of interpreting their world. For example: the schemata for a kitchen would include a room with a stove, refrigerator, counter, sink, and appropriate cooking tools and materials. Knowing a user's schemata can help documenters understand thought processes users employ when they approach work problems.

communication tasks: tasks that require the use and manipulation of information to coordinate workplace activities. Planning a meeting, evaluating employees, tracking sales data are examples of tasks with a communicative dimension.

complex tasks: tasks that require users of software to call on assumptions and understandings gained through experience, education, and training in professional work.

conceptually oriented: a type of page layout that organizes paragraphs around *ideas* that underlie software use. Researchers tell us that users with the right conceptual understanding of a task perform that task more effectively.

default user: the user who is defined as a person who uses the menu items and functionalities of a software program.

default user manual: a manual consisting primarily of step-by-step procedures based on menu items in a software program; primarily descriptive in nature and limited in application to real-world job tasks.

information anxiety: a problem experienced by computer users and others that relates to our ability to make use of information. The feeling of information anxiety comes from experiencing an overload of data (such as a computer manual or help system) but not understanding how to use it.

job deskilling: in management terminology, this term indicates what happens to a job when thinking and analytical skills get taken over by a computer. A certain job is deskilled when the skills formerly needed to perform it are no longer needed and a person possessing lesser skills can be hired, usually for less money, to perform the job.

operations: in describing software work, operations refer to units of activity usually defined by menu items, screens, or panes. "View a ruler," "Select a table," "Save a file" are operations that make up the feature set of a word processing program.

organizational existence: in end-use computing, this term refers to a person's understanding of his or her role within a corporation as part of a network of information, knowing where information comes from and where information goes. Users who understand their place in this network have a strong sense of the context of their computer work and learn to apply software to their work effectively.

routine tasks: tasks that lend themselves to easy description, are repeatable, and usually do not call for extra thought.

scenario: a kind of narrative of events that describes what a person does to perform a specific task. Often taking on the form of a story or play, a scenario tells the rich details of a person's work. Documenters use scenarios to help understand the complexities of a user's work in order to provide well-designed support. A scenario for an advertising account representative who decides on a kind of media for a client would include a description of the client and the problem and the steps the representative took to research, solve, and present the solution to the client.

semantic orientation: a type of page layout that orders or creates patterns of information on the page according to the user's task needs. Example: headings for skimming the page to find topics, and paragraphs to help the user understand concepts.

skill transfer: refers to the way skills used in one activity can also apply to the learning of a new activity. For example: a person with the basic knowledge of how to fry can learn to fry green tomatoes more quickly than the person without the basic knowledge in this area. In software work, skill transfer often refers to a person's ability to learn a new program more readily if the basics are already understood through the use of a similar program. The skills from one program transfer to the learning of the second.

tacit knowledge: the kind of internalized knowledge a person acquires as part of a work organization, a society and a culture. Often tacit knowledge is unacknowledged and part of a person's mental makeup. Social skills, experience with people, assumed rules for handling work situations: these are examples of tacit knowledge.

task-orientation: a method of organizing online and hard-copy documentation that follows the typical tasks and task sequences of the software user.

training method: in training literature, this term relates to kinds of structures of training for computer users. Training methods include: applications based (which teaches the user to apply the program to work) and construct based (which teaches the user the features of the program.) Good documentation should include both kinds of teaching for appropriate user tasks.

usability process: the process of writing task-oriented documentation that involves the user in all stages of writing to ensure an appropriate fit with workplace activities.

user community: a group of users who use the same program. They may exist within the user's organization or within the larger community, and may be organized according to various degrees of formality. Some user communities meet on a regular basis and exchange information, others exchange information only informally. Examples: Word-Perfect users in a college department, UNIX users in an R&D organization.

☑ Checklist

A manual that integrates a software program into the user's information environment has a better chance of getting used than a manual that only documents the features of the program. Often users subconsciously perform a cost/benefit analysis when considering a program: "Will this program help me do enough productive work to offset and compensate me for the time it takes to learn and operate it?"[12]

Cost/Benefit Analysis Checklist

If we translate the characteristics of the task-oriented manual presented above into a cost/benefit analysis for the user it might look something like this:

Cost/Benefit Analysis Checklist

☐ Will the manual help me use the software to solve problems?
☐ Does the manual tell me how I can control the program?
☐ Do the pages follow a logical design that emphasizes what I need to know?
☐ Does the manual clearly segmented into useful activities that I engage in at work?
☐ Is the manual designed for use rather than to describe the system?
☐ Does the manual help me connect with other users of this software?

A user who can answer yes to these questions might find the manual and the program useful, and might use it. Most importantly, your manual should function to provide access for the user to the thinking and working capabilities of the program.

Practice/Problem Solving

1. Examine a Manual or Help System for Task Orientation

Examine a copy of a manual for a software program such as a word processor, spreadsheet, or database, or examine a help system for a program you use regularly. Study the table of contents, looking at all the elements in the documents (there may be more than one document included in the printed set). Find instances where you think the manual or help reinforces the user's workplace and workplace tasks. How much of the document is in "how-to" format?

2. Examine a Computer User for Work Characteristics and Software Use Habits

Interview one or more people who use computers in their work. Find out how they use computers and what job goals software help them achieve. Compile a short report covering the following topics:

- description of the job: duties, co-workers, decisions, types of computers used, kinds of information encountered
- description of the level of computer expertise required
- description of their methods of learning software skills: user community, manuals, training sessions, and so forth
- suggestions they have for making the software manuals they use more useful

If you were writing a word processing manual for this person, which actions would you address? What about a manual for a spreadsheet or database program? A business statistics program?

3. Link Program Features with Work Tasks

Using your Internet browser, visit a web site that has shareware or free programs. Such sites include:

- http://shareware.cnet.com/
- http://www.files32.com/
- http://www.tucows.com/
- http://www.macworld.com/
- http://www.freebiedirectory.com/
- http://www.softpile.com/

You can also get samples of free software at the website for this textbook:

http://www.writingsoftwaredocumentation.com

and navigate to the software download section.

Familiarize yourself with the software program until you gain a basic understanding of it and can list some of its features. Then think of who might use the software, and complete the form below. You can add to the list of features if you like. The point is to think not only of things a program can do but how to apply the program.

PROGRAM NAME	
User workplace task	Program features
1.	
2.	
3.	

4.   List the User and Tasks for a Program

Select one of the program descriptions below. What specific tasks might the following users find for the program you select?

User	Tasks
Accountant	
Engineering consultant	
Business owner	
High school teacher	

Description: LollyDex is a document management tool optimized for correspondence control. Fast and easy-to-use data entry routines and powerful search capabilities. It adapts to your systems and is suitable for paper or electronic documents.
Search on any combination of author, recipient, date range, document number, document type, file reference, routing and subject, and print reports of results. Browse for-
wards or backwards through threads of replies. Track outstanding replies to issued
documents. View any electronic or scanned document using your default viewers.
Option to automatically generate document numbers. Optional password and authority level features. Share data across network. Basic address book features.

Description: Bids, Quotes, Estimator Software for small businesses. Easy to use
bids and quotes software allows you to enter the client's information and save it to a
database if you like or just print out the bid. The purpose of this software is so that
whoever accepts the office calls and is behind the PC can fill in all the customer
details and bid details, along with the estimator's name who will be handling the bid.
Estimators can just pick up their list of bids for the day with all of the client's information on it and the exact description of the job the estimator is supposed to bid. It is
not meant to give to the client. It is for internal office and estimator use so that each
estimator knows exactly what bids and what times he is supposed to be there. All data
can easily be backed up and it even has a mailing list feature for clients you have
already bid on. This makes it simple to do mass mailing to people you have already
quoted, to let them know of any relevant sales or new information you want to send
to your own in-house mailing list of prospective clients.

Description: PolyMap is a desktop mapping program that lets you use your own
data to customize the maps supplied with the program. Use the built-in spreadsheet
to enter data or paste it from other Windows applications. Alternatively, you can use
the import feature to bring in data from external spreadsheets, text, or database files.
The Map Presentation Wizard gives you a step-by-step process to customize your
map and the map's legend. Any data in a spreadsheet column may be utilized for
labeling. You may use a set of geographic layers in different formats (state and county boundaries, five-digit ZIP codes, U.S. and state highways, cities, rivers, and lakes)

to add details and create your new thematic map. For instance, state and county boundaries can be "extruded" to 3D prisms, shaded in color, "sprayed" with a dot-density distribution, embedded with individual pie and bar charts, or placed into a matrix for portfolio analysis. New custom point layer can be added to the maps. Maps and spreadsheets may be printed with any Windows-compatible printer. Maps can be exported to Bitmap format (.bmp) enhanced Metafile (.emf) and JPEG (.jpg). PolyMap send JPG Bitmaps of the maps as attachment, to any e-mail address using MAPI compatible e-mail clients like Outlook Express or Netscape Communicator. This trial version will function for only 30 days and comes with only a fraction of the maps available in the registered edition. Additional sample maps and a manual as pdf file are available from the PolyMap Website. New Functions: Database Import Wizard, Export to BMP, EMF and JPG format, Send Maps by e-mail (MAPI), Customize maps and layers, Add custom points to the map, Custom Shading, Measure Distance on map, Enhanced print preview.

PART I

The Forms of Software Documentation

CHAPTER 2

Writing to Teach—Tutorials

This chapter follows the organization of software documentation into three main forms: tutorial, procedures, and reference. These forms, as discussed in Chapter 1, correspond with recognized user behaviors, questions, and needs, and act as design tools. This chapter offers examples, guidelines, and discussion for design of teaching documents called *tutorials.*

Like other elements of a documentation project, the form of a tutorial grows from the purpose of accommodating information to the needs of users. Remember: People tend to associate this kind of a manual with novice users, beginners, people who want to learn the fundamentals of a program. But tutorials also can serve experienced or advanced users as quick, first introductions to new software. Tutorials follow principles of instruction that assume that users progress in skill and confidence with a program. With tutorials, as with the contact between teacher and learner, the document requires a design of text and graphical information that supports a highly focused lesson or *module,* rather than giving the reader options. Tutorials, even "open ended" *minimalist* tutorials, tend to restrict and focus the readers for productive learning.

This chapter covers the issue of how to select the right skills to teach and gives you an overview of some usual forms of tutorials that you will find popular among software documenters. It discusses two important ways to teach software—the *elaborative* and the minimalist approaches—and shows you how to design your teaching modules according to their principles.

How to Read This Chapter

As indicated earlier, this chapter forms a triad with the next two chapters, each dealing with a prominent form of writing for task-oriented product support documents. All documentation writers need to know how to write tutorials, and, increasingly, online and demonstration-type tutorials.

- If you are planning a project, this chapter can give you valuable information about tutorials as a part of the document set: how long they take, what resources they require. If you need to write a tutorial, you should at least skim the Discussion section and then follow the Guidelines. Then re-read the Discussion.

- If you are reading for understanding, the Discussion characterizes the main trends in tutorial design and applies those trends to ways to meet user task needs.

Examples

The example in Figure 2.1, from the *Essential TitleMotion,* tutorial won an STC Distinguished Technical Communication Award in 2001. It illustrates the advanced features of a document designed to teach software skills so that the user can perform them by memory. This chapter will show you the basics of design of this kind of documentation and also introduce you to a style of teaching that relies on principles of exploration of software.

Figure 2.2 illustrates the way technology can create meaningful interaction about software.

Guidelines

Writing tutorials well takes skill and experience with people and training situations, but by following these guidelines you can master the basic principles. When trying to teach skills, rely on your memories of your own learning and the difficulties you had with new software, and then combine your memories with observation of others. By following the guidelines in Figure 2.3, you can design tutorial documentation and be confident that your design follows the principles of task-oriented documentation.

1 Identify User Actions You Need to Support

You should rely on a thorough user analysis (Chapter 5) to help you identify what skills you need to teach. Any *actions* or *scenarios* that your user would participate in make good problems for users in tutorials. Decide on what skills you want to teach to support these scenarios. For a drawing program, for example, for artists you might want to teach the user to operate the freehand drawing tool, for draftspersons you might want to teach the line and box drawing tools. With an accounting program, you might want to teach the bookkeeper how to post a transaction to the general ledger but for managers focus on tracking features. Whatever user actions you select, each can only get accomplished if the user has mastered certain skills from the program: how to use menus, how to click the mouse, and, most importantly, how to combine program features meaningfully in the workplace.

Plan your tutorial around these user actions. For each action you will teach, list the program skills that the user needs. This list can serve as the basis of your lesson plans. Identify the commands that the user should associate with the skill and put workplace task and program skill together.

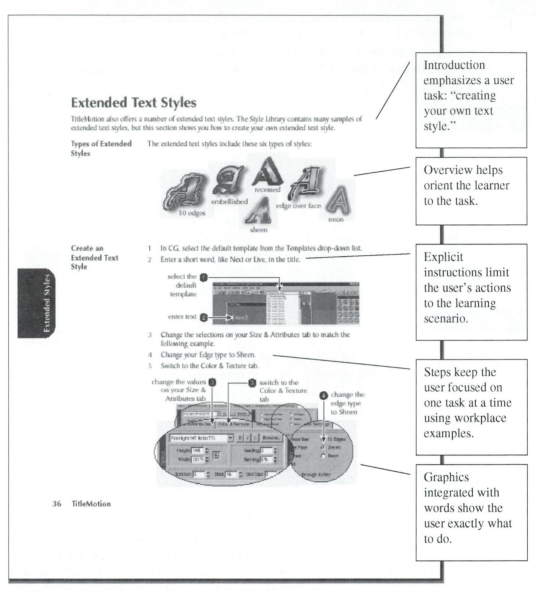

Extended Text Styles

TitleMotion also offers a number of extended text styles. The Style Library contains many samples of extended text styles, but this section shows you how to create your own extended text style.

Types of Extended Styles   The extended text styles include these six types of styles:

recessed
embellished
10 edges
edge over face
neon
sheen

Create an Extended Text Style

1   In CG, select the default template from the Templates drop-down list.
2   Enter a short word, like Next or Live, in the title.

select the default template
enter text

3   Change the selections on your Size & Attributes tab to match the following example.
4   Change your Edge type to Sheen.
5   Switch to the Color & Texture tab.

change the values on your Size & Attributes tab
switch to the Color & Texture tab
change the edge type to Sheen

36   TitleMotion

Introduction emphasizes a user task: "creating your own text style."

Overview helps orient the learner to the task.

Explicit instructions limit the user's actions to the learning scenario.

Steps keep the user focused on one task at a time using workplace examples.

Graphics integrated with words show the user exactly what to do.

FIGURE 2.1   Award-Winning Tutorial Documentation: Print
This example, from the *Essential TitleMotion* tutorial illustrates the organization of the direct instruction tutorial.

Tie Program Features to User Actions

Selecting tasks for a tutorial requires that you look over your features list and decide on which ones to treat in a tutorial. When selecting program features, you can follow these guidelines:

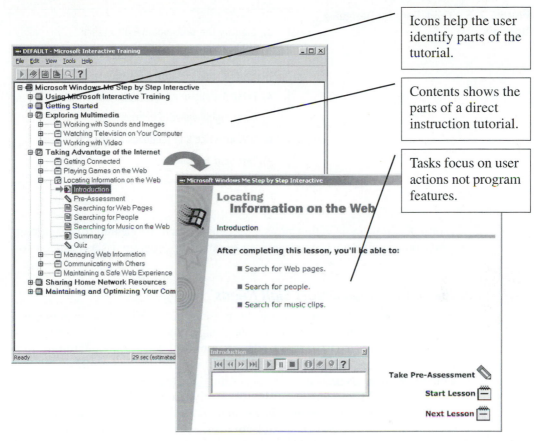

Icons help the user identify parts of the tutorial.

Contents shows the parts of a direct instruction tutorial.

Tasks focus on user actions not program features.

FIGURE 2.2   Award-Winning Tutorial Documentation: Online

This example from Microsoft *Windows ME* illustrates the way technology can create meaningful interaction about software by focusing on user tasks: searching for web pages, people, and music.

Central to job performance. Some program features will relate more directly to the user's work. Look for features like reporting or printing, or features relating to your users' communication or information transfer needs. Users may also run some other key piece of software, such as a specific operating system, that they will need to integrate with your software. Users may care more and want to learn more about certain features that they perceive as essential to problem-solving on the job.

Essential for efficient software use. Some features, like file management, security, or basic screen handling, must be taught. Teaching should acquaint users with basic concepts used in programs, such as tree structures or processing sequences.

Frequency of performance. Some features occur so frequently that you will want to teach them to your users. Which features get used hourly or daily? Also,

FIGURE 2.3   Guidelines for Designing Tutorials

1. Identify the skills you need to teach.

2. State objectives as real-world performance.

3. Choose the right type of tutorial.

4. Present skills in a logical, cumulative structure.

5. Offer highly specific instructions.

6. Give practice and feedback at each skill level.

7. Test your tutorial.

some features occur within pre-set sequences, such as opening a file, entering data, processing data, printing data, and quitting the program. Users need to know these sequences by heart.

Let the Help System Detect Skill Needs

Some help programs allow the software to present the tutorials. For example, in a drawing program, when the user moves one design object over another for the first time in the use of a piece of software, the program can detect the opportunity for a highly specific lesson. This kind of tutorial is called an *embedded tutorial.*

Embedded tutorials present a task according to some pre-determined event that triggers the tutorial program. They should clearly show choices for the user, and keep the task simple and relevant. "Next" and "Done" buttons help the user select the pace of the tutorial, and to exit at any time (illustrated in Figure 2.4 by the "close" button.) Users should clearly see how to shut down or de-select the embedded tutorial. Users going through the learning process a second time, say, after re-installation, or who for whatever reason don't see the benefit of the help system, may not need the prompting. For the user, such an intrusion may not sit well (as suggested by the humorous example in Figure 2.3,) but, on the other hand, it offers a way to deliver highly specific information at the point of user's need. Sometimes this kind of support is called an *electronic performance support system* or EPSS for short.

2   State Objectives as Real-World Performance

Write out the objectives you want your tutorial to achieve. Stating objectives in terms of performance by the user can help you plan your lessons and provide you with the outline of your tutorial project. You should state the objectives of your tutorial in your documentation plan and then again in the actual lesson or tutorial help module.

Objectives should appear as actions the user needs to accomplish and the skills that the user should learn in order to accomplish them. Often objectives sound like "In this Chapter, you will learn the following skills. . . ." Tell the user what he or she will learn from the lesson. And put the objectives in measurable terms: "This lesson will teach you to create a drawing with three colors." The example in Table 2.1 shows

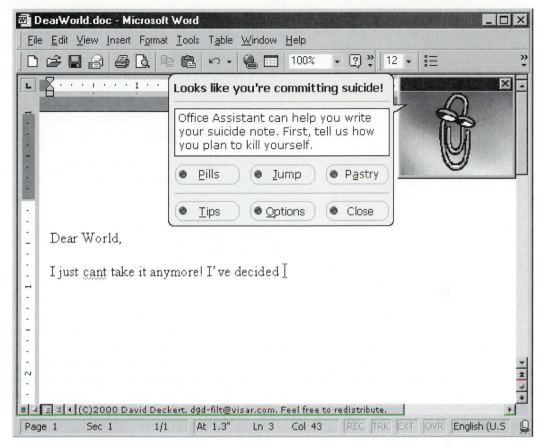

FIGURE 2.4   An Example of an Embedded Tutorial
This humorous example of an embedded tutorial shows how help can sometimes appear to intrude on the screen. Well-designed embedded help meets the user's informational needs at the appropriate time.

a summary statement that corresponds to an earlier objectives statement. See how they go in pairs?

3   Choose the Right Type of Tutorial

Tutorials come in many forms, from very brief ones to full-scale manuals, and they occur both in print and online media. I have found little consensus among writers as to terms that designate types of tutorials. Take demonstrations, for example. The example in Figure 2.1 at the beginning of this chapter comes from a 76-page tutorial guide. The descriptions below are intended to give you a brief overview of various types of tutorials. The salient features of these types of tutorial are summarized in Table 2.2: Choosing the Right Type of Tutorial.

TABLE 2.1   An Example of an Objective Statement

Document	Objective Statements
HiQ Demonstration Guide, *National Instruments Corporation,*	Overview Statement: *In this section you will use another HiQ Problem Solver, the Ordinary Differential Equation Initial Value Problem Solver (ODEIVP), and HiQ Script to perform a dynamic system analysis. Users who are familiar with differential equations will benefit most from this section.* p. 4–11
	Summary Statement: *In the final section of Module 4 you used the Ordinary Differential Equation Initial Value Problem Solver to plot the oscillation patterns of a mass-spring damper system. You also used HiQ-Script to define a forcing function to simulate an external force on the system.* p. 4–20

The Guided Tour

A guided tour presents an overview of all the program features. It focuses on the overall program capabilities and things like main screens and useful commands. The emphasis falls on introducing an overview of all the program can do, rather than just key, defining functions, as with the demonstration. Usually the tour, online or print, will follow a made-up example but provide little user interaction other than clicking for the next frame. Like other forms of tutorial, the guided tour both informs and persuades: it tells the program features, and it also helps convince the user of the usefulness of the program.

TABLE 2.2   Choosing the Right Type of Tutorial

Type	Description
The Guided Tour	Focuses on the entire program's main features and user actions Beginning and intermediate users Example: "An Overview of NetHog Plus"
The Demonstration	Focuses on specific features and user actions Beginning and intermediate users Example: "Setting up NetHog Plus"
The Quick Start	Focuses on basic features and applications Intermediate and advanced users Example: "Performing Your First Search with NetHog Plus"
The Guided Exploration	Focuses on user actions and examples Beginning and intermediate users Example: "Exploring NetHog's Sample Database"
The Instruction Manual	Useful for technically difficult software Users at all levels Example: "Learning NetHog Plus"

In form, the guided tour can occur online and in print. Print guided tours consist of a booklet or section highlighting the program's prominent features. Online, the guided tour can consist of screens and message boxes explaining the prominent and useful features of the program (see Figure 2.5). Some guided tours can include sound and animated cartoon figures superimposed on the screen that act as tour guides to the program.

The Demonstration

Design a demonstration when you want to illustrate some specific parts of a program that emphasize a real-world application of a program or an important user case. Usually you use an example of the program, often a limited version of the program (one with some features disabled). Often with demonstrations the user observes passively but can choose which features to observe. Like the other shorter forms of tutorial, the demonstration both informs and persuades.

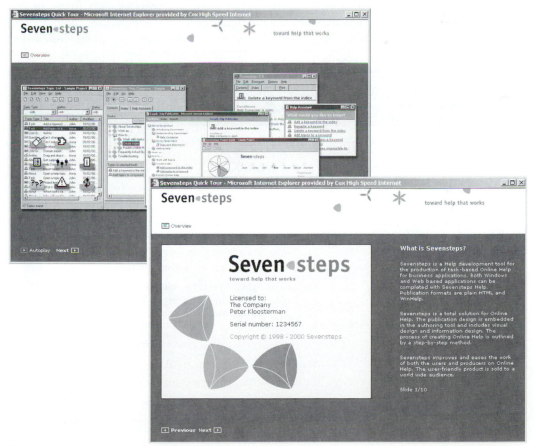

FIGURE 2.5   A Guided Tour of Sevensteps
This guided tour emphasizes the essential elements of the program.

In form, the demonstration consists of a limited version of the program and a brief print tutorial. The tutorial instructs the user in starting the program and tells the user what commands to use to perform the demonstrated procedure. Online, the demonstration shows you a procedure, allowing you to choose which procedures to see. You can link the demonstration to the help system via a button the user can press to see the command in action. This "watch while I do it for you" technique works best if you provide the user an "I've seen enough" button to allow an exit.

The demonstration in Figure 2.6 consists of an overview of one of the main user scenarios for the program—paying a bill. The screens, with example documents, appear in an overview. The user can choose which features to watch. Writing such a demonstration requires that you minimize the text and identify only key actions. However, this kind of tutorial, as well as the guided tour, allows you to highlight important workplace applications of your software.

The Quick Start

Quick-start or getting-started tutorials differ from the previous two forms in one important feature: quick-start tutorials for experienced to advanced users with *domain knowledge* who want to get going with a program, who want to explore. Quick-start

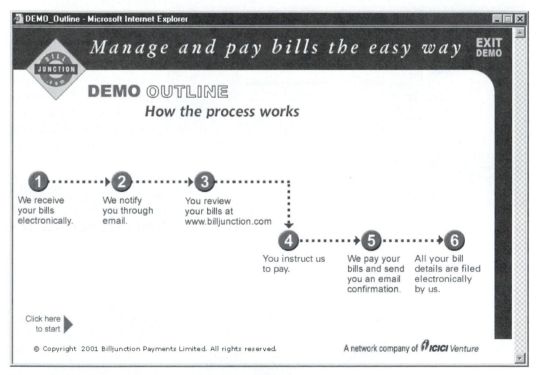

FIGURE 2.6   An Example of a Tutorial Demonstration
This demonstration walks the learner through a typical use scenario as a way of acquainting the user with the program.

tutorials involve significant user interaction with the program itself. They help users get down to work without going into complicated configuration procedures. They cover basic and advanced procedures, kind of like a survival kit for impatient users.

In form, quick-start tutorials consist of one-page or folded cards and booklets that explain how to start the program and list a sampling of commands. Often they will include a labeled main screen and tables of commands. They differ from command summary cards and quick reference cards because of the steps they include for starting the program and executing basic functions.

Unlike the two previous forms of tutorial, the quick start expects the user to interact with the program, starting the operating system, inserting the disk, starting the program, and so forth. I particularly like the way this quick-start brochure in Figure 2.7 emphasizes user actions.

The Guided Exploration

Guided explorations consist of a special, well-researched form of tutorial based on the idea that users need to control the learning experience. Like the demonstration, it may use a limited version of the program, but one some features blocked so that the user can't make serious errors. Also, these kinds of explorations contain, as you might expect, instructions for the user to "try out" commands (see Figure 2.8). Instructions like these encourage exploration of the program, but don't limit the user as to exactly what to do. Freedom from these limits helps give control over the learning experience to the user. To give the experience over to the user, the guided exploration contains little discussion. For a further discussion of guided explorations, read the section on minimalist tutorials later in this chapter.

Guided explorations usually take the form of short tutorial manuals. They may or may not provide scenarios (examples for the user to follow). They can include objectives and summaries to help give the user direction, but do not constrain the user to learning specific commands. This lack of constraint allows the user to explore functions and commands that relate specifically to his or her needs.

The tutorial encourages users to explore with the program by using statements like "Try it!" and providing examples. After the exploration event by the user the book includes a "How it works" section to reinforce the learning. By encouraging exploration, helping the user avoid errors, and keeping things brief, the manual provides a challenging learning experience.

The Instruction Manual

The instructional manual focuses on users who intend to operate a program or expect to have to learn a number of complicated commands and functions. The most traditional type of tutorial, the instruction manual consists of lessons framed by elaborate objective and summary statements. Each lesson focuses specifically on these objectives, usually tied in with specific sets of commands. This type of direct teaching also relies on the principle of accumulative learning: the idea that you have to learn one skill before you take on another, more advanced one. Like the other forms of tutorials,

Plan your video

Before you videotape that important event, make sure that you not only have the equipment you'll need, but also a plan for what activities you'll want to videotape.

1 Pack your camera bag. Make sure that you have:

- ▶ Video tapes (you may want to bring extras)

- ▶ Pencil and paper for note taking

- ▶ Extension cord (if using a power cord for the video camera)

- ▶ Extra batteries (if available)

- ▶ Tripod (if available)

2 Understand your objective. Are you trying to entertain your audience, document an important event, sell a product? What you videotape depends on your objective.

4

3 Brainstorm ideas on what activities you'll want to videotape.

- ▶ **Birthday party activities** such as: planning the party, decorating, guests arriving, playing games, lighting the candles on the cake, singing "Happy Birthday," opening gifts, and blowing out candles.

- ▶ **Wedding activities** such as: rehearsing the wedding, decorating, getting ready, guests arriving, the the arrival of the wedding party, the exchanging of vows, the wedding party entering the reception hall, and friends and relatives wishing the new couple happiness.

FIGURE 2.7   **Example of a Quick Start Guide:** *Gateway Getting Started with Your Gateway Video* :)
            Ware Package

The quick-start tutorial helps users by supplying only the beginning steps to become productive with a program.

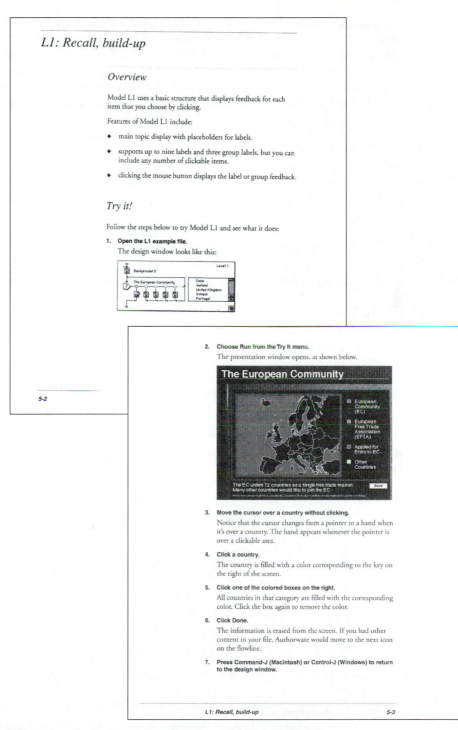

FIGURE 2.8   Example of a Guided Exploration *Authorware Models*
This guided exploration focuses on pre-set models in the software that the user can "try out."

it focuses on basic program features, at least at first, and then advanced ones. Program features take the form of lessons or modules, each about the same time length.

This kind of tutorial contains a great deal of user interaction so that the reader gets involved in and invests in the learning experience. Figure 2.2 shows this kind of interactivity and it shows the structure of events into "Pre-assessment," "Summary," and "Quiz." Often modules or lessons will contain such practice sessions and evaluations or tests to see if the user learned the material. You can find out more about direct instruction, the basic principle behind instruction manuals, by reading the section on *elaborative* tutorials later in this chapter.

In form, direct instruction tutorials take the form of a separate tutorial book or section of a book. Additionally, you may want to develop teacher's materials (overheads, lesson plans) and student's materials (worksheets, job sheets, notebooks, tests) for use in a classroom environment. The manual usually follows scenarios or presents problems for users to solve, so you may have to develop sample data sets, documents, templates, databases, or other elements the program requires for working. For instance, I once wrote a tutorial for an accounting program that used a fictitious company—ABC Lumber Co.—as an example. The learner took on the role of bookkeeper for the company.

The steps instruct the user to call up example files and demonstration *data sets* that show the program functions on data that the tutorial writer included with the program. Each chapter ends with a summary that reinforces the lesson. This kind of tutorial represents a full-blown effort to teach as much of the program as possible. It uses all the characteristics of direct instruction discussed below.

4 Present Skills in a Logical, Cumulative Structure

In Guidelines 1 and 2 you saw how you need to organize the program features in the form of instructional objectives, and how to tie these to relevant user actions in the workplace. Designing tutorials requires that you next assemble the lessons in a logical order and structure.

The most important source for your decisions about order and structure will come from the typical-use scenario, or some other user action that you think most likely resembles what most users would perform. Your tutorial will support this scenario. Examples of typical-use scenarios include a student typing a paper, a clerk calling up a record to check for payment of a bill, or a salesperson checking the computer for the availability of an inventory item. All of these scenarios or actions require different tasks or features of their respective programs. For the word processing scenario, for example, you might need these tasks: opening a document file, typing in text, editing the text, saving the file, and printing the file. For the accounting program scenario you might need these tasks: looking up a record, checking the appropriate screen for payment, closing the screen, and printing an invoice for the customer.

Some logical or cumulative structures include the following: beginning to advanced, simple to complex, generalized to specialized, input/accumulating data to output/reporting data, starting a session to ending a session, using default options to using customized options, working with text to working with graphics. See how the modules illustrated in Figure 2.10 start the user on basic tasks and then advance to more difficult tasks.

Did You Know

Learn About TitleMotion

TitleMotion Modules

TitleMotion consists of three modules: CG, Logo Compose, and FX.

- CG lets you create still, rolls, and crawls, based on blank templates or fully laid out templates. You also use CG to create the initial title for an animation.
- Logo Compose lets you clean up and add alpha to any image, and then use that image in CG.
- FX lets you add animations to any title created in CG.
- The Scrapbook is a TitleMotion feature that lets you transfer images, titles, and styles between titles and between TitleMotion modules.

About this Document

This document is designed to get you up and running with TitleMotion as quickly as possible. It won't teach you everything about TitleMotion, but it introduces you to all the major features available in TitleMotion. By the time you've looked through this document, you should be able to handle all of the basic TitleMotion tasks. From there, you can go to the online help for more detailed information about specific options.

Online Help

The online help system is designed to give you all of the information you need about every feature available in TitleMotion. If you need more details than this document provides, check the online help. You can open the online help by selecting Help→Help Topics from the menu bar.

The online help is arranged by module. It also covers all of the TitleMotion preferences and includes a complete glossary.

The TitleMotion Screen

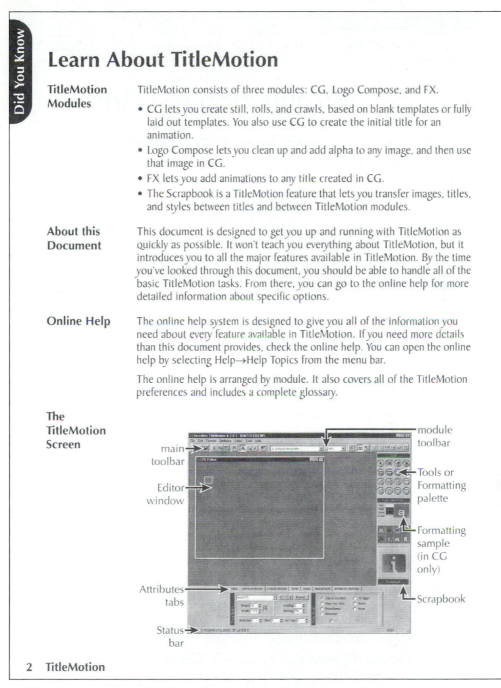

main toolbar

Editor window

Attributes tabs

Status bar

module toolbar

Tools or Formatting palette

Formatting sample (in CG only)

Scrapbook

2   TitleMotion

FIGURE 2.9   Example of an Instruction Manual
The instruction manual offers an extended example for use in learning the program features.

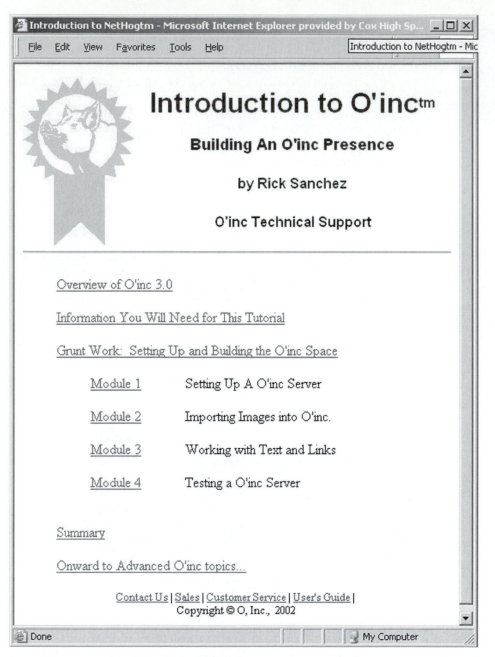

FIGURE 2.10   A Tutorial Organized from Beginning to Advanced
The structure of this tutorial follows the learning curve of the user.

Using structures of workplace tasks as organizing principles for tutorials can lead your reader into familiar territory instead of the unfamiliar. That is, the more the tutorial looks like the work your users do, the better. Figure 2.11 shows how lessons or modules can be organized using a "checklist" structure. Note also that the tutorial designer has supplied a printable version of the checklist to assist the user in actually applying the lessons.

After you have articulated the goals of the instruction and linked it to real-world task situations, you can proceed to select illustrations and write the instructions.

5   Offer Highly Specific Instructions

Your instructions or lessons should focus on a specific *scenario* or problem that the user would recognize. Exercises like these, say researchers, "can suggest typical uses

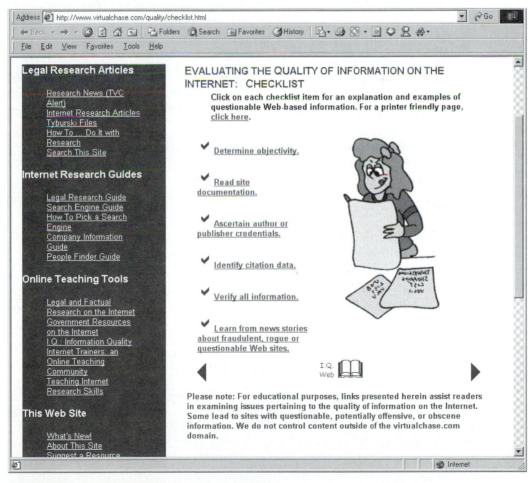

FIGURE 2.11   A Tutorial Organized by a Checklist
A checklist structure helps the user integrate the tutorial with real-world activities.

for which the software is well-suited, thus helping users to see how it could be used to advantage in their own work."[1] The example or scenario should include details, such as what data to plot, or what names and addresses to type in or look for. This way you help the user stay focused on the task. During your user analysis, look out for examples that you can use as a mock-up for particular tutorials. You may find yourself tempted to use generic instructions, such as asking your user to "Enter a name" or "Enter a number." These don't work nearly as well as "Enter 'Zachary Poole'" or "Enter '11786'." These details give a realism to your instructions that work better than phony-sounding details like "Enter 'Any Person'" or "Enter '12345'."

Often learners of software programs may feel insecure about the new program they're learning. They may think they will lose some data or they may feel anxiety because they see time spent learning software as time spent away from their job, or a general feeling of anxiousness at having so much to learn in so little time. They may feel that having to go to a class or learn a new system may make them look stupid or ignorant. Because of these anxieties surrounding learning new software, you want to do all you can to help users maintain focus. If you tell them to "type something" they may lose focus or may just get stuck trying to think of something to type. Keep them on track with specific instructions, such as the following:

- **Specific data.** Numbers, names, words, variables, formulas, search strings, customer names, client names, addresses, dates, filenames, directory names, times, printer names, printer port names, sizes, protocols, email addresses, IP addresses, phone numbers.

- **Tools.** Mouse buttons, keyboard keys and key sequences, buttons, icons, check boxes, directional arrows, hot spots, hypertext links, radio buttons, toggles, list boxes, spin boxes.

- **Screens.** Mouse selection screens, displays, panes with fields to fill in, icons, menu selections, highlighted areas, data fields.

- **Commands.** Control commands, alt commands, line commands, keyboard shortcuts, macro names, escape sequences, function keys.

Avoid distractions that could cause the learner to lose concentration. Make sure you carefully edit your screens to eliminate any extraneous images, extra files, extra menu items, other disk icons—all the stuff that can distract your user. Do what you can to not draw attention to details of page design.

6   Give Practice and Feedback at Each Skill Level

Like all good teaching, the completion of a lesson by the student should result in praise and reward. Do this in a tone of what interviewers call an "unconditional positive regard." Imagine that the user always has patience, imagination, and a pleasant disposition. State goals positively and avoid controversial jargon. Remind the user of the reward for understanding and correct performance (a new skill or job capacity), or use the conclusion of the lesson to help you establish a feeling of goodwill. The TC8215 Sectra Management System for Windows *User's Guide* establishes a feeling of goodwill in the following way: "This is the end of the Console program guided tour. Thanks for coming and we hope you enjoy the rest of your trip!" (p. 2–18).

You may use the "we" pronoun—as well as the "you"—liberally, to reinforce the performance orientation. Also, contractions add a colloquial tone. They can help a novice user relax, and they give breathing room around difficult concepts the user encounters for the first time. Again, the TC8215 Sectra Management System for Windows *User's Guide* (explaining how to create program objects manually) states: "The advantage of this method is that it lets you create objects you know about (and that's your job, right?) so you can get started managing them right away" (p. 2–8).

Imagine you can lean over the user's shoulder and point to the screen. Say things like "Notice that the text has turned gray, indicating that you have already chosen it" or "See how the icon changes from open to closed when you click on the *close* box." Let this imaginary posture of helpful teacher help you find ways to guide the reader's attention on the screen as he or she follows the steps on the page or examines a screen to see if the steps worked.

Build a Pattern of Exposition

Remember, too, to build in a pattern of exposition, whereby you repeat the following rhythm (or something like it):

1. **Give action to take.** "Select Open . . . from the file menu."
2. **Explain the result.** "The program will display an empty file."

Spend your time explaining the result, and avoid giving alternative advice, as in "You could have also used the keyboard to. . . ." Alternative advice, as an *elaboration* in software tutorials, tends to distract the user from the task. Key the user to the screen and where to look.

Sometimes you can give practice and feedback by including exercises in your lessons. These exercises, if you design them realistically, can give your user the kind of freedom to experiment that adult learners like. Remember that most software learners are adults and so their orientation is toward problem solving instead of accumulating facts. So for feedback, you might try including a quiz or question-and-answer list that reinforces the application of the lesson.

Pace the Tutorial

When you put your tutorial together, consider that you don't want to waste your user's time. Neither do you want to waste yours, given the high cost of developing tutorials. Try not to take the user away from the job for more than an hour at a time, and expect that they won't have even that much time to give. Keeping the lessons down to about ten to twelve minutes each enables the user to maintain concentration during a lesson. Also, consider that busy professionals may get called away during a training session or may only have a limited time to devote to learning each day. So give them a chance to quit during the tutorial and show them how to quit the program without losing data and having to restart later.

7   Test Your Tutorial

Your tutorial, like other documentation products, should get a thorough session in the testing lab, whether you use a fancy, well-designed lab with recording equipment and a coffee machine, or simply the user's environment (where possible). You should base your testing on the objectives of the tutorial. (Chapter 8, "Conducting Usability Tests," has information on specific ways to design your test.) But you should keep testing in mind during your planning stages and watch out for points you will want to verify through testing.

Design your test by determining whether you will test the entire tutorial, or just parts. Probably you can only test parts: besides, you get diminishing returns if you spend too much hard work testing. Design the test, also, to focus, as much as possible, on the design elements: the *cuing system,* the effectiveness of the graphics, the pace, or style of writing steps. Get feedback on the tone if you allow yourself some humor. Above all, find out if you can count on your lessons to get the user from point A of ignorance to point B of skill mastery, in the allotted time using the instruction you provided. When you can, try out the tutorial and revise it based on results. If you don't have a real user of the program at your disposal, do your best to mock-up the situation with someone of similar background as the users.

Discussion

This section will present the elements of tutorials, followed by a close examination of two philosophies of teaching: the *elaborative* approach and the *minimalist* approach. Finally, we will examine design guidelines for each approach.

Designing Tutorials

When you set out to design and write tutorial documentation, you should start with the knowledge of how tutorials work and when your particular users need them. Because not all *documentation sets* contain tutorials, you should know when to use this form of documentation, and when to apply others. This section will help you make that decision by examining some of the basic elements of tutorials.

Intention to Teach

With tutorial documentation you should intend for the user to internalize certain skills or concepts about a program and apply them in their work. You want the user not only to gain a familiarity with skills but also to remember them and perform them later from memory—a tall order. Documentation that accomplishes this operates on the teaching level of task orientation, meaning that often you must create a close relationship between the *persona* of the writer and the reader.

But most of all, you try to limit the awareness of the user, so that he or she can focus only on the problems discussed in the lessons. In other forms of documentation, you try to expand the user's awareness—of options, alternatives, short cuts—but in tutorials you limit the user's awareness to one way, one option, one problem, one activity. This takes a great deal of control and structuring of the user's interaction with the material.

Selectivity in Choosing Material

Clearly you can not *teach* all the functions of a software program. To do so would take many books, given the fact that tutorial documentation takes up more space, usually, than documentation at other levels of task orientation. This need for selectivity means that you must know your users very well. You should know which essential tasks need learning and which don't. Usually you can only afford to teach the essential ones: the others, well, the user has to get them somewhere else.

To select material wisely, you should first do a thorough user analysis. The user analysis first narrows down the field of all users to potential users, then to user types, then to usual scenarios of use of the program, and finally to the *scenario.* The typical-use scenario represents the fundamental tasks of them all—the ones that probably would get performed most often. If you have done your user analysis well, then you can design a tutorial around these tasks.

Tutorial Users Need Special Care

A number of studies of tutorial users show us that they often require special considerations, mostly because, as adults, they have special learning styles. For example, most adult learners are oriented toward goals: They want to know why they have to learn something and what good it will do them. Also, they like to have control of their learning. Adults like to think of themselves as self-motivated and self-assured, not as ignorant bumblers. They do not like to make mistakes and often do not realize the value of making mistakes in the learning process. Stephen Lieb, Senior Technical Writer and Planner, Arizona Department of Health Services notes that in the case of adults, "Typical motivations include a requirement for competence or licensing, an expected (or realized) promotion, job enrichment, a need to maintain old skills or learn new ones, a need to adapt to job changes, or the need to learn in order to comply with company directives."[2] Many of these factors motivate software learning.

The more we know about these styles and motivations, the better we can design effective documentation for them. This presents design challenges for the software documenter, because to build task analysis into tutorial documentation means that you have to accommodate the learning styles of a specific population. The designer needs to know how to build tutorial modules that avoid public display of a user's mistakes, limit the lesson times, give positive feedback and reinforcement, and also imbue a sense of self-direction in the steps. You can accomplish this by studying carefully how you yourself learn programs, and how others do, and by applying the principles of task analysis to the documentation situation.

The first step in your study of tutorial design begins with an awareness of the two trends in tutorial design that have grown in the United States during the last decade: the elaborative approach and the minimalist approach. In the area of document design, these two approaches fall at opposite ends of a spectrum of information design ideas. In many ways, document design is a matter of determining more or less information in a document, and these two approaches represent major trends based on either of those poles. Each of the two approaches described below represents different philosophies of learning. The discussion attempts to bring out the task-oriented characteristics of each so you can make informed design decisions about the approach that works best for your situation.

The Elaborative Approach

Research supporting the elaborative approach answers "yes" to the question "Does elaboration improve retention of skills in software manuals?" Elaboration includes summaries, explanations, examples, articulations of goals and objectives. Elaboration also includes elements of good storytelling, the ability to describe a scenario carefully, and the ability to pace, in measured steps, the user's progress through highly technical material and to make it stick.

The elaborative approach responds to the needs of the new-to-computers user: not the engineer who needed referential documentation, but the person from a non-computer background who needs a highly structured, skill-oriented approach. Indeed, the skill orientation of the elaborative approach should make the designer pay attention to its principles. Task orientation, the emphasis taken in this book, highly values any structure that assists the designer in building performance elements into documentation. We have something to learn from the elaborative approach.

The elaborative approach borrows elements of instructional principles from the field of instructional design, as a way of approaching the problem of teaching foreign, often abstract and highly technical material. Other researchers have studied the effectiveness of the *direct instructional* or *elaborative* approach to manual design and have also discovered how computer manual users learn.

Among the foremost researchers of elaborations in tutorials, Davida Charney and Lynn Redder[3] have studied the effects of elaborative elements in software manuals and found that while sometimes summaries and overviews distract the user from focusing on information, these elaborative elements helped them *apply* their tasks to real-world situations. Elaborative elements help the readers see how the program could help them perform more efficiently in their jobs. They also found that people learn skills in complex ways. Some characteristics of their learning of skills, such as how they understand procedures versus how they understand concepts, makes summaries useful at certain times and not as useful at others.

Elaboration serves your purpose in tutorial documentation when you have abstract concepts to teach and the user is encountering a basic idea for the first time. Elaboration helps users learn to apply certain functions of the program. On the other hand, when you must concentrate on teaching procedures—steps for performance—elaborations in the form of examples work best. In particular, research seems to indicate that when you use elaborations, lots of examples, tables of commands, and so forth, you should use them in conjunction with accurately designed steps.

Finally, you should always consider using the elaborative approach with novice users who know little about computers or your program. These users have a much more difficult time experimenting and need more guidance than more advanced users. Besides, they need specific guidance in applying tasks; they may lack the experience to make the connections themselves.

The design of the elaborative manual follows the traditional principles of lesson design:

1. Instruction results in articulated skills.
2. Skills transfer capability to real-world performance.
3. Steps should present skills in a logical, cumulative structure.

4. Highly specific instructions work best.
5. Give practice and feedback at each skill level.
6. Master one skill before going on to the next.

The Minimalist Approach

We noted earlier that the minimalist structure takes what some see as a *realistic* view of human behavior. In the research that supports this approach, we find this realistic view reflected in the kinds of sobering observations made about user behavior. Minimalist principles assert that people learn on a concrete plane. In this approach, less means more: out go the introductions and the reviews. Let's explore this approach more closely to see how minimalist ideas can contribute to the design of a task-oriented manual.

Observations of Software Users

Researcher John Carroll[4] explored the ways people learn software programs. He makes the following observations about user behavior:

Users jump the gun. From the work Carroll and others have done, it appears that users of computer manuals like to get started right away with a program and will resist reading information designed to introduce or orient. They want to see results from a new program, and will not read the manual first. They will use the program first.

Users will skip information. Users will rarely read the introduction to a manual. Carroll relates an interesting anecdote in which a researcher observed that a user flipped quickly through the first pages of a manual. The user decided that part could be skipped "because it's just information." Such a casual approach seems incongruous with life in the "information age." But on the other hand, we want relevant information and have learned to sort useful information from what doesn't seem immediately relevant. So in that case, the casualness of the user makes sense.

Users like to lead. Users like to create their own perspectives on their training. Researchers found that when you ask learners of a word processing program, for example, to type whatever sample document they want, they may not always pick the job-related option of a memo. Instead, he found that many computer learners would prefer to write a letter to their mom. Users, adult learners most of the time, like to take charge of situations, they like the control and don't like manipulative instructional strategies.

The Principles of Minimalist Design

These and other observations about how users react to traditional learning materials makes common sense, and they present design challenges to software documenters. Rising to the occasion, Carroll has devised the minimalist approach, which is very popular among software documenters.

The minimalist manual teaches by following four basic principles:

1. Choose an action-oriented approach
2. Anchor the tool in the task domain (workplace context)
3. Support error recognition and recovery

4. Support reading to do, study, and locate

The psychology dictates that these three principles will help the user focus better on information, allowing him or her to try out the program and get out of trouble when needed.

1. CHOOSE AN ACTION-ORIENTED APPROACH. The need to focus on real tasks and activities may seem obvious to you. But the reason for that focus comes partly from the observation that users prefer to do something with a piece of software rather than learn about it. If they perceive that the tutorial focuses them on the system of teaching, the highly structured nature of the traditional elaboration, they often will try to subvert the tutorial. They want to type the letter to mom or supply their own example for the tutorial. For this reason you should provide immediate opportunities to act, and you should encourage exploration.

People have a natural human tendency to want to try things out, so software use is unpredictable. Computer users, especially those who have to take time away from their jobs to learn a new program, like to go their own way. The minimalist model capitalizes on this explorative and unpredictable impulse. Minimalist tutorials, instead of having a practice session at the end of a lesson, encourage practice as the main way of learning all the way along. They suggest that the user "try it out." People often try out products before they read the manual anyway. "As a last resort," they say, "read the instructions." They want to know what's inside the box, what they can do with their new toy or tool, what happens if they press *this* key. To the extent that your users would feel this natural curiosity, this drive to try out the computer program, you should consider encouraging exploration with program features.

But the documenter must make sure that the exploration leads in the right direction. Real-world tasks can guide the user in this effort. The user analysis proves invaluable here, to the extent that it includes descriptions of what users really do in the workplace from a non-automated perspective. The designer of tutorials should study the work which the program will eventually support. The tasks that make up this work also make up the goals, the direction, provided by guided exploration.

2. ANCHOR THE TOOL IN THE TASK DOMAIN. Software is not an end in and of itself, but a tool to accomplish workplace tasks. This fundamental point guides you toward identifying tasks that come not from the software itself (operations performed by manipulating the interface) but from the user's workplace. In Chapter 1 (Figure 1.6) we examined some of the kinds of actions users associate with software use. Some of these are also indicated in Figure 2.7, where you can easily see that users are not as interested in using their camera software as they are in taking pictures of little kids at a birthday party or guests at a wedding. According to minimalism, "Users will be able to recognize an activity as genuine only to the extent that they have had adequate prior experience in the task domain to underwrite such a judgment."[5] By task domain we mean the workplace situation or other areas of expertise. Therefore centering instruction around real-world activities works best for software instruction.

3. SUPPORT ERROR RECOGNITION AND RECOVERY. It should come as no secret to you that we learn by making mistakes. The progress from ignorance and ineptitude to awareness and skill necessarily seems to involve trying out a solution, failing, and trying again. This happens no matter how carefully we plan our actions or choose a possible solution. The direct, elaborative approach to teaching accommodates this tendency to make mistakes by minimizing it.

Directly instructional materials do not allow us to make mistakes, but instead they carefully guide the learner around the mistakes to the desired goal. On the other hand, the indirect method used in minimalist manuals, by following the strategy of exploration, leads to mistakes, almost encourages them. For many learners, exploration leads to a delightful serendipity—a learning of unexpected things. Learning like this sticks with the student. It also leads to mistakes: a necessary part of exploring.

Why is this so? The reason is that making mistakes is a key part of problem solving. Indeed when you reflect on it, making mistakes is a kind of "thinking in action" that works something like this. You take an action and then learn the results. You make a direct connection between action and learning in this way. So then you act again, but this time it's based on your experience *plus* the new experience you gained from your results. And whether you made a mistake or not, results are results and knowledge is knowledge, right? In this way, interactive thinking—thinking that's based on action—leads reinforces learning. The trick is, as Carroll puts it, to make sure that the information you get from your mistakes can easily feed back into your learning. Unfortunately most error messages, such as the one illustrated in Figure 2.12, generated by computer programs are insufficient in explaining what went wrong in terms of the user's current activity.

If you design minimalist approaches to information, you need to not only support error recognition, but make it easy for the user to get out of trouble. Study the user, and learn where mistakes can and probably will occur. Whatever the cause of errors, in writing the minimalist tutorial, you should find out the kind of errors a user most likely will make (or which ones the procedure may lead to) and include information for recovering from mistakes. Turn the user loose, but give the steps to recover.

Carroll calls this technique the "training wheels" technique. On your first bike, the training wheels allowed you to take off down the street, but they caught you in case of a mistake. Usually you can catch and avert potential errors simply by including a statement like this: "If you make a mistake typing, use the backspace key." You can also give reassurance: "You can always restart the system without damaging the data."

4. SUPPORT READING TO DO, STUDY, AND LOCATE. As far as the *minimalist* designer of tutorials cares, the elaborative manual resembles the long-haired military recruit

FIGURE 2.12   A Circular Error Message
This error message doesn't promote learning because it's so confusing.

TABLE 2.3   Comparing the Elaborative and Minimalist Approaches

Criterion	Elaborative	Minimalist
Uses	Programs with highly abstract concepts, complicated procedures, large systems	Getting started booklets, guided tours, demos, programs with intuitive interfaces, programs requiring creativity by the user
Advantages	Good for users who like structure, first-time users, traditional	Cuts writing time, document length, interesting
Disadvantages	Limits documents to one or two scenarios, boring	May frustrate first-time user, may backfire, increases testing time

getting a haircut at the induction station. The introductions, overviews, illustrative examples, statements of objectives, double-checks, exercises, and practice sessions get swept out like curls on the barber's floor. As far as language goes, the minimalist manual sounds lean and mean. But it may gain in brevity. Some have suggested that all introductions go, because the user doesn't read them anyway. A minimalist manual may have as few as three pages in a chapter. This economy of language accommodates the impatient user.

The reason for this economy lies in the observation that users read to locate necessary information rather than from front to back like you would a novel. So you should indicate first what the user should do (using a simple command). Then, when needed you can present the explanation. Notice that this practice inverts the usual "explanation followed by command" structure found in direct instruction.

An example from a manual we examined earlier in this chapter bears this principle out. The writers of the *AuthorWare* manual wanted users to try their models first, so they organized the chapters of their manual to support this.

Notice in Figure 2.13 that the "How the model works" *follows* the command to "Try it!" In this way the user has the experience with the model and can better understand the explanation that follows.

Depending on your user's needs and learning preferences, you may decide to use either an elaborative or minimalist tutorial, or to create a hybrid of the two. Table 2.3 lists some of the issues you might consider in such a decision.

Glossary

actions: tasks that require a combination of various menu functions and program features to accomplish, but that grow out of the user's actual work environment. Actions associated with a fitness tracking program would be getting into shape, controlling one's diet, deciding on an exercise program. Actions arise out of the user's workplace or activity context and consist of one or more *operations.*

cuing system: refers to the pattern you establish of formatting or other noticeable change to signal a specific type of information. Usually you use boldface, italics, or all caps to cue items such as steps, numbers, commands, menu items, and so forth. In this book, for example, glossary entries are cued with boldface italics *(like this)*. When you see a word in boldface italics you know you can look it up in the glossary.

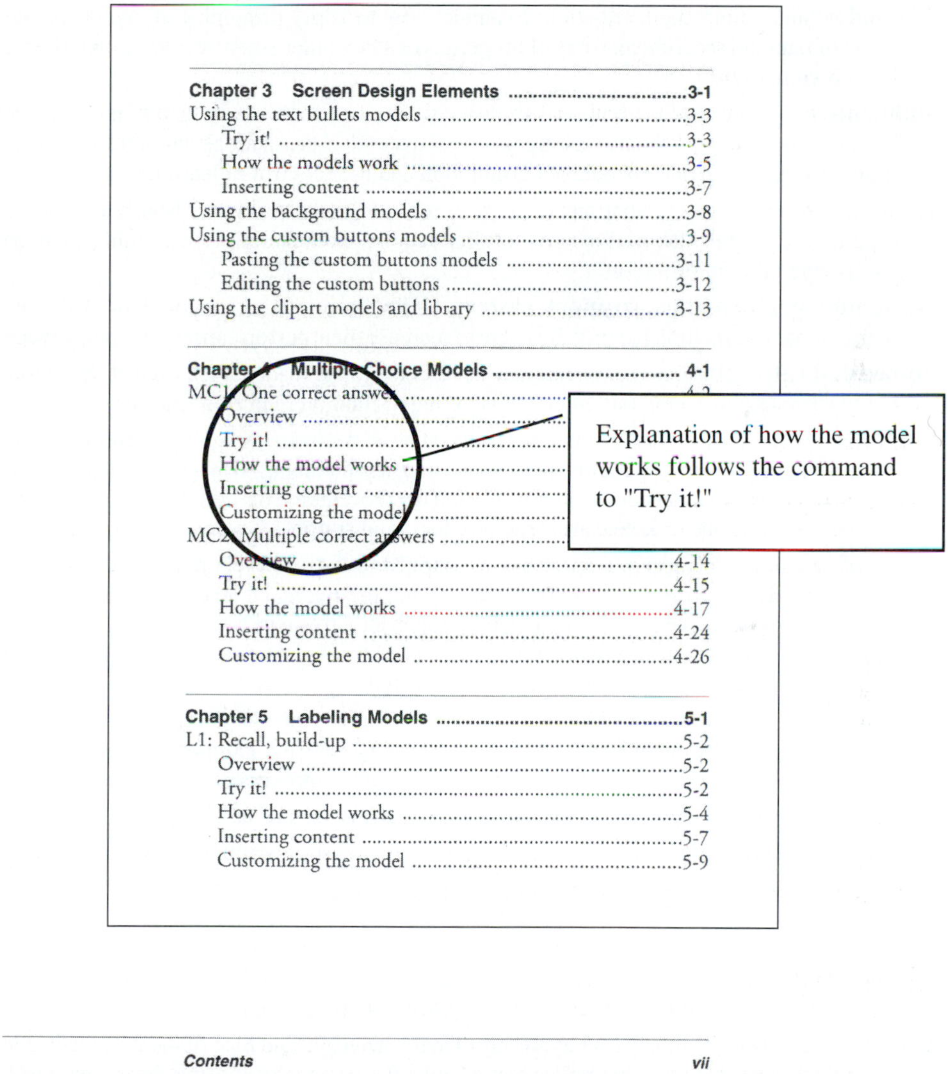

Explanation of how the model works follows the command to "Try it!"

Contents   *vii*

FIGURE 2.13   Putting Explanations *After* the Command
Software features can be reinforced after the user has had a chance to explore the program.

data sets: Examples to be used in the learning of a program. For example: author, title, and publisher information would form the data set for learning a library cataloging system.

direct instruction: an approach very similar to the elaborative approach in teaching. The direct instruction approach determines the knowledge the user needs and then designs lessons that focus on that knowledge. It contrasts with teaching approaches focusing on activities and user experiences.

documentation sets: groups of printed or online manuals and help that a software company provides to software users. Documentation sets usually include: a getting started booklet, installation guide, user's guide, reference manual, and a maintenance manual.

domain knowledge: knowledge specific to the user's workplace, professional field, organization, and employment situation. Example: how to apply principles of chemistry, elements of banking security, analysis of properties of steel under stress would be classified as domain knowledge.

elaborative: a more-or-less traditional approach to teaching software skills relying on a strict focus on a mocked-up scenario and tight control over the user's actions. It contrasts with the minimalist approach that encourages exploration and user control of learning.

elaboration: refers to explanations of steps in procedures or tutorials. Usually one or two sentences in length, elaborations give further details, explanations of why things happen, results, and other information.

electronic performance support system (EPSS): an online method of delivering instructional information to software users to increase their performance with the software.

embedded tutorial: a kind of tutorial that is presented to the reader at the time of need based on the user clicking a help button or encountering a situation requiring learning.

guidance level: a type of documentation designed to lead the user through a procedure one step at a time from a designated starting place (such as a certain menu) to an ending state (such as a printed report). Guidance level documentation (or procedures) defines the task for the user, but does not teach the task. See also **teaching level** and **support level.**

levels of support: categories of information supplied to users. Levels relate to the *teaching* level (tutorial), the *guidance* level (procedures) and the *reference* level (reference). Levels differ in terms of purpose: to teach, to walk through step-by-step, and to provide data. They also differ in the relationship of the writer to the user: from very close and controlling (with the teaching level), to distant and business-like (with the reference level).

minimalist: an approach to teaching software skills that relies on encouraging exploration and giving control of the learning to the user. It contrasts with the *elaborative* approach that emphasizes a focus on a mocked-up scenario and a tight control over the user's actions.

module: a unit of instruction in a tutorial document. Contains a lesson and is often self-contained.

notational conventions: conventions relating to how terms, commands, menus, and other interface elements appear in a manual. For example, often manuals will use italics (as in *dir, copy*) as the notational convention for commands.

operations: in describing software work, operations refer to units of activity usually defined by menu items, screens, or panes. "View a ruler," "Select a table," "Save a file" are operations that make up the feature set of a word processing program.

persona: the character of the writer as portrayed in the language and tone of the documentation. In teaching documentation the writer may assume the persona of a counselor or teacher. In guidance documentation the writer may assume the persona of a colleague. The writer often does not assume a persona in reference documentation, depending on it does on the orderly presentation of data more than a relationship between the writer and the reader.

scenario: a story or narrative describing the kinds of *actions* a user would undergo in using a program. Example: Sloane would open the word processor, open the daily report, edit it for new transaction, save the daily report, and print the daily report. These events make up a scenario.

reference level: a type of documentation intended to provide the user with a piece of information needed to perform a task. Reference documentation does not define the task for the user, but provides the necessary data the user needs to complete a task. See **guidance level** and **teaching level.**

teaching level: a type of documentation intended to instill a knowledge of how to use a program feature in the memory of the user. Teaching level documentation (tutorials) aims to enable the user to perform a task from memory. See **guidance level** and **support level.**

typical-use scenario: a description of the most usual task or tasks that a user would perform with a program. It often forms the core of a tutorial project. For example: a typical-use scenario for a word processing program would entail opening a file, typing, formatting, saving, and closing the file.

tags: tags refers to words or phrases inserted into a computer program that relate to specific help topics. When the user calls for help at a point within the program, the program reads the tag and presents the appropriate topic to the user as a help screen.

☑ Checklist

Use the following checklist as a way to evaluate your tutorial design. Depending on the kind of tutorial you have and your users, some of these items may not apply.

Tutorial Design Checklist
Identifying Skills to Teach

- ❏ Do the tasks you wish to teach relate closely to the users' critical job tasks?
- ❏ Do the tasks you wish to teach relate to effective use of the program?
- ❏ Do you have the option of letting the context sensitive help system detect skill needs?

Identifying Objectives

- ❏ Do you state the teaching objectives in terms of real-world performance?

Choosing the Right Type of Tutorial

Which of the following types best fits your users needs for efficient and effective software use?
- ❏ The Guided Tour
- ❏ The Demonstration
- ❏ Quick Start
- ❏ The Guided Exploration
- ❏ The Instruction Manual

Presenting Skills in a Logical, Cumulative Structure

Which of the following orders best suits your users?
- ❏ Beginning to advanced
- ❏ Starting to using to ending a session
- ❏ Using defaults to using custom options
- ❏ Working with [topic 1 of 2] to working with [topic 2 of 2]

Specificity of Instructions

Which of the following specific details do you intend to include in your tutorial exercises?
- ❏ Screens
- ❏ Specific tools
- ❏ Commands

Practice and Feedback

- ❏ Do you give practice and feedback, where appropriate, at each skill level?
- ❏ Have you paced the tutorial to match the users' concentration level and work requirements?

Testing

- ❏ Have you chosen a test site (lab, users' environment)?
 Have you decided on the most relevant test points for your tutorial?
  - ❏ test the cuing system
  - ❏ test for suitability of details (scenario, tools, screens, commands)
  - ❏ test the pace
  - ❏ test for users' familiarity with the form of the tutorial

Elaborative Tutorials

If you choose to design an elaborative tutorial, does it follow the principles of this type of teaching?
- ❏ Instruction that results in articulated skills
- ❏ Skills that transfer capability to real-world performance
- ❏ Steps presented in a logical, cumulative structure
- ❏ Highly specific instructions
- ❏ Practice and feedback at each skill level

Minimalist Tutorials

If you choose to design a minimalist tutorial, does it follow the principles of this type of teaching?
- ❏ Focus on real tasks and activities
- ❏ Encourage exploration
- ❏ Slash the verbage
- ❏ Support error recovery

Practice/Problem Solving

1. Analyze a Tutorial

You work for a medical office that wants to buying a new word processor for use by secretaries, doctors, and nurses at the office. The committee working on the choice has a number of problems to face in choosing just the right system, including the problem of training. How will they learn the new system?

The committee has turned to you, the resident expert on training, for help. They would like you to analyze two word processing packages (you pick which two) in terms of their training. Analyze the tutorial material accompanying both programs. Compare the differences between the tutorials you find, and recommend the one that you think will provide the least difficulty of learning. Remember: Justify your choice in terms of the tutorial, not the inherent ease of use of the program.

2. Analyze a Program Operations List

Examine the following program operation's list and identify three elements for each operation: 1) importance to job performance, 2) importance to efficient software use, and 3) frequency of performance. Use the grid below to analyze the operations list.

Operation	Importance to Job Performance	Importance to Software Efficiency	Frequency of Performance
Create new file	3	2	1
Open File	3	3	1
Save File			
Update record			
Import data from a spreadsheet	2	1	3
Add new record			
Find existing record			
Scroll through records			
Exit update			
Change record			
Print all records			
Print individual record			
Delete record			
Sort records			
Pack records			
List records			
Analyze records			
Set up customized search macros			
Customize *StampView* screen			
Exit *StampView*			

- Put a "1" for highly important to job performance, a "2" for useful but not critical, and a "3" for not important to job performance.
- Put a "1" for highly important to efficient software use, a "2" for useful but not critical to efficient software use, and an "3" for not important to efficient software use.
- Put a "1" for used very frequently, a "2" for used frequently, and an "3" for used infrequently.

Use your analysis to reorganize the list of operations from most likely candidates for a tutorial to least likely candidates for a tutorial. In a brief report, describe the contents of the tutorial you would create for the program and justify your design. What operations will you include, and which would you leave out? What purpose does your tutorial fulfill? What does it offer users that makes it worth their while to use it?

> Program: *StampView*—a program to record, sort, and maintain large stamp collections for professional collectors and stamp store owners.

> User: A manager/owner of a hobby store specializing in stamps and rare baseball cards. The owner uses the program to record new trades and purchases, to find specific stamps for customers, and to calculate the value of the total collection for tax purposes.

3. Analyze Elaborate versus Minimal Methods

Imagine that you work for a publications department in a software development organization called Software Associates. They have come out with a new line of intermediate user software and they want to do some predevelopment thinking on how to handle training for the new programs. To this end, they have asked you, as a person who would possibly develop the materials, to do some thinking about one of the programs, and tell the committee what you think. Should they go elaborate or minimal?

   Follow these directions to prepare your thinking, then put together a brief recommendation report based on your findings.

1. Identify three and *only* three job-critical operations that you would support in tutorial for *one* of the following programs:

   - A modem/fax/voice-mail program called ModemMaster that operates on a business PC server. User: Traveling sales associates with laptops.
   - A dialup tracking program for a national chain of rental trucks called Road-Warrior. User: Franchise customer representatives at remote offices.
   - A program to manage timed and scored athletic events called RodeoBoss. User: Officials at collegiate and professional rodeo events.

2. Pick one of the programs and users and briefly write down your justification of your choice of tasks.

3. Then brainstorm how you would develop the tasks into a tutorial using both the elaborative approach and the minimalist approach. You may have to make up or imagine some of the details of the programs, their users, and the environments in which you want to make them effective.

You can use the worksheet provided to record your ideas.

Elaborate vs. Minimalist Analysis Worksheet

Program name:

Three most job-critical operations:

Justification for picking these operations:

Elaborate Method (intro. and lessons with mock data)

List the *program skills* the user will need to know to perform these operations.

List the *objectives* of the lessons.

Evaluate the elaborate approach (for this case).

Strengths: Weaknesses:

Minimalist (hands-on exploration)

List the skill objectives the user should attain.

List the points the tutorial would explore:

Evaluate the minimalist approach (for this case).

Strengths: Weaknesses:

Recommendation

Which treatment (elaborate or minimal) do you think would work best in this case?

---

4. Revise the Objectives Statements

The following objective statement contains all the information required to introduce a tutorial lesson for the program PhotoBase, used by scientific users to record and analyze photographs of museum specimens. Rewrite it using a user orientation, emphasizing the terms the user should know by the end of the lesson, and the commands the user will master. Make whatever format and content revisions (you may have to make up some details) that you need in order to make an effective overview.

The program requires familiarity with registration, threshold, stability points, quartertone calibration, and dither tables. This chapter presents several methods of data input for developing a photobase database. Data development (mostly digitizing) represents between 75% and 95% of the time spent using photobase. The first exercise will be focused on digitizing. For maximum digitization accuracy several digitizing methods discussed in this chapter should be used. The sub-program *ddriver* must be mastered. The commands used in digitizing are: PB.STAB, PB.QUART, PB.IMPORT, PB.EXPORT, PB.VECTsORSET, and the combination of P.DOT and P.LASER.

5. Write Practice Problems for Users

Often you will have to describe problems that users would have to overcome as scenarios in tutorials. For example: with an accounting software program, the users would have to delete a transaction to a ledger after discovering a mistake, or a manager would have to extend an existing database by adding new entries. Each of these user tasks requires knowledge of certain program tasks that you can provide from your task list. Your skill is in writing the task in a way that reflects your analysis of the user.

Choose one of the following users and write a description of a task or problem suitable for the introduction to a tutorial. Then describe what software functions you would need to design the tutorial.

A business traveler in a hotel room . . .GlobCon connection software

A sales accountant . . . . .Client database software

A clerk in a Pack 'n' Mail Outlet . . . . .Client billing software at the point of sale

A city planner . . . . .Urban resources allocation tracking software

CHAPTER 3

Writing to Guide—Procedures

This chapter follows the organization of software documentation into three main levels, called *levels of support:* teaching, guidance, and reference. These levels tie in with recognized user behaviors, questions, and needs, and act as design tools. The chapter offers examples, guidelines, and discussion for design of guidance level documentation.

Guidance information, also known as step-by-step instructions, or how-to instructions, or the best-known term, *procedures,* makes up the heart of all task-oriented documentation systems. Much of the documentation you write will consist of procedures. Guidance documentation gets its name from the characteristic way that procedures guide the user from step to step through the task. All procedures share the characteristic of guiding the user, as opposed to teaching a task by memory (tutorial) or supporting a user-defined task (reference). *Guidance* means that the user temporarily forfeits a certain amount of control to the manual or help system in order to get help in performing a discrete task. Then he or she resumes control again, possibly forgetting the actual steps, as one might do when following a map to a hotel in a strange city.

Such procedures consist of a mix of explanations and steps, as you will see. That is, procedures consist of how-to-do-it explanations, but they also require how-it-works and why-it-works overviews. Your job as a designer of procedures is to balance these elements to meet users' informational needs and to make them efficient and effective software users in their workplace.

This chapter covers formats you will find among manuals and online help: standard, prose, parallel, and context sensitive. The chapter then discusses the elements of a procedure, breaking it down so that you can see how to combine information in ways to offer your user the maximum in usability and efficiency of use.

How to Read This Chapter

This chapter forms a trio with Chapter 2 and Chapter 3, each dealing with a form of writing for task-oriented product support documents.

- The Procedures Checklist and the Procedure Test Form in Chapter 8, "Conducting Usability Tests," can assist the project-oriented reader in managing the checking and testing of procedures.

- Those reading with a project in mind might skim the Guidelines first, but for the most part all readers should read the Discussion section (containing an analysis of the parts of a procedure) before reading the Guidelines.

Example

Most task-oriented documentation takes the form of procedures, also known as step-by-step or how-to documentation. Procedures work in practically all media and fall naturally into a chronological order. The procedure in Figure 3.1 follows this order, as indicated by the step numbers. It also indicates a number of other elements that you must carefully design to maximize the user's efficiency and effectiveness in the workplace.

Guidelines

1   Relate the Task to Meaningful Workplace Activities

A procedure is a step-by-step series of commands for accomplishing a meaningful operation with a software program. Figure 3.1 and Figure 3.2 both show examples of single procedures. But what makes a procedure meaningful does not necessarily reside in the operation itself, because users don't use the draw program documented in Figure 3.1 just to "add cues," nor do they use the program documented in Figure 3.2 to "borrow and item." They use these programs to do other workplace actions: to create multi-media presentations for an advertising firm or to manage the computer resources in a company. The "meaning" or meaningfulness of a procedure comes from its application to work.

The goal then of writing procedures is to see them as part of larger activities, as part of the activity/action/operations model that we examined in Chapter 1. Procedures occur at the "operations" level of that model. Your job as a writer is to clarify how they fit into the larger picture of actions and activities in the user's workplace. They act as building blocks for these larger actions. Thus the "skill" in a procedure lies not in the procedure itself, but in the tacit, understood knowledge that it evokes in the user.

The knowledge that the user brings to the procedure comes from the user's context, as illustrated in Figure 3.4. That context is composed of a web of different resources. First, the reasons and goals the user brings to the task make up a context of meaning. The photographer, the advertising layout expert, the chemical engineer, the manager, the writer bring their work aims and reasons to the operation in the form of workplace projects. Such projects involve other professionals and workers, materials, data collected from surveys and web sites, and other software and hardware tools. They use the program, your procedures, and then they communicate about it to others. Their awareness of this larger context of their work constitutes their vision. It drives and makes meaningful such limited operations as opening a file, adding cues to a file, searching for numbers, configuring

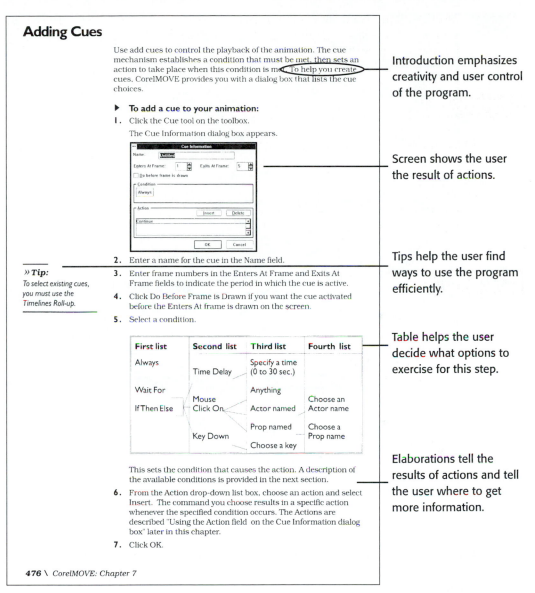

Adding Cues

Use add cues to control the playback of the animation. The cue mechanism establishes a condition that must be met, then sets an action to take place when this condition is met. To help you create cues, CorelMOVE provides you with a dialog box that lists the cue choices.

▶ **To add a cue to your animation:**

1. Click the Cue tool on the toolbox.

   The Cue Information dialog box appears.

2. Enter a name for the cue in the Name field.

» *Tip:*
To select existing cues, you must use the Timelines Roll-up.

3. Enter frame numbers in the Enters At Frame and Exits At Frame fields to indicate the period in which the cue is active.

4. Click Do Before Frame is Drawn if you want the cue activated before the Enters At frame is drawn on the screen.

5. Select a condition.

First list	Second list	Third list	Fourth list
Always		Specify a time	
	Time Delay	(0 to 30 sec.)	
Wait For		Anything	
	Mouse		Choose an
If Then Else	Click On	Actor named	Actor name
		Prop named	Choose a
	Key Down		Prop name
		Choose a key	

This sets the condition that causes the action. A description of the available conditions is provided in the next section.

6. From the Action drop-down list box, choose an action and select Insert. The command you choose results in a specific action whenever the specified condition occurs. The Actions are described "Using the Action field on the Cue Information dialog box" later in this chapter.

7. Click OK.

476 \ *CorelMOVE: Chapter 7*

Introduction emphasizes creativity and user control of the program.

Screen shows the user the result of actions.

Tips help the user find ways to use the program efficiently.

Table helps the user decide what options to exercise for this step.

Elaborations tell the results of actions and tell the user where to get more information.

FIGURE 3.1   Example of an Effective Procedure
This procedure from the CorelDRAW *User's Manual* indicates the many kinds of information you need to provide in an effective procedure.

passwords and usernames, and all the otherwise directionless operations of computer software.

These users also solve problems in their work by applying the software tools you provide, and they do so in a certain period of time, hoping not to make too many mistakes and hoping to use the operation meaningfully in conjunction with other procedures. So when you define an operation with a software problem, keep in mind these larger actions and do your best to situate the operation within them.

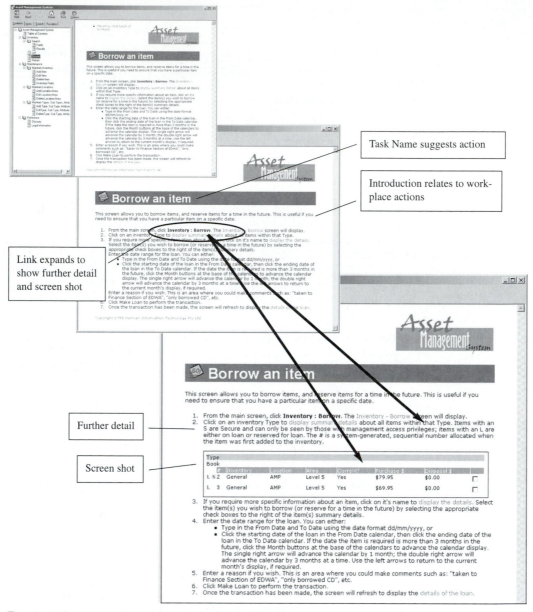

FIGURE 3.2   Example of an Effective Online Help System
This procedure from the Asset Management System shows how an online system can allow the user to adjust the amount of detail a user needs.

2   Determine How Much Information Your User Needs

You may design procedures to contain varying amounts of detail, sometimes rich with detail, at other times more sparse, depending on the difficulty of the task or the

FIGURE 3.3   Guidelines for Designing Procedures

1. Relate the task to meaningful workplace actions.
2. Determine how much information your user needs.
3. Choose the appropriate instructional format.
4. Follow a rhythm of exposition.
5. Test all procedures for usability.

reader's experience. A richly detailed procedure needs more visuals and a greater amount of explanation. It requires you to state more options and describe more results than a sparse procedure. Some rich (highly detailed) procedures will contain a note for each step, pointing out all the "what ifs" and all the other alternatives: basically, more information. Sparse (less detailed) procedures, on the other hand, because of the nature of the task and the reader's needs, often require only the stating of the steps in chronological order.

Your user analysis (Chapter 5) will indicate whether your user needs a lot of detail or not so much. The paragraphs that follow outline some of the elements of a procedure that you can vary according to the amount of detail you wish to present. Also, as you can see from Figure 3.2, you can use features of electronic presentation to allow the user to get more detail if he or she needs it at the time. This technique is called *layering* and it simply means that you provide both levels of

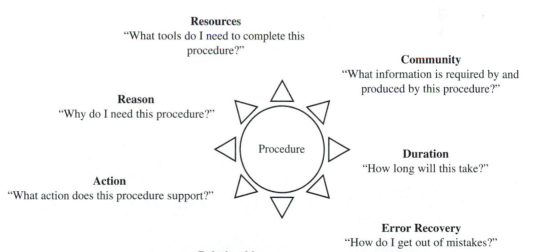

FIGURE 3.4   The Activity Context of a Procedure
Asking these questions can help you focus on workplace activities. You can use these questions to help you write introductions to procedures.

details on the same screen. You can layer information on pages too, as in using one column for fast-track users and one column (with more detail) for slow-track users. Layering techniques for both online and print documents are covered in Chapter 10, "Designing for Task Orientation," and Chapter 11, "Laying out Pages and Screens."

Figure 3.2 illustrates the difference between a step in a procedure with sparse detail and one with rich detail. Notice that the step with rich detail contains both more text (explaining what appears in the screen shot) and the screen shot itself (illustrating what the actual interface looks like.)

Other details you can include to enrich procedures include the following:

- Screen shots
- Cautions and Warnings
- Tips for efficient use
- Tables showing options the user can take with a specific step
- References to other sections of the manual or other resources
- Explanations

Screen Shots

Screen shots show the actual user interface, what menus to display, and what choices to make. The screen shot in Figure 3.5 is from a program called WordNet.app, a graphical interface program for a lexical database of words in the English language. In this example, the screen shot shows how the user selects one of the results of the database searches.

The following three examples are for cautions, warnings, and notes and tips (Figure 3.6). They all come from the *Oinc User's Guide*. This guide covers how to set up an Internet web environment for accounting and information storage. The examples illustrate how to use these elements in a procedure.

Cautions and Warnings

The cautions and warnings cover occasions where the user needs to be careful of possibly damaging the product, losing data, or in the case of the example in Figure 3.7, compromising the performance of the system. Warnings, as illustrated in Figure 3.8, involve possible harm to human beings as a result of errors with software or equipment.

Notes and Tips

Notes and tips, illustrated in Figure 3.9, offer you the opportunity to suggest alternatives, workarounds, or helpful applications to the user's workplace activity. The note in Figure 3.9 helps guide the user in using profile setup wizard software.

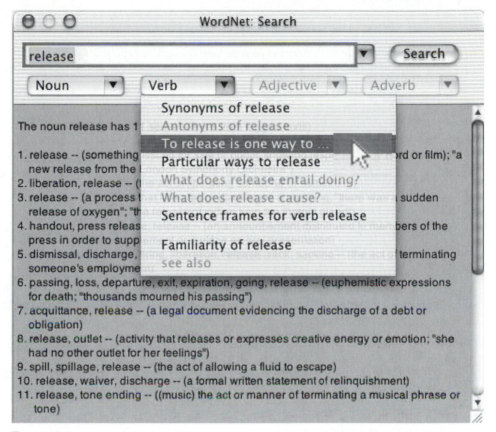

FIGURE 3.5   A Typical Screen Shot
This screen shot of a web page application can show users an overall view of an interface.

Notes and tips convey information that elaborates on a step or command. For example, in Figure 3.9 the note follows an instruction to use a wizard to create time saving wizards. Now the writer could have left it at that, but instead, considered that the wizard contained a number of steps and that the result could or could not help the user in a communication skill ("good quality communications"). So the writer added some advice about leaving some values "default" and changing others to help guide the user at the point of making decisions about the wizard.

Tables

Tables abound in user guides and procedural manuals and help because they allow you, the writer, to organize information and present it efficiently. Tables consist of categories, columns, and rows, lines and spaces and cells that make for very useful overviews, summaries, collections and other information structures. The table

• To select a hidden tool, hold down the mouse button on the related tool with the triangle until the additional tools appear, and then drag to the tool you want. Or hold down Shift, and press the letter key showing in the tool's tip to cycle through the group of tools.

Press on a tool with a triangle to open a hidden group of tools.

FIGURE 3.6   A Partial Screen Shot
This partial screen shot from a manual can help users focus on just part of the information.

in Figure 3.10 contains just such elements: a column that lists the choices a user has (in this case for deciding on quality of a copy of a compact disc) and the columns explaining the possible solutions. Notice also that to enhance the user's understanding, the writer has included a footnote to explain what one of the column headings means.

You should use tables whenever a procedure involves a number of choices in a problem-solving sequence in a procedure. Chapter 11, "Laying Out Pages and Screens," contains a number of guidelines for building tables. In general you can use them to present the following kinds of information:

• Features and uses
• Terms and definitions

Automatic Setup

Oinc will arrange for a number of default settings (name, password, account number, field format, filing system, etc.). These settings follow accepted and tested methods and usually work without you having to worry about them. However, you can specify settings if you want, as in the case of multiple accounts using the same database (see Appendix on page 150 for a listing of automatic setup options.)

⚠ **Caution:** Do not change the default settings manually unless you are sure how each setting affects the performance of your *Oinc* environment. Your record retrieval speed may decrease significantly if the wrong settings are used.

FIGURE 3.7   Cautions
Notices of caution can help users solve problems when using the software.

Preferences Settings

On the Preferences page you can set properties which are common for all your user profiles. These settings will remain unchanged even if you switch to using another profile.

Preferences tab

Country selection

You must always configure the country setting to match the country where the program resides (even though users may reside and use the program in different countries.)

- Select the correct country from the list and click **Apply.**

⚠ **Warning:** Use only the country setting appropriate for the area where your server environment will reside (as in "USA" or "United Kingdom"). Using the server in any other country than the one specified may be illegal.

The Oinc system operates under the tax laws of the country in which the server resides and retrieves tax information automatically from government web sites. Therefore you need to specify the legal residence of the server site.

FIGURE 3.8   Warnings
Use warnings when users face a risk of losing data, damaging equipment, or, in this case, breaking the law.

- Setting names and options
- Users and program applications
- Section tables of contents

Other opportunities for adding detail to manuals and help systems include these:

- Cross references and links
- Icons
- Graphics showing program concepts
- Keystroke combinations
- Examples
- References to other sections of the manual
- Footnotes

You should follow the guidelines for tables presented in Chapter 11, but also observe some specific guidelines for using them in procedures:

Keep tables simple. Start with columns and bold style headings for a simple table. For more complex tables, add a rule under the headings. Next add a rule under

Creating user profiles

By creating profiles for different users you can easily switch from one set of data to another without having to remember the personal settings for all users.

1. To create a new profile, click New. This opens the Profile Wizard. This wizard will guide you through creating a new profile. To continue, click **Next.**

     Note: The program will load the default settings, most of which you can just leave as is. Changing the name and database ID (see Section 2 on "Database IDs") will do. However in some cases you will have to alter the default settings.

2. Type a name for the new profile. Check your user list for names or let the system derive names for each person who logs into your Oinc site. To include a person who hasn't signed up simply type the name and the database ID in the appropriate text boxes. Then click **Next**

    Tip: When you create new profiles, add a nickname to it that describes the profile and sets it apart from the others. This will help you quickly select the right one when you are running your maintenance procedures or troubleshooting user inquiries.

FIGURE 3.9   Notes and Tips
Notes and tips allow you to share wisdom and help users negotiate through difficult steps in a procedure or tutorial lesson.

the columns: the bottom rule. After this you could add a complete box around the table for more complexity. Next, add vertical rules separating the columns; next, add horizontal lines separating column entries, with the boxes sized to the largest entry. Whew! But a simple table works best.

*   Cite the table in the text. Citing the table makes it clear when the user should consult it, and for what purpose.
*   Use descriptive, performance-based column titles.
*   Use visual cues for keys or commands, or menu selections presented in tables.

3   Choose the Appropriate Procedural Format

The well-designed procedure should follow one or two accepted formats for instructions or a format of your own design. The formats in this section give you a basic outline of the formats you will encounter in most manuals. It's a good idea to stick to one format for the sake of consistency, however you will find a great deal of variation and inconsistency in manuals today. For the most part these formats cross over between online and paper formats, however the functionalities of the online formats (links, pop-ups, and so on) are not available on a printed page.

Select quality level

Click this option if you want to specify the audio quality level of the music that is copied to your portable device. The following table lists the range of quality levels:

Quality	Bit rate	Disk space*
Smallest size	32 Kbps	14 MB
Medium	64 Kbps	28 MB
Best Quality	128 Kbps	56 MB

*Refers to the amount of disk space required if you copy an entire CD according to quality level.

FIGURE 3.10   A Table
A table allows you to arrange sets of numerical information and text information, or to organize text to support decision making.

Standard Format

So far we have described the standard format of instructions. The standard format consists of steps, notes, screens, and other elements aligned on the left margin and continuing in either one or two columns, in a numbered sequence, from first step to last. Figure 3.11 is a good illustration of this format.

The advantages of this format include the fact that the user will most likely recognize it. It follows the steps clearly to the end. The convention of putting each step on a separate line helps the user follow the step and retain it in short term memory long enough to execute it on the computer without interrupting the flow of work.

Advantages of the Standard Format

The overall advantage of the standard format lies in the fact that users have seen it in software manuals and other instructional writing, and that transfer of recognition carries a lot of weight. The advantages of this format follow in the list below:

- Recognizable by users
- Easy to flow from one page to another
- Easy to re-number and test
- Easy to see the steps using hanging indent

Disadvantages of the Standard Format

The standard format works best when you don't have a variety in complexity from task to task, and your users become familiar with it. It has the following disadvantages:

- May take a lot of space for really simple, brief procedures.
- Confusing if you have to mix complex steps with simple steps. The user in this case can loose track of the chronology while handling a difficult interface item.

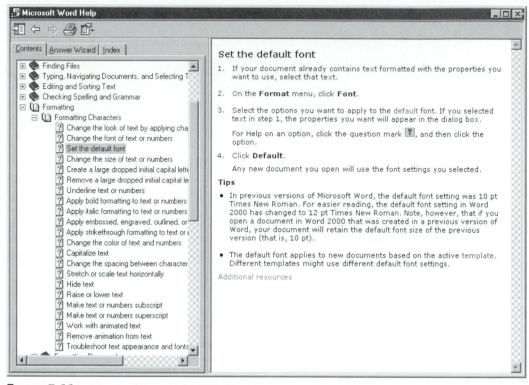

FIGURE 3.11   Standard Format for Help Topics
Standard format for help topics consists of navigation tools, a contents pane, and a topic pane. The contents pane also has tabs for the Answer Wizard and Index, which also display topics in the topic pane.

Prose Format

The prose format for instructions puts the steps in sentences and paragraph form instead of the command-oriented, numbered pattern found in the standard format. The prose format gives a conversational tone because of the use of sentences rather than a list of command verbs. It allows you, as a writer, to make asides, qualify ideas, and give a relaxed rhythm to sentences.

The prose format occurs fairly commonly in programs with relatively simple tasks (about three or four steps per task) and a simple interface. Prose format instructions also work well in reference sections, where you want to include only abbreviated steps. Because many manuals use this form, the user will probably recognize it, especially the experienced user. Finally, it conserves precious space where you simply need to cover the basic steps. You might use it, for instance, in the cramped space of an error message box to give users a way to correct a problem. Figure 3.12 shows an example of prose format.

The prose format often uses **bold** or *italics* to indicate command verbs, function keys, buttons, and text that the user is supposed to type in—as you might imagine.

Logging Tasks

Logging Tasks

When a task on your list has been completed, you are ready to check it off and move it into the program's task log so that you will have a record of its completion. The buttons on the toolbar used for the logging function are the following:

- **Check/Uncheck** Used to check off completed tasks or uncheck ones that were thought to be completed but now need to be re-done. Follow these simple steps:

  Click your cursor in the small box at the far left of the task in your list that you wish to check off. Click the **Check-mark** button on the toolbar, and your task will be labeled "completed" and ready to be logged. If you want to uncheck an item, place the cursor in the checked box and then click the **Check/Uncheck** button to remove the checkmark.

  Note: You can view your task history at any time using the **View Task History** button.

FIGURE 3.12   **Procedure in Prose Format**
The prose format saves space when you write for experienced users.

Also, some prose instructions keep the sentences short (under twenty words) and the paragraphs short (fewer than five sentences). This brevity guarantees that the user will find it comfortable to transfer the instructions from the page to the keyboard or mouse. No rules exist for how many screens or other illustrations you may use with the prose format, but usually you would include no more than one screen per task, if that.

Advantages of the Prose Format
- Uses a conversational, relaxed tone
- Saves space
- Clarifies simple, basic steps
- Accommodates experienced users

Disadvantages of the Prose Format
- Buries steps in the paragraph
- Precludes lengthy explanation of individual steps

- Can't accommodate graphics for individual steps
- Doesn't offer much support for novice users

Parallel Format

The parallel format comes in handy when you have a program that uses complicated data fields or dialog boxes. Examples would include database programs, address or Rolodex type programs, invoice or order entry programs—in general, programs that require the user to move from one field on the screen to another, filling each one along the way. The example in Figure 3.13 shows how the writer used this format to assist a user in filling out a dialog box. The parallel format shows the screen with the fields empty, and parallels the field names in the steps that follow. Each of the steps indicates to the user what kind of information to include in each field—characters or digits or both—gives examples, and cites special cases. There is one step for each field, usually. Figure 3.13 shows a procedure using letters (A, B, C, and so on) to correspond with the fields.

When you find yourself confronted with forms and dialog boxes, the parallel format can help your users stay organized, as long as you keep the correspondence between the steps and the screen clear. The format helps keep the information centralized, and helps the reader see the filling out of the dialog box or screen as a single task. On the other hand, the parallel format can break down if the procedures get so long as to take up numerous pages because the user has to refer back to the illustration on a previous page—an awkward situation that allows for plenty of mistakes. Should you decide that your users require the parallel format, you can set it up easily by following the directions below.

1. Keep the terminology consistent. If the screen uses the terms "Employee number" and "Employee name" then use the same terms in the explanatory steps. Slight variations, such as "Your Employee Number" or "An Employee Number" increase the user's thinking load and thus should be avoided.

2. Cue the terms to the screen. Keep the same type font, size, and style in your steps as the screen shows. Even more efficiently, reproduce the screen element, showing the box, say around the field, or the line the user would write into, in the steps. This increases the user's ability to recognize what steps go with what screen elements.

3. Discuss one screen item at a time. Usually when you set up parallel instructions you cover only one field at a time, even though two fields may be related. For example, your user may have to fill in the "employee number" field before the "employee name" field will take any data. In this case, resist the temptation to discuss both fields at a time. Simply make filling out the first field a prerequisite for filling out the second, and mention this in the discussion of the second field.

4. Use plenty of examples. Tell the user whether the fields require characters and digits, and give examples of each in your notes to the steps. Note the example in field "E" in Figure 3.13.

5. Make sure you explain to the user how the parallel format works. Introduce the idea, and, in complicated tasks, explain the cueing scheme and other conventions used.

Using Custom Scaling

Scalable analog modules can be set to show something other than the actual inputs or outputs. For example, you could scale the readings of a -10 to +10 VDC input point to measure its input as zero liters per second when the real-world reading is zero VDC, and 1000 liters per second when the real-world reading is five VDC.

NOTE: Custom scaling has no effect on the resolution or accuracy of the module.

1. In the Add Analog Point dialog box, click the Custom button in the Scaling area to open the Scale Analog Readings dialog box:

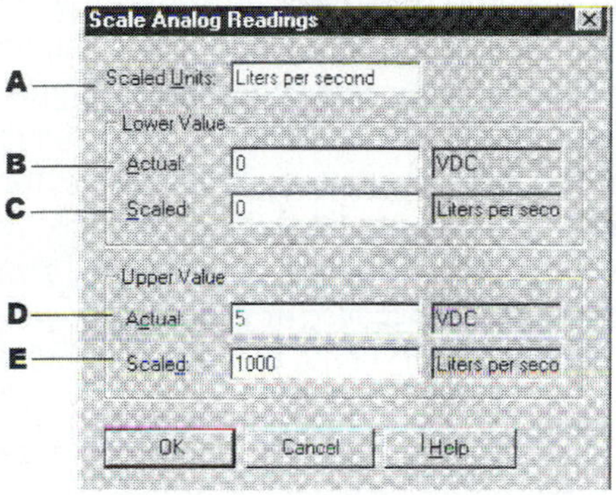

2. Complete the fields as follows:

   A Enter new engineering units for the module. The example uses liters per second.

   B Enter the actual real-world lower value that the scaled lower value corresponds to. Note that inputs typically have under-range capability, which means you can specify a lower actual value that is less than the zero-scale value. Outputs do not have under-range capability.

   C Enter the new scaled lower value. This value can be any floating point value.

   D Enter the actual real-world upper value that the scaled upper value corresponds to. For inputs, you can specify an upper actual value greater than the full-scale value.

   E Enter the new scaled upper value. This value can be any floating point value greater than the scaled lower value. This example uses 1000, which scales the output to 1000 liters per second when its actual reading is 5 VDC.

3. Click OK.

FIGURE 3.13   Procedure Using Parallel Format
The parallel-format procedures work extremely well when the user needs to fill out a form or dialog box.

Advantages of the Parallel Format

- Can help your users stay organized
- Works best with shorter procedures
- Good for filling out complicated screens and dialog boxes

Disadvantages of the Parallel Format

- Does not present information in a step-by-step, task-oriented manner
- Specialized: can't use it for all procedures
- May confuse user who can get lost moving between steps and screen
- Has to fit on one page

Embedded Help

Remember those times when you had a dialog box open, probably in a Windows program, and you just pressed the F1 key to call up a help message? The help program "knew" your location in the program using *embedded help* (Figure 3.14). Sometimes this kind of help is called "interactive assistance."[1] No matter what menu or dialog box you have open, the help program will display information appropriate for that location. The help program recognized the dialog box you have open or the item you clicked on (using the "What's This?" prompt) and responded with information. Such kinds of responses occur because of information called *tags* that the writer, working with a programmer, put into both the program and the help system.

An innovative approach to embedded help is the "coach" or help that guides the user through a complicated procedure. With the coach feature turned on the novice user of the software gets assistance that is highly focused on tasks. Consider the example in Figure 3.2. In this example, the user sees not a static set of steps for performing a procedure, but a well-organized representation of actual workplace tasks. Each of the gray panes of help is associated with a field in the interface and the user can click on the "Cancel," "Back," or "Continue" buttons to control the flow of the operation. The "Related Links" panel on the left allows the user to situate the immediate process within the larger, decision-making context.

This method of representing tasks does not depend on constricting descriptions of procedures, but instead reflects the dynamic nature of actual workplace decision making.

Some systems use a design called "fly out" help that provides procedures at the time of need. Help systems of this kind open a panel on your screen and allow you to follow the procedure while you're working with the program. Alternately they can interrupt your work temporarily and provide a demonstration of a concept or procedure. Embedded help can offer a number of types of procedural information.

- **Tips for efficient use.** Reminders of keyboard shortcuts, suggested filenames
- **Cue cards.** Brief explanations of buttons and fields, done in a memorable way
- **Short, animated demonstrations.** User-paced procedures showing movement, where the computer program performs a function for the user, who clicks on a "next" button or a "close" button when done.
- **Trouble-shooting tips.** Procedures offered when users perform the same, nonproductive keystrokes over and over.

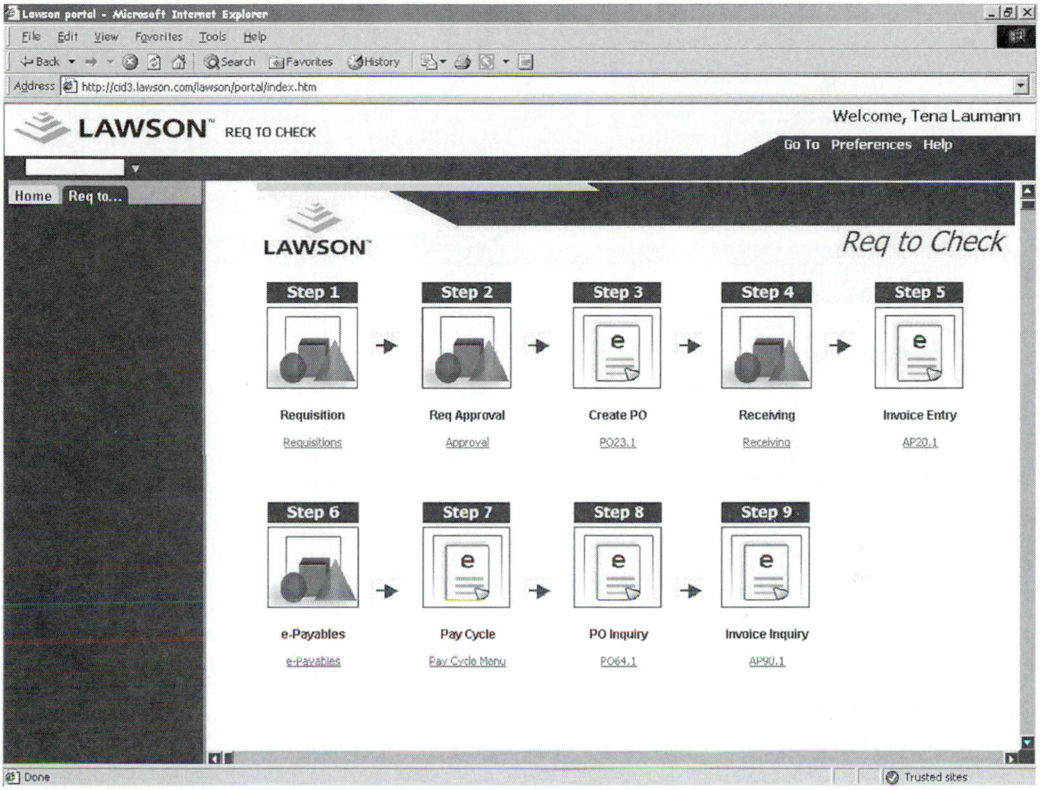

FIGURE 3.14   Embedded Help
Embedded help provides help at the time of need in the field or interface object where the user is working.

Embedded help can come in a number of formats and types. The following is a brief overview of some of these types.

- **Flyout help.** Help that appears in a box or panel on the screen at the user's request.
- **Interactive flyout help.** This form of flyout help monitors the user's progress in filling out a dialog box and highlights one step at a time until the procedure is complete.
- **Do it for me help.** This form of help contains links within the online help procedure that activate the screen element or dialog box described in the procedure.
- **Field-level help.** Help that provides information on how to enter information in fields. This form of help is illustrated in Figure 3.14.
- **Interface help.** Help information (brief instructions) provided in a designated section of a screen.
- **Pop-up definitions.** Pop-up definitions provide brief definitions of interface elements activated by a mouse click.
- **Roll-overs.** Definitions of interface items that appear when you move a mouse over the item and (often) pause for a second or two.

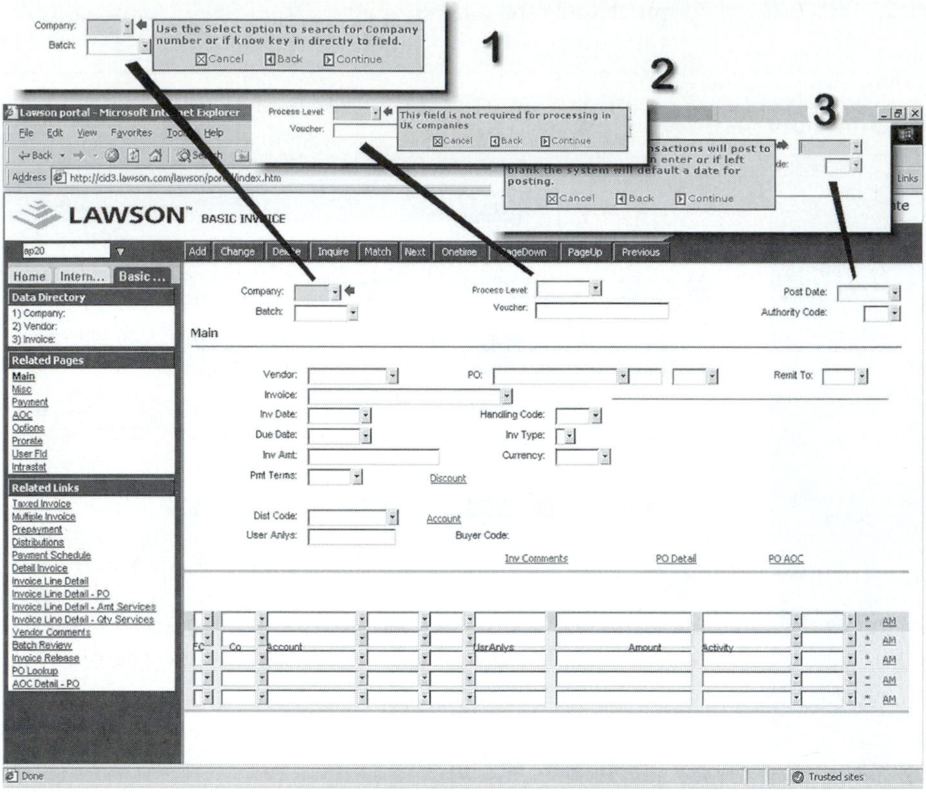

FIGURE 3.15   An Example of a Coach

Coaches assist users by providing both instructions and tools for complicated procedures.

4 | Follow a Rhythm of Exposition

By *rhythm of exposition* we mean a pattern of step, note, and illustration. Think of your procedure as occurring in this way:

- First I give command for the step.
- Then I say how the program will respond.
- Then I illustrate what happened.
- Then I tell the next step.

The basic idea of a rhythm of exposition lies in the action/response pattern. Computer programs work in that way: Take an action, the system responds. These two events get repeated over and over with incremental progress toward the goal of the whole procedure. Technically, then, each step should have a note to explain the result of the action. But not always. Often, with simple steps or more advanced users, the results do not need explanation. Thus, with more sparse procedures—depending on your users' information needs—you would just give the steps.

Whether your procedures contain lots of notes or few notes, they should be compact enough so that the user doesn't get lost between steps. The eye needs to follow easily from step to step. If your procedure contains extra information—other options, definitions of terms, or complicated interpretations of results—then put the extra information after the steps, so that the reader sees them clearly.

5   ## Test All Procedures for Accuracy

During the developmental phase of your projects you will most likely test your procedures to gauge whether the pacing and the format conventions you follow have the desired effect. Once you settle on a format, however, the testing does not stop.

As a designer and writer of procedures, you must see that your descriptions accurately reflect the program. To do this, you need to test every procedure you write. Tests of this type are called *evaluative tests;* which means that after you finish the procedure, you have an actual user, or a prototype of the user, or yourself as a last resort, perform the steps. Get ready to have your eyes opened to all the conditions, alternatives, options, and other details you left out.

As part of your review of your procedures, you should double-check them, to make sure that the screens represent the program accurately, that all the options you need get included, that your statements of syntax, field content (digits versus characters), and field size are accurate and complete. (Chapter 8 gives further details on kinds of tests and test methods.)

Discussion

This section examines the structure of procedures to determine how you can design them to guide the reader effectively. We will look, first, at the users' psychology when regarding procedures, and how they need to focus heavily on user actions. In fact, of all the documentation forms you will design, those offering procedures will most closely resemble the context-free operations of the program. Next, we will see how the parts of a procedure, examined analytically, can help in orienting the user toward productivity in the workplace.

What Constitutes a Procedure?

Procedures are often rooted in the features of the software program. But the *features* of the program can differ greatly from the *uses* of the system in the office or business to perform meaningful actions. In fact, you should see the features of the program—whether the program processes words, or creates graphic images, or calculates numbers—as constituting the tools that the program provides for the user. Thus, the procedure you design might support any number of actions. Conversely, any action—writing letters, laying out advertisements, calculating crop yields—can involve any number of procedures organized in creative ways to solve problems and get work done.

Nor do procedures derive merely from descriptions of the components of the software program. These components—menus, panels, toolbars, and so forth—do little (usually) to help the user because they work in the background. Procedures result directly from your putting the functions of the program into usable sets of steps that do the user's work.

Most of your writing in software documentation consists of writing procedures. All procedural documentation fulfills the user's simple purpose: "how do you use the program?" Procedures get used when the user is actually *doing something* with the program. He or she may have undergone training in the basics of the software using a tutorial or class, but when the user follows a procedure, he or she is actually *at work* with all the urgency, importance, anxiety and stress of a work situation.

In the midst of a work situation, a procedure functions on the **guidance level.** The reader of procedures needs to know what keys to press, what reports and screens will look like, and how to get out of trouble. But mainly, the task of writing procedures consists of giving guidance, of leading by the hand, being an assistant rather than a teacher. The **teaching level** of task orientation, which you accomplish through tutorials, has as its goal the internalization of concepts: you try to get the user to remember the features after the lesson finishes. But procedures are different. Procedures focus much more on what to do *at any given moment.* As such, the stakes are often higher because errors can cause lost data, mistakes can waste time, and getting stuck half way through can stop the larger workplace action in its tracks.

Guidance-level documentation also differs significantly from **support level** or reference documentation, in that procedures tend to follow a beginning and ending: a chronological sequence. With support-level or reference documentation, the *user* defines the task and goes to the documentation to get an essential tidbit or chunk of information needed to perform the task. A reference user might only need to know the *name* of a menu, the function of a tool button, the *meaning* of an error message, or the *type* of data to put into a specific field. Probably this difference explains why reference documents (for example: lists of error messages, ASCII codes, command summaries) usually consist of smaller units of information than procedures.

With what kinds of information would the user need guidance? With installation, for one thing, because installations vary from system to system. With installation the guidance is often produced by a software program called a **wizard:** the personification of a guide or assistant. Users also need guidance in maintaining and repairing systems: open this file, check this variable, close the file, and so on. But for the most part procedures concentrate on the actual operation of the program, the step-by-step of manipulating the program interface.

To help enhance the day to day step-by-step use of a program interface, each procedure can contain a description of how and why to use it. And each scenario should indicate the user's role, and the goal of the procedure. For example, you might point out that: "You perform these tasks every day after you have posted your last transaction and before you turn the computer off." User's roles might require you to cast operations in terms of office roles—sales clerk procedures, accountant procedures, front office procedures—or in terms of program roles—programmer procedures, maintenance programmer procedures, end user procedures, installer procedures. The goal of the procedure should indicate what action it relates to: improving printing capabilities, solving a sales problem, finding a client, organizing information in a report, and so on.

How Does a Procedure Work?

A procedure that guides the user through a series of tasks to a designated end works because you design each of its parts to do a specific job in measuring time productively. The sections below discuss how those parts each contribute to the overall task-orientation of the procedure.

Task Name

The task name identifies the program function in *performance-oriented* language. You can design most task names based on the following model: "Opening a file," or "Recalling a Record from the Client Database." The important point when naming tasks: the task name should describe what job the user performs, not what functions he or she uses. For example, the task name of "Using the Open . . . option" indicates the use of a program function. You should describe the task as "Opening a file." Other examples include the following:

Program-Oriented Task Name: Weak	Task-Oriented Task Name: Strong
Using the Print Function	Printing a Card
Selecting the "List All" command	Listing All the Disk Functions

Overview

The overview serves as an introduction, and orients the reader to the use of the procedure. It reminds the reader what the task will allow him or her to accomplish in a work setting. It should indicate, using informal language, what the user should be able to accomplish with the procedure. The overview or introduction is a bridge or link between the operation of the program and the action the user wants to accomplish with it. To write overviews well you need to analyze not just what steps the user needs to follow, but why.

The overview should set the user up to perform the steps. If the user needs to have certain skills to perform the task, then mention them, as in "To perform this step you should be familiar with raster-formatted images." Likewise, if the task has conditions for performance, mention them in the scenario, as in, "You should only perform the end-of-day posting after you have closed down the general ledger file."

Study examples of overviews and introductions to tasks, starting with the ones shown in Table 3.1.

You might be tempted to omit the introduction, thinking that the user may have read the beginning section of the manual and will know how the procedures apply to work activities. But this is not always the case. Often specific actions in the workplace, where the user is responding to a new problem or has taken on new workplace tasks, require him or her to seek out unfamiliar procedures and operations. For example, if my boss informs me that the home office has changed the format requirements for the yearly reports (to which I contribute) then I may have to employ software operations and functions that I have never used before. I can easily look those functions up using the table of contents, but that doesn't necessarily mean I'm going to understand exactly how the procedures apply once I locate them on the page or in the help system.

TABLE 3.1   Examples of Introductions to Tasks

Task Name	Introduction/Scenario
Changing Default Settings	"Picture Publisher lets you save defaults in the tool ribbons and in some dialog boxes. You will find these features extremely helpful when working with Picture Publisher." *Picture Publisher Reference Guide,* p. 4–11
Creating Groups	"*Groups* are containers; they can contain objects and/or other groups. A group window can also contain links to other groups to form a *group structure.* The group structure is useful to represent the structure of your network hierarchically or functionally." *TC8215 Sectra Management System for Windows User's Guide,* p. 4–9
Including Graphics in a Document	"When you prepare reports, manuscripts, or other types of technical documents, there may be times when you want to include GCG graphics. The Wisconsin Package lets you save files in Encapsulated Postscript (EPS) format, which you can include but not edit in most commercial document processing programs." *Wisconsin Sequence Analysis Package User's Guide,* p. 5–36
Controlling the Screen Display	You can use the Display command on the Options menu to change the colors in the MS-DOS Editor window, display or hide scroll bars, and set tabs. *Microsoft MS-DOS User's Guide and Reference,* p. 221

The second use of an introduction is to help direct the user once he or she gets started with the procedure. Consider this example: the user knows ahead of time that the filling out of a configuration panel will result in values being typed into every single field. But when he or she gets toward the end of the panel there is a temptation to skip a field. The task gets tiring and boring. But if the introduction pointed out that every field requires some kind of value, the user might resist that temptation. What you say in the introduction gives you some degree of assurance that the user, when faced with a decision in the completion of a task, will make the *right* decision.

Steps

Steps make up the most important part of the procedure because they embody the segment of time during which the procedure directs the user's activity. The steps constitute the time actual people devote to following your procedure. However, often the user will skim the steps, either to avoid having to read the explanations, or to pick out the essential step and try it without reading further. Even so, you should take care to write them well.

Steps tell the user what to do, and in so doing, accomplish two things: give the user tools to use and actions to take with the tools. But while steps can include both the tool and the action, as in: "Use the mouse to select 'Open' from the file menu," you may limit the step to only the action. This is the general rule. The decision to include the "use the . . ." part of a step depends on the user's familiarity with the task and the program. Often users will not read the notes and explanations that go along with steps, so you should make the steps as self sufficient as possible. Imagine that your user only read the steps: would they contain sufficient information to perform the task?

Following are listed four versions of a "Step 1." Note how they increase in elaboration.

Step 1: Open a file.

Step 1: Select the **Open** . . . option in the **File** menu.

Step 1: Use the **Open** . . . in the **File** menu to open an existing file.

Step 1: Using the mouse, select the **Open** . . . option in the **File** menu to open an existing file.

Needless to say, steps should always occur in chronological order. Putting them in chronological order ensures that the user will not get lost, or get the sequence mixed up. For these reasons, use numbers instead of bullets for steps. You can also include the word "step" as in "Step 1 . . . , Step 2 . . ." to help the reader note the need to take action.

If your procedure contains smaller actions, remember to keep all the main steps in one continuous sequence. In other words, do not renumber under the other actions required for a step because this can cause confusion as to which step to take next. Consider the following weaker example:

WEAKER

Step 1: Choose Groups from the Maintenance Menu.

Step 2: Choose an action from the Groups dialog box.

    1. Select a name for the group.

    2. Select a directory name for the group.

    3. Set the access code to either Open or Restricted.

Step 3: Choose Close from the Groups dialog box.

This sequence of steps risks confusing the user because it contains two sequences of steps. In other words, you have two step 2s, and so on. You can avoid this confusion by trying a format like the following stronger example:

STRONGER

Step 1: Choose Groups from the Maintenance Menu.

Step 2: Choose an action from the Groups dialog box.

Once you have opened the Groups dialog box, you need to select a name for the group, then select a directory name for the group, then set the access code to either Open or Restricted.

Step 3: Choose Close from the Groups dialog box.

The second example puts the smaller actions in a prose format and doesn't number them. This way the user runs less risk of confusing the step sequence. As indicated in the example, it's also a good idea to separate the smaller steps from the main steps by putting them in a paragraph on the next line.

Also, you should avoid giving *commands* in paragraphs where you offer notes and explanations. Reserve commands for steps and use imperative verbs. Technically, if the user needs to perform an action and that action has some result, then the action should appear as a numbered step. Putting actions in explanatory notes that accompany your steps begs for the user to ignore them—which may happen.

WEAKER

1. Chose Open from the File Menu
2. A new document appears.
3. Choose Font from the Format Menu.
4. The Font dialog box appears.

STRONGER

1. Chose Open from the File Menu
   A new document appears.
2. Choose Font from the Format Menu.
   The Font dialog box appears.

Elaborations

Performing the steps will get the task completed, but not without explanations. Here, elaborations come in. They explain the steps, commenting on them *as they get performed.* You learn a lot about how to perform the procedures when you study the program, so share that experience with your users when you write the procedures. In elaborations you share the following kinds of advice with your users:

• Possible mistakes and how to avoid them
• How to perform procedures efficiently
• Alternatives such as keystrokes, toolbars, or function keys
• Definitions of terms
• Ways to tell if a step has been performed correctly
• Where else to look for additional information

When you write elaborations, always try to use the active voice and refer to the program. For example, instead of saying "The control panel will be displayed on the screen," use the active voice and say "MarketMaster will display the control panel."

Options

Often when you describe a procedure you will have to include a list of optional commands or keystrokes needed. Put these in an easy-to-read table. Tables in procedures give the user options and save time and space. Consider the following example:

3. Adjust the color of your image.
   At this point you can adjust the color of your image by using the following commands.

To do this . . .	Use these keys . . .
Set colors to black and white	Ctrl-M
Revert to default colors	Ctrl-D
Adjust the brightness	Ctrl-B
Adjust the tones	Ctrl-T

Screens

Include screens in your procedures when the user needs either to see the tool in use or the goal or results of an action. Rich procedures would include a screen for the starting state of the task, one or more screens illustrating how to use interface tools, and a screen indicating the result of the task. Depending on your constraints of time and budget, and depending on your user's level of expertise, you will include more or fewer of these screens. Use a box or active white space around your screens to make sure the user can distinguish them easily from the surrounding text. Finally, if you sense any chance that the user may not associate the screen to the appropriate step, give the screen a name, or descriptive caption.

In Figure 3.16 the screen shot comprises almost the entire procedure, as the commands are indicated in the callouts.

Usually you will find yourself using screens to do the following:

- Give an **overview** of the main panel of an interface.
- Show the **partial result** of a procedure (a stage in the process) to help the user keep on track.
- Show the **final result** of the procedure to let the user know where the procedure ends.
- Show **dialog boxes** where the user has to make choices.
- Show **toolbars** indicating which tools the user needs.
- Show **menus** indicating what commands the user needs.

Chapter 13 deals in greater length with the subject of screens and other graphics used in procedures.

Glossary

elaborations: elements of a procedure that give extra information, usually about the application of a procedure to the user's larger workplace purpose.

embedded help: refers to help that displays procedural information as a part of the actual interface of the system. Embedded help differs from conventional help in that conventional help is a separate, stand-alone program accessible via a Help button or menu.

evaluative test: a test of a document's usability, done after releasing to the user.

guidance level: a type of documentation designed to lead the user through a procedure one step at a time from a designated starting place (such as a certain menu) to an ending state (such as a report printed.) Guidance level documentation (or procedures) defines the task for the user, but does not teach the task. See also **teaching level** and **support level.**

layering: a formatting technique that allows for different levels of information (beginning and advanced) on the same page.

screen shot: an image of a screen from a computer program electronically recorded in the form of an image file that you can use in a manual or help system to illustrate what the screen looks like.

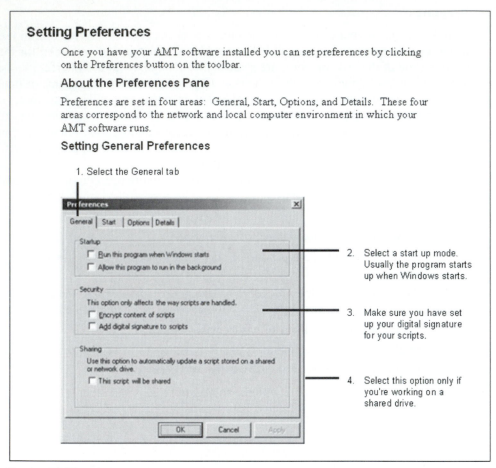

Setting Preferences

Once you have your AMT software installed you can set preferences by clicking on the Preferences button on the toolbar.

About the Preferences Pane

Preferences are set in four areas: General, Start, Options, and Details. These four areas correspond to the network and local computer environment in which your AMT software runs.

Setting General Preferences

1. Select the General tab

2. Select a start up mode. Usually the program starts up when Windows starts.

3. Make sure you have set up your digital signature for your scripts.

4. Select this option only if you're working on a shared drive.

FIGURE 3.16   A Screen Shot Using Commands

Putting commands on a screen shot can be an alternative to listing steps in procedures. The technique has the advantage of showing the user exactly where interface items are.

support level: documentation designed to provide reference information for a user.

performance-oriented: language emphasizing performance of a task, as opposed to emphasizing the features of the software system. All good documentation should focus first on performance—using the program, then on the features of the software system.

teaching level: a type of documentation intended to instill a knowledge of how to use a program feature in the memory of the user. Teaching level documentation (tutorials) aims to enable the user to perform a task from memory. See **guidance level** and **support level.**

wizard: an online software program that follows the steps of a procedure and performs the tasks for the user. Wizards typically handle procedures for installation and those requiring advanced knowledge.

☑ Checklist

Use the following checklist as a way to evaluate the efficiency of your design for procedures. Depending on your users and the level of detail you choose for your procedures, some of these items may not apply.

Procedures Checklist
Determining How Much Information the User Needs

Your users require what level of detail in the procedures?
- ☐ Sparse
- ☐ Moderate
- ☐ Rich

Specific details to meet your users' need include which of the following?
- ☐ Screens shots
- ☐ Cautions and warnings
- ☐ Notes and Tips
- ☐ Tables

Format

Which format suits your users best?
- ☐ Standard format (step after step)
- ☐ Prose format
- ☐ Parallel format
- ☐ Embedded help

Rhythm of Exposition

- ☐ Have you reviewed your task design so that each one follows a similar pattern?

Testing

Have you identified the appropriate test points for your procedures?
- ☐ Accuracy of steps
- ☐ Accuracy of details
- ☐ Pacing/rhythm
- ☐ Format conventions
- ☐ Inclusions of options/conditions/alternatives

Elements of Procedures

Have you double-checked your procedures for the suitability of the following elements?
- ☐ Performance-oriented task names
- ☐ Overviews relating to user actions
- ☐ Steps including tools, actions, and results
- ☐ Steps in an orderly sequence

❏ Elaborations emphasizing user performance
❏ Tables to conserve space

Practice/Problem Solving

1. Write a Basic Procedure

In Figure 3.17 you will find a portion of the interface for Microsoft Word for Windows program Using it (or some other interface element of your choice) to practice writing a procedure for one of the following users.

• A student organizing a list of classes and times

• An assistant in a veterinary hospital organizing appointments for doctors

• A sales associate organizing clients and phone numbers

• A musician listing songs and times for performance

Choose what you think would be an appropriate format, level of detail, and elaborations.

2. Fix this Procedure

The following procedure has problems in format and organization. Rewrite it in a way that makes it more logical and consistent.

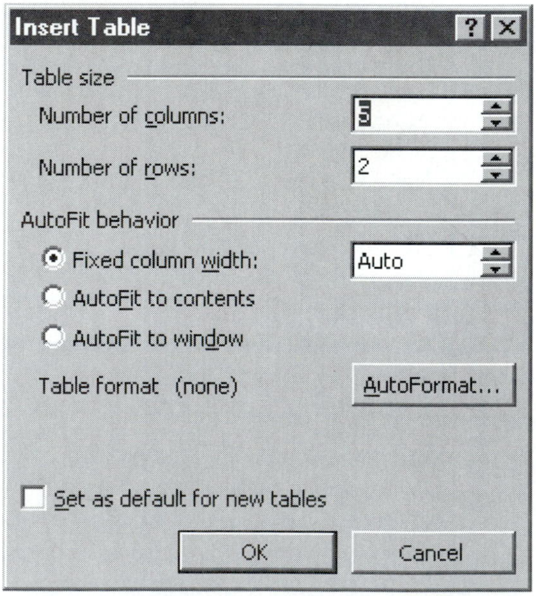

FIGURE 3.17   The "Insert Table" Dialog in Microsoft Word for Windows
Write a procedure to help the user set up a table using this dialog box.

From the **CardMaster** *User's Guide*

Erasing a Card

CardMaster allows you to write customer information onto magnetic cards using the CardWriter module. You can erase the existing information on a card and update it with new information. The CardWriter uses both write only and read write cards. To erase a card insert the card into the CardMaster recorder.

1. Choose Erase Card from the toolbar
2. Check the radio button for the type of card in the recorder
3. Cards with magnetic IDs and cards with magnetic IDs and UPI scanners are available. Click on Erase.
4. You have two options: Number change and Full erase.
5. Choose the kind of erase you want to do.
6. Click Yes when prompted to verify the Erase procedure.
7. If you do a Full erase all data will be erased from the card. You can not recover this data. Number changes just deletes the customer number but leaves the other information on the card.

3. Rewrite Prose Format into Standard Format

The following example shows a procedure done in prose format that probably should appear in step format. Rewrite it in the different format. Correct any errors you find and reorganize the passage.

Getting a Chart

To get a chart you should follow this procedure. Select Option #4, file from the Main Menu. Then, select Option #1, Get, from the File Menu. A listing of files in the default directory will appear. (If necessary, TAB to the DIRECTORY field to change the default directory.) Highlight the file you would like to retrieve and hit Enter. As an alternative you can use the CTRL+G speed keys to get a chart from any screen in ZQ. This allows you to get a chart without having to back out to the Main Menu.

CHAPTER 4

Writing to Support—Reference

This chapter follows the organization of software documentation into three main levels, called *levels of support:* teaching, guidance, and reference. These levels correspond to recognized user behaviors, questions, and needs, and act as design tools. This chapter offers examples, guidelines, and discussion for design of print and online reference information.

Reference documentation, also referred to as *support* documentation, includes all the look-up sections and elements of your manuals and help. It takes forms such as command descriptions, menu overviews, lists of definitions, function descriptions, examples, and error messages. Often reference documentation is associated with advanced users: users who know the software well, but need to look up specific elements of it that are too complicated to memorize and don't necessarily relate to step-by-step procedures.

So often, we think that these sections don't require design, or that they consist of merely alphabetized lists of commands. Often they do, but, with the rapid growth of computer skills among all workers, we should study ways to design reference information in a task-oriented way. Additionally, electronic and hypertext-based interfaces allow writers of reference documentation to design online documents that use powerful and useful technology for looking up information and relating it to workplace actions and purposes. These technologies include electronic contents pages, user profiling capabilities, and searching capabilities that allow users to find reference information fast and accurately. The designer of reference documents should know about the opportunities to build task orientation into look-up documents.

To this end, this chapter covers how to select the right form of support documentation (online and print) by examining both usual and special forms of reference. The chapter also discusses the idea of parallel patterns in reference documentation, emphasizing methods of organizing: alphabetical, menu by menu, and context sensitivity. The chapter then offers you a discussion of why users consult support documentation, looking specifically at the psychology of the reference user and how reference entries parallel the user's informational needs.

How to Read This Chapter

- For the project-oriented reader, the examples in the Figures will provide enough background to get you started on the design of your project as outlined in the Guidelines. But before you finish your project, you should familiarize yourself with the principles in the Discussion section.
- For the reader reading to understand, the Discussion section offers a basic background for which to read the Guidelines.

Examples

The example in Figure 4.1 shows a number of elements of a typical, structured reference entry for a programming language manual. The entry reflects the questions that reference users have about software, and it illustrates how to present answers to those questions in a highly accessible way.

These reference pages from the Ceridian Source Quick Reference manual (Figure 4.2) show how graphics can orient the reader to the overall process and supply information for each of the steps in the cycle. The graphic helps the reference user situate the information within the overall workplace activity.

Guidelines

Figure 4.3 lists guidelines to follow as you design support documentation.

1 Choose the Right Form of Reference

Reference users probably don't come to the manual naive; they know the interface and usually come looking for a specific piece of data to complete some task they themselves have defined and probably know how to do already. For this reason the structures in reference document facilitate fast look-up of information. This section gives you an overview of some of the forms of reference documents.

The Forms of Reference

With some programs you will design information in all three of the forms below. With still others you will just supply "a quick reference," along with the procedures in a user's guide. When you design a documentation set with all three kinds (and others), you gain the advantage of using the same information in each form. Using the same information in each form helps you as a writer because you can see the consistency among the forms. Consistent information helps the user, who begins to learn

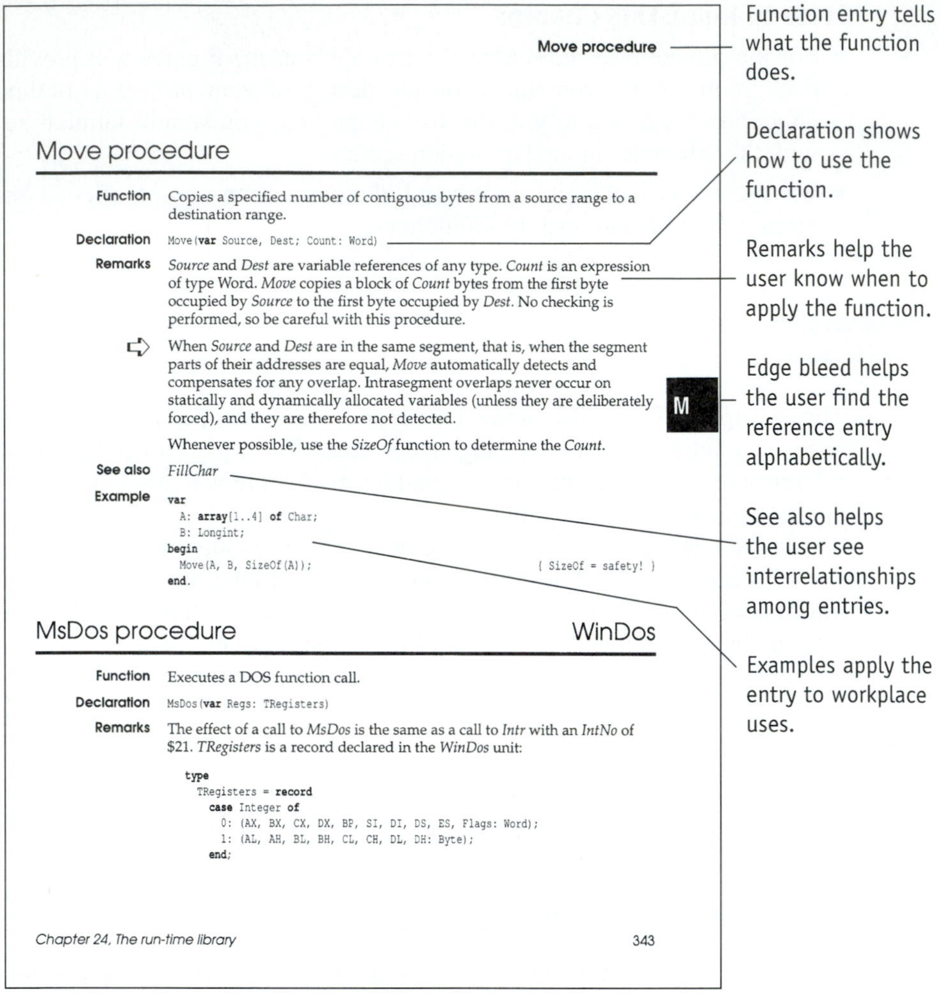

Function entry tells what the function does.

Declaration shows how to use the function.

Remarks help the user know when to apply the function.

Edge bleed helps the user find the reference entry alphabetically.

See also helps the user see interrelationships among entries.

Examples apply the entry to workplace uses.

Move procedure

Move procedure

Function   Copies a specified number of contiguous bytes from a source range to a destination range.

Declaration   Move(**var** Source, Dest; Count: Word)

Remarks   *Source* and *Dest* are variable references of any type. *Count* is an expression of type Word. *Move* copies a block of *Count* bytes from the first byte occupied by *Source* to the first byte occupied by *Dest*. No checking is performed, so be careful with this procedure.

⇨   When *Source* and *Dest* are in the same segment, that is, when the segment parts of their addresses are equal, *Move* automatically detects and compensates for any overlap. Intrasegment overlaps never occur on statically and dynamically allocated variables (unless they are deliberately forced), and they are therefore not detected.

Whenever possible, use the *SizeOf* function to determine the *Count*.

See also   *FillChar*

Example
```
var
    A: array[1..4] of Char;
    B: Longint;
begin
    Move(A, B, SizeOf(A));                          { SizeOf = safety! }
end.
```

MsDos procedure   WinDos

Function   Executes a DOS function call.

Declaration   MsDos(**var** Regs: TRegisters)

Remarks   The effect of a call to *MsDos* is the same as a call to *Intr* with an *IntNo* of $21. *TRegisters* is a record declared in the *WinDos* unit:
```
type
    TRegisters = record
        case Integer of
            0: (AX, BX, CX, DX, BP, SI, DI, DS, ES, Flags: Word);
            1: (AL, AH, BL, BH, CL, CH, DL, DH: Byte);
        end;
```

FIGURE 4.1   A Structured Reference Entry
This reference entry from the Turbo Pascal for Windows *Programmer's Guide* shows how elements of a reference entry create a structure for explaining program functions.

how to cross-reference information in various elements of the documentation set. For instance, the example in Figure 4.2 contains references to keywords included in the online help document.

Appendices

The appendix in a software manual often contains some of the most valuable information relating to the use of the program. Appendices allow documenters a place to put all the highly detailed, technical information that highly detailed, technical per-

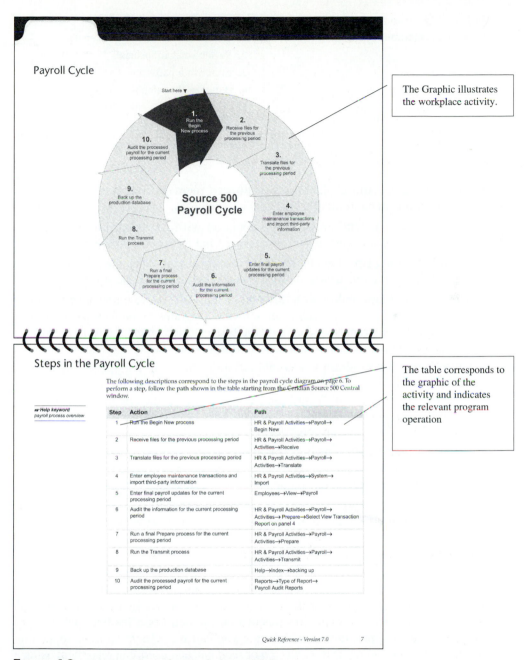

The Graphic illustrates the workplace activity.

The table corresponds to the graphic of the activity and indicates the relevant program operation

FIGURE 4.2   An Overview of a Workplace Activity
This reference source links program information to workplace activity cycles.

sons would want and use in the workplace. The print appendix contains information that's relevant and useful, but not *essential,* to all users. Examine some appendices to software manuals and you will find the kinds of information listed below.

FIGURE 4.3   Guidelines for Designing Reference Documentation

1. Choose the right form of reference.
2. Decide what to include.
3. Establish a pattern.
4. Organize the reference section.
5. Show how to use the reference information.

- **Error messages and explanations of how to recover from them.** Usually this information is in the form of tables with explanations.
- **Filenames and extensions of files associated with the program.** For program installers and maintainers this information is vital for troubleshooting.
- **Troubleshooting tips.** Usually this kind of information takes the form of a problem/solution structure, describing first the problem and then the solution.
- **Matrixes of compatibility with other programs.** Many programs have functions that are similar to or work with other software programs in the workplace. In fact, it's rare that a program is used entirely on its own. Supplemental information about how to transfer files or use information in other programs usually goes in an appendix.
- **Charts showing program key-combinations.** Almost all commands that appear on toolbars and menus have corresponding keystroke combinations. For example, to save a file in Microsoft Word I can choose Save from the File menu, or press the keystroke combination Shift-F12. The appendix is where users often look to find tables of these time-saving combinations.
- **Printer driver charts showing capabilities with various printer brands.** Many programs provide printer driver programs for use with various printers, or with other kinds of equipment like plotters, network connections, and so on.
- **FAQ.** Frequently Asked Questions are usually in alphabetical order and are a shorthand way of orienting reference users to information to solve their problems.

Update Information Sections/Readme Files

Readme files consist of text files that accompany programs on the distribution disk or come when you download the software from an Internet site. In a way these readme files a kind of primitive online documentation. Sometimes a readme plays a very important part in a program, in which case you call it a *rdmefrst.doc* or something like that. Usually you name it *readme.txt* or *read.me,* and users can read them using a notepad or word processing program.

Readme files contain the following kinds of information:

- Installation details
- Last-minute changes too late to be included in the manual or online help system
- New features in a recent release of a program

- Revision histories
- Errata in manuals and online help
- File descriptions
- Contents of directories
- Installation tips
- Compatibility requirements with other programs

Job Aids

Job aids, sometimes called *job performance aids,* are shorter forms of reference documentation that include keystrokes, definitions, brief processes, command summaries and other information useful to computer users as they are working. Rossett and Gautier-Downes, experts on job aids, define a job aid as "a repository for information, processes, or perspectives that is external to the individual and that supports work and activity by directing, guiding, and enlightening performance."[1] As reference documents for software, they comprise one of the most important elements of effective and efficient work. These authors also point out that job aids are not "instructional," that is, they do not intend to take the place of tutorials and are designed for users who have already received basic training in software. Their job is to provide information useful at the point of need.

Job aids also differ from minimalist forms of guided exploration cards in that they do not contain information specifically relevant to the overall tasks and goals of the user.[2] That is, they are not geared toward orienting the user to the overall activities within which the software use occurs. So how can they be task oriented, one might ask? The reason, I believe, is that while the task may not be embodied in the document (as it is, for instance in the example of the "payroll" process in (Figure 4.1), the task is nevertheless real to the user. The difference is that the task resides in the user's consciousness, and the user has performed it enough times that he or she knows it by heart. While they may not have clues as to the user's activity, they are often very specific to the user's workplace, sometimes written by persons who work there. What the reader doesn't know, often, is *which* task to perform, or the user needs a reminder of details too lengthy to memorize, such as codes associated with forms, or complicated menu items. This need for detail creates a performance gap: a desired workplace functionality of a worker with a lack of information to perform. Job aids containing lists of procedures (and often procedural information) can fill a need in this performance gap. (Figure 4.4)

Job aids consist of a number of forms, including keyboard templates, cheat sheets, laminated program overview cards, and quick reference cards, and other forms. Keyboard templates usually consist of very brief reminders that attach to the keyboard. Usually limited to defining keys, they can stick to the keyboard or overlap the keys. Once more popular than now, they help the user remember how a specific program uses the ten-key pad on the right-hand side of the keyboard, or what Shift-, Alt-, and Ctrl- keys the program uses. Usually keyboard templates contain a subset of the most frequently used keystrokes.

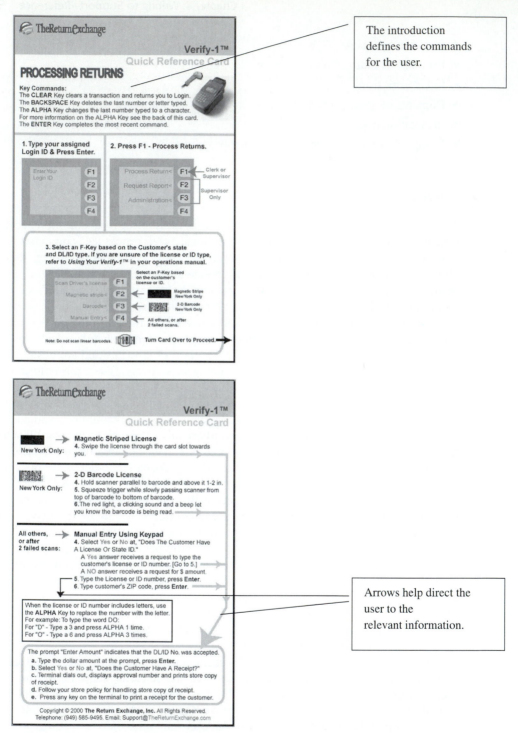

The introduction defines the commands for the user.

Arrows help direct the user to the relevant information.

FIGURE 4.4   An Example of A Job Aid

This quick reference guide won an award for its careful design for usefulness in the workplace.

Innovative Forms: Flipcards

Innovative forms make up a special class of reference information. It's very easy to put command information on foldouts, posters, and the back covers of documents. One such form is the flipcard, (Figure 4.5) a specially designed form that mounts a card on the monitor of the computer and contains brief overviews of program information.

The flipcard is an interesting adaptation of the job aid because it provides such useful, highly focused reference information right at the user's fingertips.

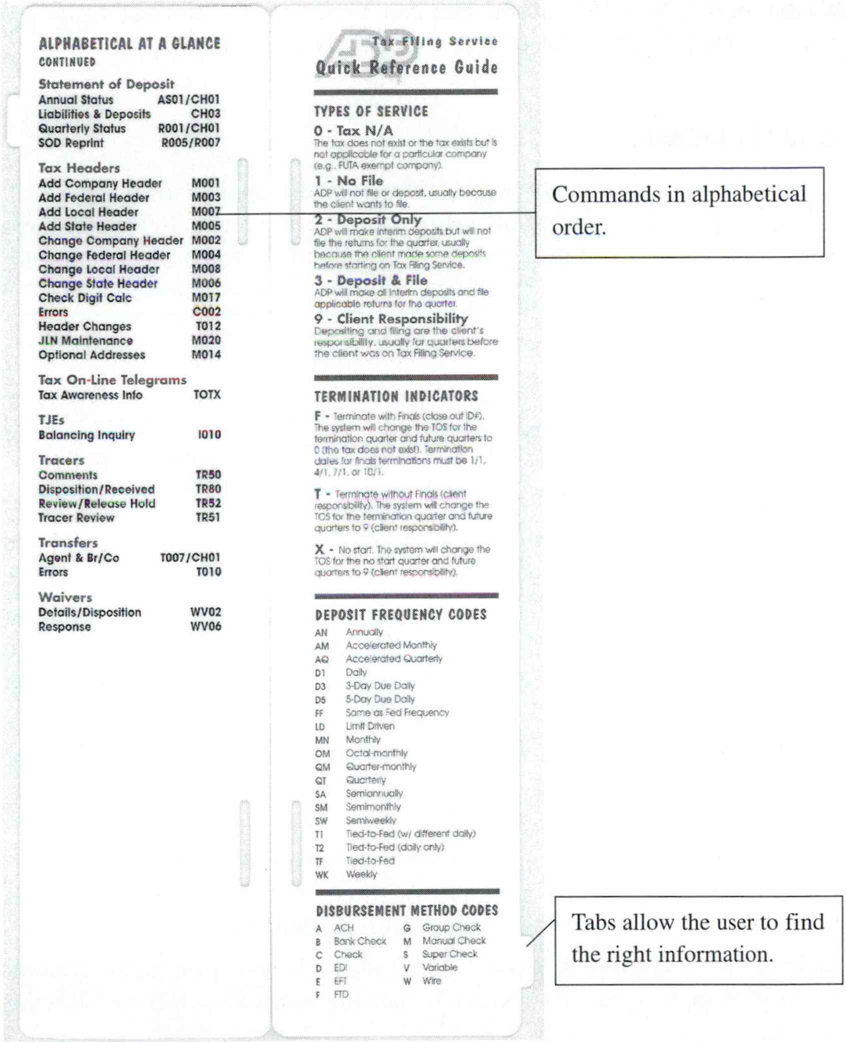

FIGURE 4.5   A Flipcard for Data Entry Reference Users
This document attaches to the edge of the computer screen and helps data entry persons at Automatic Data Processing Tax and Financial Services remember transaction codes and other information.

- **Easy to read.** You can turn easily flip to the kind of information you need.

- **Contains a surprisingly large amount of information.** Because of the size of the pages and the small type size you can put a lot of information in a small space.

- **Colored headings.** The headings in red make the sections of information easy to find. Also, icons assist in identifying information.

- **Unique and interesting.** The flipcards attach easily to the edge of the computer screen, thus delivering information where the user needs it without him or her having to shut down the program.

- **Affords easy access to information.** You can easily get to information on the cards simply by flipping from one to the other. It's always available.

2   Decide What to Include

When you look over your program or program *development specifications,* you will want to begin identifying topic areas you want to cover as reference. Below, you will find an overview of the kinds of information to include in a reference section, categorized into three groups: commands, interface elements, and terminology.

Commands

In reference, *commands* refers to all the instructions that the user employs to put the program to work. Information about commands would include the following:

- **Meanings of special function groups.** These groups contain commands that the program assigns to specific function keys, such as CTRL-C or ALT-Shift.

- **Explanations of set commands.** In some programs you can set certain things at the prompt. For example, in the VMS operating system, you can use "set host" to identify another host for your computer to connect to. Users need to know what kinds of things they can use the set commands for.

- **Definitions and uses of format commands.** These groups contain commands or tags, such as one might find in an HTML document, like *<fontsize=z>text</font>* to set the font size.

- **Special procedures for using utilities.** Programs have utility functions that do things like transfer files to other programs or filter images from different draw programs. Usually you provide a list of these with definitions.

- **Explanations of toolbars.** Almost all Windows-based programs contain toolbars: a series of icon-based commands that the user can select by clicking on the mouse.

- **Definitions of macros.** *Macros* refers to pre-identified and prewritten sets of commands that do certain things with programs that you need to list for the user. Most programs allow users to create sets of commands, but some provide them already. Manuals often describe them in a reference section.

Interface Elements

Interface elements refer to the parts of the screen that the user sees and has to read and manipulate in order to put the program to work. Information about interface elements would include the following:

- **Explanations of menus.** Menus contain most of the commands of a program and most reference consists of detailing what the menu items do.
- **Definitions of keys.** Here you define all the keys that the program assigns to special functions, like CTRL-S for save or ALT-T for tab, and so forth. With some programs, like large PC word processors, this group of interface elements gets very large, because they use the ALT, CTRL, SHIFT, ALT/SHIFT, CTRL/SHIFT, and ALT/CTRL key combinations along with letters, numbers, and symbols to allow keyboard input for just about every command in the program.
- **Labels of screen regions.** These refer to things like toolbars or scrolling regions of which you need to inform the reader.
- **Explanations of rulers.** Word processors and draw programs have different kinds of rulers to help the user measure and draw on the screen. These elements of the interface need referencing, often with a labeled screen.

Definitions of Terms (Glossary)

Glossaries contain definitions of terms that the user finds in the manual and needs to understand in order to work the program. Basically, you have two kinds of glossary items: concepts in the software, and concepts in the subject matter of the software.

- **Concepts that underlie the software.** Examples include *masks, shells, routers,* and so forth. Some will get defined with the procedures they relate to, but you'll collect them all in the glossary.
- **Terms relating to the subject matter of the software.** For example, *general ledger, connectivity, gutter.*

What to Include in a Single Reference Entry

Imagine that you have a reference section in a manual or help document that covers the commands of a program in alphabetical order. You will still have to answer the question of what to put in each command description. Clearly, the simplest response would involve simply listing the command and a brief definition. But you could go further. In fact, you will find a bewildering array of things to include in each reference entry. Consider the following list of possible elements of reference entries. For the sake of convenience, these information elements are divided into groupings relating to the kinds of information each presents to the user.

Conceptual information (emphasizing the idea of the command and its function)

- The command itself
- Definitions and descriptions of the command and what it does

- Explanations of how the command affects the user's work
- Notes to explain the command further
- Sample reports showing what the command produces

Structural information (emphasizing the relationship of the command to other commands)

- Access sequences to tell the user how to get to the prompt where the command works
- Screens or menus showing where to find the command
- Alternative commands (such as keyboard or mouse alternatives)
- Cross references to tutorials, procedures, other entries where the command gets further explanations
- Tables showing variations of the command and how it relates to other commands

How-to information (emphasizing the use of the command)

- Steps for executing the command in tasks
- Examples showing the command in a syntax statement
- Tips for when and how to use the command efficiently
- Error messages for when you use the command incorrectly

Technical information (emphasizing the software programming associated with the command)

- File specifications for what files the command uses
- Input requirements (characters, data, or both) telling what kind of data to use with the command
- Warnings telling how not to lose data when using the command
- Syntax diagrams telling how to use the command in statements
- Switches that allow the user to tailor the command to various needs

Clearly, you cannot include all the items in the list above in each reference entry. Your task will be to include or omit items that match your particular user's characteristics and workplace needs. In fact, the task orientation of your reference section may depend on your ability to build a set of elements that meshes with your user's expectations and needs.

The next guideline demonstrates how you can select from the list of all possible reference elements to create a pattern for your reference entries.

3   Establish a Pattern

The key to reference material lies in patterns: repeating the same set of elements over and over again so the user learns to identify how each element of an entry functions. Like other methods of building patterns into reading material, you want this one to assist recognition through regularity, yet allow for enough flexibility to present variations in information effectively. The following selection of topics works well as a recognizable yet flexible pattern for reference entries.

- **Definition:** Tell what the command or function does.
- **Explanation:** Tell how to apply the command or function.
- **Example/syntax:** Give an example of the command or function in use.
- **Step-by-step:** Present abbreviated steps for using the command or function.
- **Warnings/cautions:** Let the user know what problems might arise.

4 Organize the Reference Section

As part of your design of reference sections, you will have to decide on an organizing principle. Unlike documentation on the teaching and guidance level of task orientation, the reference does not come with a built-in sequential organizational scheme. You will have to decide on what comes first, what comes next, and so on. And you will have to make a decision that supports, overall, the task orientation of the manual. In general, you have two basic choices for organizing your reference section: alphabetical or menu-by-menu.

Alphabetical Organization

As the name implies, when you set up the reference section alphabetically, you usually heap all the functions of the program together, regardless of the menu structure, and go through them one at a time, starting with the a's. The MS-DOS 5.0 *Reference Guide* follows this method, starting with the *append* command and ending with the *xcopy* command. Like other documents, it divides the commands into groups: MS-DOS commands, batch commands, *config.sys* commands, and so on. Each command grouping starts the alphabet again. Likewise, you can organize your reference section according to sections of like commands, like **command sets,** or other topic areas. Notice that in Figure 4.6 the order is based on the actual commands (in parenthesis) but each entry begins with the name of the command.

In the case of topic areas, or command sets, the question becomes how to organize the sections. You may, for example, put them in a simple-to-complex order, including the basic ones at the first and the advanced command sets at the end. Or you might choose to start with the more abstract, concept-oriented information and progress to greater and greater levels of concrete, procedural information. The Lotus *Manuscript* manual, for example, starts with "A View of Manuscript" and progresses to various categories of conventions, file names, interface conventions and other features. The Adobe Postscript Language *Reference Manual* begins with definitions of terms that are:

> essential to understanding the problems the PostScript language is designed to solve and the environment in which it is designed to operate. Terminology introduced here [in the second of eight chapters] appears throughout the manual. (p.11)

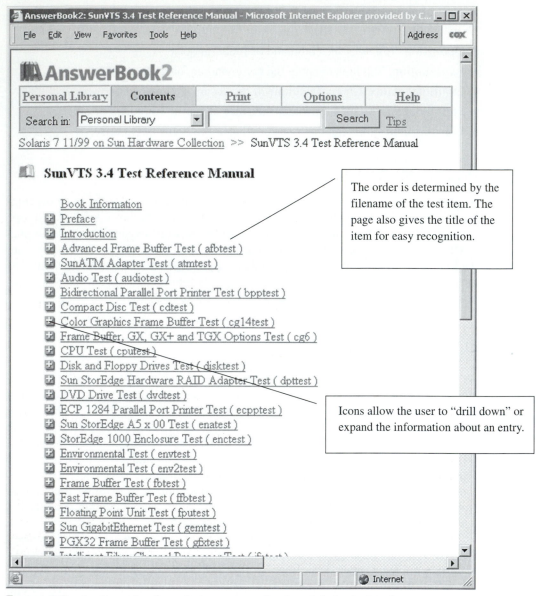

FIGURE 4.6   An Example of a Reference Manual Using Alphabetical Order
This manual organizes topics according to the alphabetical order of commands.

Glossaries of functions follow an alphabetical organization because of the great virtue of predictability. As long as your user knows the name of the function, he or she can find a description of it easily. It also strongly emphasizes the features of the program over the functionality. It lets the user know what features the program contains, and so appears comprehensive. Indeed, organizing this way helps the writer by allowing you to make sure you explain all the features.

The drawbacks of the alphabetical method often outweigh its advantages. While it may reassure some users to see a familiar alphabetical list of features, it does little to support the task orientation of your manual. For instance, since the items appear alphabetically, you always have to include access information for each item, telling what menu it goes under or how to get there. The user could get this access information from the surrounding text, but not if the next item occurs on a menu from another part of the program.

Menu-by-Menu

In many ways, the alphabetical organizational structure does not enhance the task orientation of a manual as much as a menu-by-menu organization. Using this scheme, you set up your reference section by menus, according to how the user sees them in the program. You could alphabetize the menus, but it makes more sense to arrange them as they appear on the program's main screen or menu bar, as does the example in Figure 4.7. Often such menu groupings reflect workplace tasks anyway. Using the menu-by-menu method, you start with the main menu, then secondary menus. Present each menu, and then, in the subsequent pages, describe each of the commands in the order they appear on the menu. Present each command in a standardized format. You can also include dialog boxes, confirmation boxes, and other elements the user sees when using each command.

The very strong advantage of the menu-by-menu system lies in its reinforcement of the task orientation of your work. The user sees the information in the same shape as he or she does when using the program. Thus, the document and the screen reinforce one another. The user's experience with the program itself helps with the understanding of the manual, and vice versa.

5 Show How to Use the Reference Information

In many cases, your reference section or document will require no instructions. Maps of menus or one-page summaries of commands represent this kind of self-explanatory reference page. However, with reference entries containing multiple elements, you should tell the user what pattern you intend to follow. This establishes the pattern in the user's mind, sets up the right expectations, and serves as a handy reminder of how you organized each entry.

Usually you instruct your user in the organization of the reference information in an introduction. In some instances the introduction takes the form of a paragraph or so explaining each of the entries, along with other items. Such an introduction should explain the following:

- **Who should use the information.** Remind the user that the information will best serve the advanced user, or users with special tasks, such as supervisory tasks, or transferring data from other programs.
- **How you organized the information.** Point out whether you used alphabetical order or menu-by-menu order, or some other category.

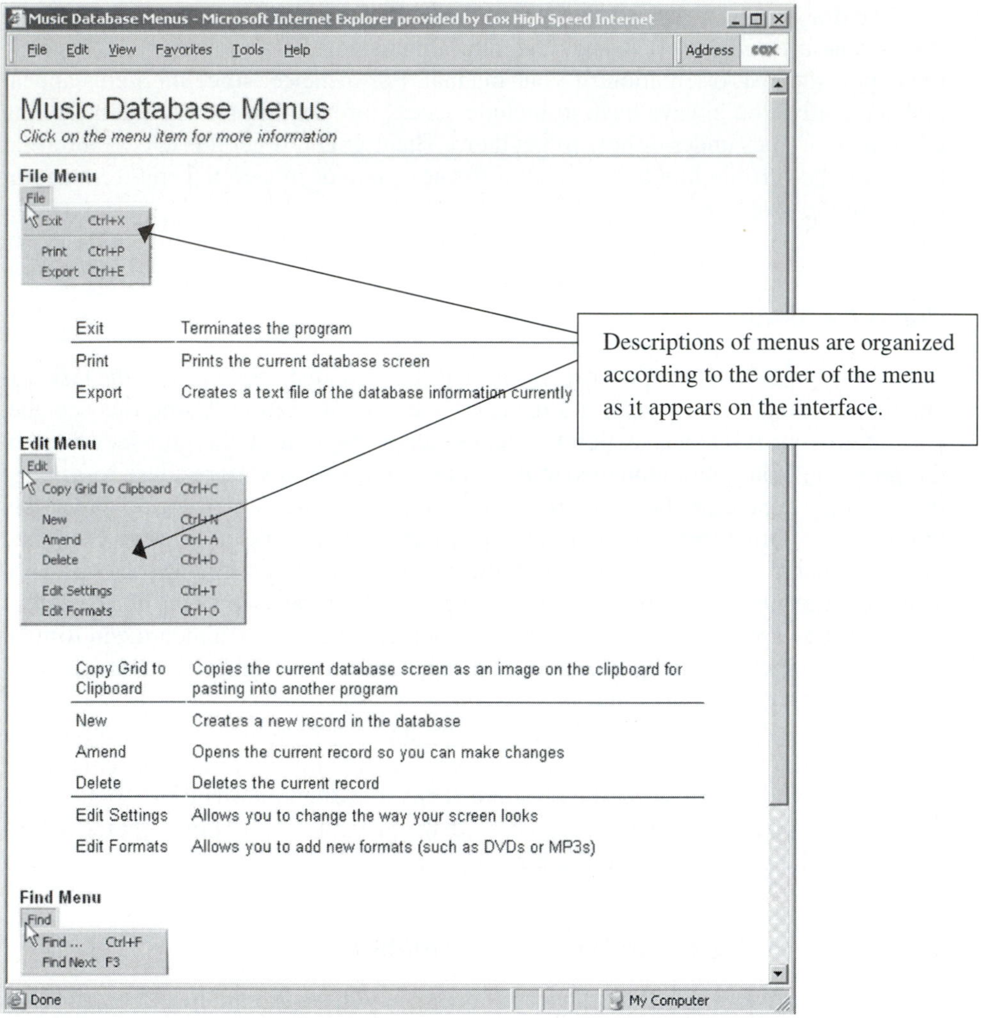

Music Database Menus - Microsoft Internet Explorer provided by Cox High Speed Internet

File　Edit　View　Favorites　Tools　Help　　　　　　　Address　cox

Music Database Menus
Click on the menu item for more information

File Menu

File
Exit　Ctrl+X
Print　Ctrl+P
Export　Ctrl+E

Exit	Terminates the program
Print	Prints the current database screen
Export	Creates a text file of the database information currently

> Descriptions of menus are organized according to the order of the menu as it appears on the interface.

Edit Menu

Edit
Copy Grid To Clipboard　Ctrl+C
New　Ctrl+N
Amend　Ctrl+A
Delete　Ctrl+D
Edit Settings　Ctrl+T
Edit Formats　Ctrl+O

Copy Grid to Clipboard	Copies the current database screen as an image on the clipboard for pasting into another program
New	Creates a new record in the database
Amend	Opens the current record so you can make changes
Delete	Deletes the current record
Edit Settings	Allows you to change the way your screen looks
Edit Formats	Allows you to add new formats (such as DVDs or MP3s)

Find Menu

Find
Find ...　Ctrl+F
Find Next　F3

Done　　　　　　　　　　　My Computer

FIGURE 4.7　An Example of a Reference Manual Organized by Menus
Organizing by menus has the advantage of allowing users to look up menu items that they find confusing.

- **Elements of each entry.** List the elements of each entry, telling what kind of information each contains.
- **Relations to other sections of the documentation.** Indicate cross-references to other parts of the documentation set, like the procedures or tutorial or help. This helps reinforce the idea of the documentation system in the user's mind.

Figure 4.8 illustrates this kind of simple introduction to a reference section.

As an alternative to an introduction, many reference manuals and sections include a sample *library entry* showing categories and telling what each does. The

---

AccountMaster User's Manual

Reference

This section contains information for users who have some familiarity with AccountMaster, and who want to look up specific commands.

Commands are arranged alphabetically, and each entry, where appropriate, will contain the following elements:

Name: (the name of the command)
Access: (the keystrokes for getting to the menu containing
                    the command)
Usage: (what the command does)
Exceptions: (special qualifications in the way you use the
                    command)
Notes and Tips: (advice for increasing your efficiency using the
                    command)
See Also: (page and screen numbers of related
                    AccountMaster commands.)

You can find additional information in the AccountMaster User's Guide and Help.

5-1

FIGURE 4.8   **Introduction to a Reference Section**
This section helps orient the user to the pattern used in the reference section.

advantage of using this technique lies in the capability to show an entry when you have a highly structured, visually-oriented format. You give the user a chance to see an entry, labeled clearly. In one example, the writers of the Codewright *Programmer's Reference* included a sample entry with numbers for each of the items (i.e., "Function Name," "Usage Syntax," "Icons," and so on). They then explained what each numbered entry meant in a list of definitions taking a few pages.

The example in Figure 4.9, from a Turbo Pascal *Programmer's Guide,* illustrates the use of a sample library lookup entry.

Discussion

Reference documentation means those parts of your documentation product that you design for the reader to be able to look up information about the program. Almost all software manuals contain reference sections describing details such as

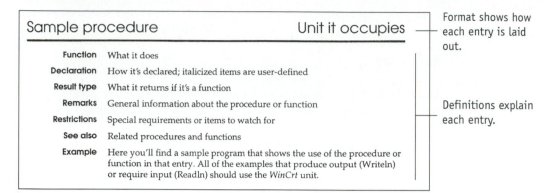

FIGURE 4.9   Library Entry to Orient Reference User
This example from the Turbo Pascal for Windows *Programmer's Guide* allows users to know the pattern to follow in each entry.

what printers the program supports, or what settings the program needs to perform certain functions. Ordinarily users do not read reference sections completely; they only go there for help.

Reference documentation functions on the ***support*** level of task orientation. By support, I mean that much of the motivation for success and guidance in the use of the program comes from the user, not from the document. So you don't really push the users, or constrain their focus to your carefully pre-selected path; instead, you try to respond "on call" and let the users take the initiative. Unlike in ***teaching*** or ***guidance*** levels of task orientation, the reference document contains very little "how to" information. If a reference entry does contain steps, it functions merely to jump-start the user, who then can take off on his or her own power. The reference entry contains primarily data (commands, keystroke lists, menu item definitions), with well-designed but relatively limited access and tracking devices (such as elaborate headings or routing information).

Also, unlike documentation that functions on the other levels of task orientation, reference documentation establishes, through its style and tone, the least engaging relationship with the user. The relationship between the writer's persona and user more formal, almost businesslike. The business like nature of the reference section reflects the special needs of the reference user.

Understanding the Reference User

Often the reference user knows the program well, at least better than the novice or casual user. Put yourself in the reference user's place and consider the following scenario. You have been using a desktop publishing package for a few months and have done one or two projects with it. But you have never tried to import graphics files from, say, PC Paintbrush. Now, however, a client has some graphics done in PC Paintbrush and wants them in the manual. You have to consult your program to see what utility to use to import these specially formatted files. At the last minute, after you have already taken on the job and have much of the document put together, you turn to the reference manual that accompanies your desktop publishing program and start looking for the utility that will do the trick.

Given this description, what generalizations can we make about you as a reference user? You do not like to waste time looking things up. Probably, when the solution shows up, it will have to reside in your short-term memory only long enough to do the job of importing the files. The scenario above also indicates something of the psychology of the reference user: He or she dislikes leaving the screen to go looking for information. Thus, the well-designed reference section should cater to the values of efficiency and, above all, immediate usability.

Another important characteristic of reference sections relates to what we have discussed above: the structure of the page. In reference sections, structure, often parallel or repeated structures, abounds. To an extent, a reference entry resembles a card in the card catalog of a library or a record in a database. You fill in a form. Sometimes you structure entries with relatively simple headings, and sometimes you put information into tables and matrices. At the heart of it, you have to establish a pattern and then follow it, so the user quickly understands where to look. Generally, the more structure you can build into your reference entries without overdoing it, the more usable your reference entries. We will discuss the structuring of reference entries below.

Large systems, network programs, programming languages, operating systems—these kinds of software programs have lots of look-up commands to make them work. These system get used by computer professionals whose advanced information needs differ from those of the novice or intermediate user. These users often know a lot about a particular system and they know a lot about computers in general. If your user analysis shows that you need to support these kinds of users who know both your program *and* computers in general, then you can increase the proportion of reference material.

Understanding a Reference Entry

The elements of your reference entries deserves some special consideration. To understand a reference entry, such as the one in Figure 4.10, you need to look to the idea behind the repeated categories, column heads, or other user-oriented reference elements. These introduce, orient, inform, and direct the user in the search for a solution. But what exactly makes them work? They work because each one answers a question your user might have about a function or command. Figure 4.10 illustrates the psychology behind using structured reference entries.

Notice how the elements of a reference entry respond to the needs of the reference user. These elements come from a program called FormMaker, used by managers to attach multiple notes to forms during the review process for developing forms. The authors of the manual chose to alphabetize the functions—a strategy that bets that the user will know the right names for things or get set straight by the index. At any rate, each entry includes the following:

- **Access information.** Access information tells what menu contains the function, and shows the chain of commands the user needs to issue in order to get to the function.

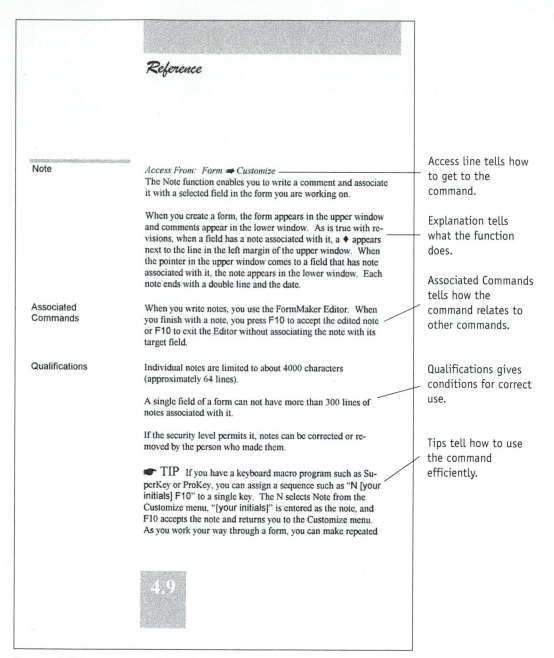

Reference

Note

Associated
Commands

Qualifications

Access From: Form ➡ Customize
The Note function enables you to write a comment and associate it with a selected field in the form you are working on.

When you create a form, the form appears in the upper window and comments appear in the lower window. As is true with revisions, when a field has a note associated with it, a ✦ appears next to the line in the left margin of the upper window. When the pointer in the upper window comes to a field that has note associated with it, the note appears in the lower window. Each note ends with a double line and the date.

When you write notes, you use the FormMaker Editor. When you finish with a note, you press F10 to accept the edited note or F10 to exit the Editor without associating the note with its target field.

Individual notes are limited to about 4000 characters (approximately 64 lines).

A single field of a form can not have more than 300 lines of notes associated with it.

If the security level permits it, notes can be corrected or removed by the person who made them.

☞ TIP If you have a keyboard macro program such as SuperKey or ProKey, you can assign a sequence such as "N [your initials] F10" to a single key. The N selects Note from the Customize menu, "[your initials]" is entered as the note, and F10 accepts the note and returns you to the Customize menu. As you work your way through a form, you can make repeated

4.9

Access line tells how to get to the command.

Explanation tells what the function does.

Associated Commands tells how the command relates to other commands.

Qualifications gives conditions for correct use.

Tips tell how to use the command efficiently.

FIGURE 4.10   A Well-Structured Reference Entry
Each part of this entry from the FormMaker Reference Manual answers an important question posed by the reference user.

- **Function definition.** The function definition tells what the function does and what it enables the user to do.

- **Associated commands.** Associated commands tell what other keys or commands the user needs to use the function effectively.

- **Qualifications/special cases.** Qualifications and special cases offer information about *field* lengths, maximum or minimum allowable inputs, time limits, or pre-requisites (i.e., you can't save a file that you have not created). It also tells what to do in special cases, such as hardware limitations or dealing with earlier versions of the program.

- **Tips.** The tips section of the reference entry tells ways to use the function efficiently, plus shortcuts and potential problems. It represents software "wisdom": good advice from an old pro.

As you can see, even a simple, five-part structure of a reference entry contains plenty of useful information for the experienced user. To help see clearly how such an entry works, think of each of the items as answering some need or question of the reference user.

- **How do I get to the function?** Probably the experienced user needs to know simply how to find a function more than any other type of information. Many advanced users need only this to get them going. Once they have the sequence of commands, many users will close the book and return to the program to try them out.

- **What does the function do?** After checking the chain of commands to get to a function, the experienced user would want a very brief, performance-oriented explanation of what the function does. This allows him or her to double-check whether this or some other function really applies.

- **What other commands do I need to know about?** The user wants to know how to use the command along with other commands, as well as how to get out of trouble. The reference entry should indicate what other commands to use and how to back out of trouble.

- **When can I use the function?** In case the user had difficulty using the command, he or she needs to know that special conditions might exist, such as disk drive incompatibility or file size limits. Certain functions will not work unless the user has set the program up just right. Certain menu items stay dimmed until the user loads a file or creates an entry.

- **How do I use the function well?** The experienced user wants to make the most out of the system and needs to know any short cuts or efficiency measures that apply. Here one might include an example of the function in use.

As we can see from the discussion above, the elements of a reference entry do much of the work of establishing the task orientation of your manual. They make up the structural elements that you use to analyze and present each reference entry—a command, a menu option, or a definition. In your study of reference documentation—look-up documentation on the support level of task orientation—make sure to study any reference sections you come across to see what elements the writer has used to help structure the entry information to the user's needs.

Glossary

commands: refers to that part of a software program that allows a user to tell a program what to do. Example: "print," "compile," or "open a file." Most commands are run by menus. Often reference documentation consists of descriptions of program commands and what they do.

command sets: refers to groups of commands that resemble one another, or use the same control keys on the keyboard. Example, "CTRL-S," "CTRL-Z," and "CTRL-Y," belong to a command set.

development specification: descriptions of a software program that describe the users and functions of a program. These documents guide the work of programmers and writers while the program is being developed. They contain lists of program menus and features.

guidance level: a type of documentation designed to lead the user through a procedure one step at a time from a designated starting place (such as a certain menu) to an ending state (such as a report printed.) Guidance-level documentation (or procedures) defines the task for the user, but does not teach the task. See also **teaching level** and **support level.**

interface elements: parts of a program that the user sees. These elements include menus, toolbars, dialog boxes, and buttons.

keyword searches: electronic and automatic searches of the topics in a help system to find pre-identified words relating to certain topics. Example: the keyword "File" might call up the following topics: "opening files," "saving files," and "deleting files."

levels of support: categories of kinds of information supplied to users. Levels relate to the teaching level (tutorial), the guidance level (procedures) and the reference level (reference). Levels differ in terms of purpose: to teach, to walk through step-by-step, and to provide data. They also differ in the relationship of the writer to the user: from very close and controlling with the teaching level, to distant and businesslike with the reference level.

support level: a type of documentation intended to provide the user with a piece of information needed to perform a task. Reference documentation does not define the task for the user, but provides the necessary data the user needs to complete a task. See **guidance level** and **teaching level.**

teaching level: a type of documentation intended to instill a knowledge of how to use a program feature in the memory of the user. Teaching-level documentation (tutorials) aims to enable the user to perform a task from memory. See **guidance level** and **support level.**

☑ Checklist

Use the following checklist to evaluate the efficiency of your design for reference.

Reference Checklist
Choosing the Right Form of Reference

Have you chosen the right form of reference to match your users' needs?
- ❏ Reference section (tables of date, definitions, command descriptions)
- ❏ Quick reference (card, booklet, brochure)
- ❏ Appendix (tables, messages, troubleshooting)
- ❏ Update information/ ReadMe files (disk file containing update highlights)

❑ Job aids (reminders, command lists)
❑ Innovative forms: flipcards

Deciding What to Include in the Reference Section

Which of the following elements meet your users' needs?
❑ Commands
❑ Interface elements
❑ Definitions
❑ Concepts
❑ Relationships among commands
❑ How-to information
❑ Technical information

Deciding What to Include in Each Reference Entry

Identify which of the following elements you want or need to include in each reference entry to meet your user's support needs.
❑ Definition: Tell what the command or function does.
❑ Explanation: Tell how to apply the command or function.
❑ Example/syntax: Give an example of the command or function in use.
❑ Step-by-step: Present abbreviated steps for using the command or function.
❑ Warnings/cautions: Let the user know what problems might arise.

Establishing a Pattern

❑ Have you established a pattern for each reference entry?

Organizing the Reference Section

Which of the following organizational patterns do you intend to use?
❑ Alphabetical
❑ Menu by menu

Showing the User How to Use the Reference Information

Have you written a section that tells the user the following things?
❑ Who should use the information
❑ How you organized the information
❑ The elements of each entry
❑ Cross references
❑ Sample library entry

Practice/Problem Solving

1. Write a Tool Bar Reference

Write a brief reference document describing the commands on the toolbar in Figure 4.11. You can use a tool bar from another program if you like.

FIGURE 4.11   The "Standard" Toolbar from Microsoft Word
Describe the tools on this toolbar.

2. Analyze Reference Formats

According to an analysis of this information, "This online guide targets engineers and engineering technicians who are familiar with general-purpose interface bus instruments and who want to control the instrument remotely."[3] Figure 4.12 and Figure 4.13 both show the same information, but in different formats. Compare and analyze the formats according to the guidelines for reference design in this chapter.

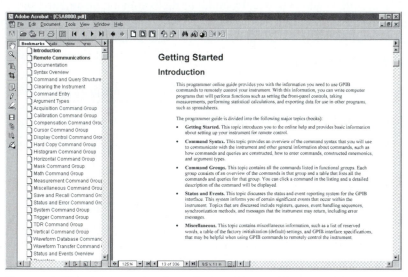

FIGURE 4.12   An Example of a Reference in Portable Document Format
This version of the document shows the features of the Adobe Acrobat document reader.

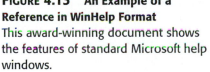

FIGURE 4.13   An Example of a Reference in WinHelp Format
This award-winning document shows the features of standard Microsoft help windows.

3.   Analyze a Reference Section

The example of a reference system in Figure 4.14 operates allows the users (students, faculty, and staff at a university who need to use the system to set up their email accounts) shows a number of features desirable in a help system. Study the sample screen in the figure (or visit the site for the system at the following address:

http://www.exchangepop3.com/ep3/help/frameset.ref.html

Write a brief summary of the features you see in this system that you think make it an effective reference system for both beginning and advanced users.

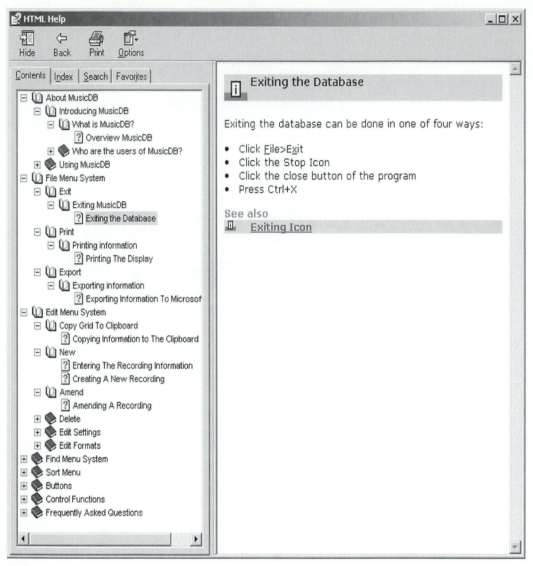

FIGURE 4.14   A Sample Reference Guide
What features can you identify that make this reference entry effective for users of a database system?

PART II

The Process of Software Documentation

CHAPTER 5

Analyzing Your Users

The user analysis makes up the basic research phase of the documentation process. In preparing a user analysis you use interviews, questionnaires, and surveys to gather information about users. As an activity, the user analysis requires you to make contact with persons who might use the software that you want to document. It helps you find out how your software fits in to the pattern of workplace activities of potential users in productive and appropriate ways. The analysis itself consists of inquiry into eight areas:

1. Tasks and activities the user will perform
2. User's informational needs
3. User's work motivations
4. Level of the user's computer experience
5. User's knowledge of the program's subject matter
6. User community
7. User's learning preference
8. User's usage pattern

Information gathered during the user analysis informs the goals you set for the documents and the features included in them. You will use information from your user analysis later in the development process when you design documents.

How to Read This Chapter

- If you're reading this chapter to *understand,* read the Example, then the Discussion section, then the Guidelines.

- If you're reading this chapter to *do,* read the Example section, then the Guidelines section to find some techniques for forming focus groups, creating questionnaires, and conducting interviews. Also check out the section "How to Design Documentation for Intermittent Usage" in Chapter 10, "Designing for Task Orientation."

Example

User analyses can get very detailed because the more you can immerse yourself into the workplace world of your users, the more you will learn how to help them integrate a software program into it. The example in Figure 5.1 illustrates how you can summarize information from interviews, observations, and surveys using the analysis checklist provided at the end of this chapter.

Degree Audit User Analysis

Project: Degree Audit, v. 1.0
Client: University Department of Engineering
Description: The Degree Audit program was created to help advisors to prepare accurate degree plans for students in the engineering program and to help advise students with existing plans. It also builds a knowledge base of existing plans for future reference and problem-solving.
Users: *Advisors* (who meet and work with students in building and maintaining degree plans prior to graduation) and *Administrators* (who coordinate the work of student advisors)

Advisor

Job description	The advisor keeps up with a student's degree progress and determines when they will graduate. The advisor helps students shape their schedules to meet their educational needs.
Work objectives	The advisor must understand the program and know how it works so he or she can give the correct information to the students. Also, the advisor must be able to communicate with the system administrator in case of any errors in a student's file.
Information needs	The knowledge of the program requirements, classes, scheduling principles, and pre-requisites should be high. Advisors should have a clear understanding of which classes a student needs to take for a particular degree emphasis in order to graduate. Additionally, advisors should have a high degree of knowledge about computer programs. They use computers now to access class information, and should not have difficulty adapting their existing advising processes to this new system.
Software environment	The advisors will have the software installed on their computers. They will learn the program through the quick start guide and online help. They will then use the program as a more efficient way to advise students, cutting down on paperwork.
User tasks	• Consult with students about degree plans • Evaluate students' degree plans • Display student degree plans • Update degree student degree plans • Simulate changes to student schedules

Continued

FIGURE 5.1   An Example of a User Analysis
This analysis demonstrates how the writer distills information about users of a program into an overview of computer use in the context of workplace activities. As part of an information plan (Chapter 6) it would form the basis for document design.

Continued

Administrator

Job description	The administrator oversees the program and advisors and maintains and updates the knowledge base for the Degree Audit program.
Work objectives	The administrator must learn how to use the program to create accounts for advisors, log in and use the system to update degree plans, delete user logins, and make changes/updates to the system.
Information Needs	Knowledge about the subject matter is high. Administrators understand the advising process and can serve as resources to individual advisors. Administrators should also have a high knowledge of computers and basic computer software (scheduling type software, especially). The user would use the program on the job to create accounts for new advisors and make updates to the system when degree requirements change.
Software environment	Users will access the Degree Audit program through the Web browser on their personal computers. The user would then learn how to use the program by reading job aids and online help. The user would then use the software as a tool to manage the advisement process efficiently. This software would cut down on paper-based processes of degree auditing.
User tasks	• Coordinate the work of student advisors • Maintain the Degree Audit program • Add and modify degree plans • Change and update the knowledge base • Introduce the degree audit process to new advisors

FIGURE 5.1   An Example of a User Analysis

Guidelines

Figure 5.2 lists the steps involved in a user analysis. Each step is explored in the paragraphs that follow.

1   Choose Users Carefully

When you approach the task of analyzing your users, you may have to use your imagination in identifying whom to select as examples. For starters you should list as many types or groups of users as you can, no matter how improbable a user might sound at first. Table 5.1 shows some examples of user groups for programs.

Study the list of possible users you have brainstormed and ask yourself: "Which users would most probably use the program and which users can I interview most easily?" The answers to these questions will help you narrow your search. Look for the most typical users and users with whom you can develop a working relationship.

When you choose your readers you should try to show sensitivity to the culture they come from. "Culture, " or the patterns of behaviors and attitudes that persons characterize everything from nations to workplaces, can affect how people react to

FIGURE 5.2   Guidelines for Conducting a User Analysis

1. Choose users carefully.
2. Anticipate transfer of learning: Study users before and after using a program.
3. Research professional behaviors.
4. Write use cases.
5. Plan interviews carefully.
6. Involve users in all phases of the project.
7. Identify document goals.
8. Tie the user analysis to document features.

TABLE 5.1   Examples of User Groups for Programs

Program	User Group	Tasks
Network Program	System Administrator	Set up user accounts and passwords.
	User	Log in, run programs, store files.
Jogging Log	Coach	Enter data about athletes, create reports, track progress.
	Athlete	Enter data about workouts, create reports, track progress.
Math Tutor	Teacher	Set up tutorials, track progress.
	Student	Run tutorials, score tests, record progress.
Family Financial Planner	Financial Consultant	Create customer profiles, reports for customers.
	Head of Household	Enter data about assets, income, investments, create tax reports.

your asking them questions about their work, examining their software habits, and, in general, using them as subjects in order to do your work. They may represent different national cultures than yours (Chinese, French, etc.) or they may represent different work cultures, such as the visual cultures of engineers,[1] or the organizational culture of bankers. You may need to get authorization to interview people in the workplace, or you may need, in the case of Japanese culture, for example, to have an introduction ahead of time to help your users understand your presence and accept your intrusion.

Once you have narrowed the list you can conduct user interviews. If you don't work with the users, locate some people with similar jobs in your own organization. Call them, explain your documentation project, and ask if they would let you discuss their work with them. As you build your list of job tasks—activities users will use the program for—pay particular attention to information-related tasks: communicating, storing, sharing. Identify users with these concerns; they should be able to tell you a lot about how your software could be implemented in task-relevant ways.

2   Anticipate Transfer of Learning: Study Users Before and After Using a Program

The user interview should provide you with a wealth of information about a particular user. But remember, when it comes to job duties, you want to describe the activities the user does, *without the benefit of your program.* You can then evoke these activities in your tutorials and procedures. For example, talk to a club president about the kinds of documents he or she writes, as well as about how he or she would use your word processor. Another example: if your user conducts mathematical research, find out what computational tasks he or she performs, as well as hypothetical information about what he or she *might* do with your graphing software.

Study the example in Figure 5.3. This description of user tasks was prepared during the documentation planning phase for some database management software (DBMS). Research shows that skills learned in the workplace can transfer to skills using software. The more you understand these skills in your user, the better you can transfer this valuable learning to the use of your software.

Notice even small facts about users, such as their attitudes toward computer programs or their habits when developing computer skills. Look for artifacts like notes stuck on bulletin boards, diaries of skills, or third-party computer manuals that users employ in their daily struggle to get the most out of their existing technology. Learn what other types of employees the user interacts with. Find out what values— efficiency, team orientation, environmental concerns, ethical issues—confront your

Every Friday, Charlie had to do the weekly sales summary for his boss, Adam. He downloaded raw data from the mainframe into a comma-delimited ASCII file his spreadsheet could read. Next, he used the spreadsheet's database functions to extract the information for each salesperson, one person at a time, summed the figures, and manually transferred the total to another part of the spreadsheet. He manually maintained weekly totals since January 1, and adjusted the line chart to include another week.	Describes the user before using the program
With his new DBMS, he creates a single query that requests the sales figures since January 1 from the mainframe, cross-tabulates them by salesperson by week. A one-page report with a 3-D line graph is automatically updated.	Describes the user after using the program

FIGURE 5.3   An Example of Users Before and After Using a Program
A description of a user task should describe the non-computerized work. This will help the documentation designer find ways to promote transfer learning.

user, and how he or she responds to them. Such information, called *tacit knowledge,* can help you understand the way users see their work: attitudes and understandings that can provide the energy to get them to use a tutorial or follow a procedure.

3 Research Professional Behaviors

In some instances you don't have actual users at your disposal or you need to research professional behaviors with which you're unfamiliar. In such cases you can construct a mock-up of the user, a kind of model to use as a resource in making design decisions. Do this by focusing on the user's occupation as described in easy-to-find library materials.

Occupations tend to have a degree of generality about them. You can find out about what tasks people perform in various occupations by consulting occupational guides such as *Occupational Outlook Handbook (http://www.bls.gov/oco/home.htm),* or O*Net Online (http://online.onetcenter.org/). In addition, you can consult of the many industry-specific guides listed at the University of California, Berkley University Health Services Career Occupational Links page. (http://www.uhs.berkeley.edu/Students/CareerLibrary/Links/occup.cfm). Placement services at your local college or university often have computerized career information systems containing descriptions of job titles.

Beyond this, you can look over some of the job descriptions found in publications put out by companies (usually found in placement services, personnel agencies, or from the companies themselves. Figure 5.4). The descriptions you find in these kinds of publications may provide the best information about jobs and tasks, because they show how different organizations define different jobs. A systems analyst, for example, gets mentioned in just about every guide to job titles, but the definition of a systems analyst varies greatly from company to company. To find out what a systems analyst would do in a bank as opposed to a communications firm, you can consult the literature produced by specific companies found on their websites.

4 Write Use Cases

You should always remember that your goal is to uncover motivations, behaviors, values and knowledge pertaining to your users that might not be visible on the surface. Such information is called tacit knowledge and it often embodies the "pay dirt" of a user analysis. It informs the activities, not the operations, that your software manual or help system supports. For example, you may discover that a user provides a sales report each week and use that activity as a focus for procedures in your database program that help create sales reports (data gathering, sorting, printing, etc.) These are the computer operations that the user would perform. But what you may miss is that the report needs to be prepared by Monday morning each week, or else it is useless because that's when the sales associates meet. In other words, the *timeliness* of a report is often what makes it valuable. So a help system that incorporates this value into explanations, overviews, and so on, will naturally sound more relevant to the user than one that ignores this key bit of *tacit knowledge,* this unspoken "trick" of the profession.[2]

One way to express the unspoken rules and values of professional activities is the case or use case. Use cases depict the actual work flow tasks that users of the software would perform. For example: "On Friday afternoon Debbie begins to

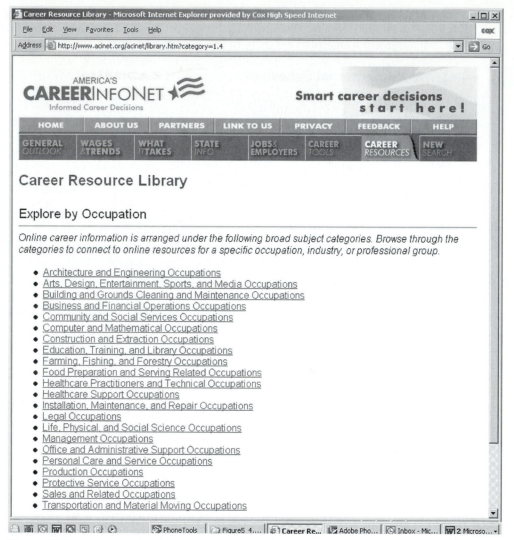

FIGURE 5.4   An Example of Occupational Information Sites
Sites like this one at America's Job Bank can help you understand the higher-level activities
of your users.

prepare the sales report in time for the meeting on Monday . . ." Doing this takes
time, but it helps you visualize users and creates a useful document that you can
share with clients, potential users, or other members of the writing team. You can
employ use cases in your documentation to provide role models for users and in your
documentation plan, to illustrate the activities you will support.

You should prepare one or more use cases for each user type that you envision
for the software you are documenting. The documentation needed for each user
group may vary: you may have special sections for each type of user or, if their needs
differ greatly, each type of user may require a separate document. The major sections

of the use case illustrated in Table 5.2 should provide you with enough general guidance to prepare one easily.

Sometimes it is a good idea to draw out a work flow diagram to accompany a user scenario. The example of a work flow diagram in Figure 5.5 shows how the "system" the software package being documented, interacts with the persons who use it (the "manager," the "deliverer," and the "customer.")

When you decide what documentation is needed, you should study the chapters on documentation forms—Chapters 2 (tutorials), 3 (procedures) and 4 (reference)—and document design, Chapter 10. When you describe a form, whether as a section or an entire manual, think of it as a commitment to creating that form. Also, consider that you want to provide the minimum documentation you can. Don't think up an array of many different document forms if you only have the resources to produce a limited number.

Review your user scenarios carefully. Ask your potential users to read them and comment on how correctly you have described the overall and typical-use descriptions. Figure 5.6 shows a scenario that potential learners would easily be able to identify with. How easily can your users identify with the role model they find in your scenarios?

5   Plan Interviews Carefully

User interviews provide the most important source of information for planning your documentation project. In an interview, you encounter potential users of the

TABLE 5.2   Elements of a Use Case

Topic Area	Writer's Question	How to Write It
Professional Role	How do my users function in their work environment? What objectives must the users achieve to perform their jobs successfully?	State the job title of the person who would actually use the program. For example "Manager in charge of scheduling a fleet of school buses," or "Temporary secretary needed to fill in for vacationing employees."
Profile	How much do my users know about the subject matter of the program and about computers in general?	Describe how the user would use the program on the job, and how much the user knows about computers in general. List the user type.
Overall Use of the Program	How will my users integrate the program into their work environment? Which program tasks relate to which professional goals?	Describe the series of actions the user would take in integrating the program into a work situation. Give realistic life examples of what a person would do with a program.
Typical Use of the Program	What tasks will the user perform most typically?	Describe the **most typical** use (no more than 5 or 6 tasks.) This will be a subset of the tasks in the program task list.

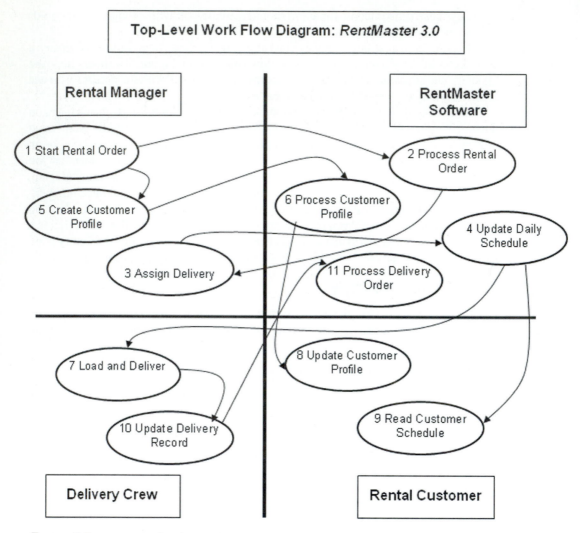

FIGURE 5.5   An Example of a Work Flow Diagram
This work flow diagram looks chaotic and complicated, but all it does is record the flow of orders and information among users of the software system.

program and, often in one or two settings, learn enough to design a documentation system that will satisfy your goals as a writer: to encourage them to learn the features of the program and put the program to useful work. The advice under this guideline pertains to the process of interviewing users: making site visits, spending an hour or two shadowing a potential user, asking questions, and so on. A full user analysis should not only study users in one or two episodes, but should involve

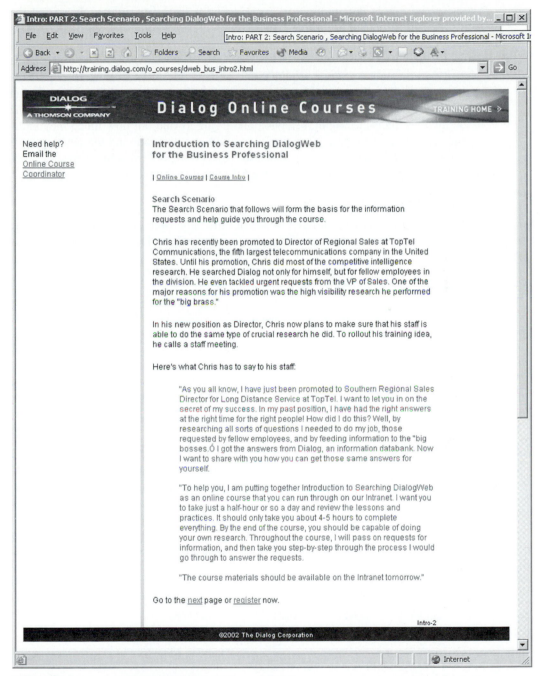

FIGURE 5.6   An Example of a Scenario for a Tutorial

This scenario allows the tutorial writer to identify which program operations to teach and also sets expectations for the learner.

users in the whole documentation cycle, using a *usability approach,* so see the activities described below in that larger context.

What Support Issues Does Your Program Raise?

Not all documentation projects require the same kinds of questions of users, or pose the same set of challenges to the writer. For example, with a customized messaging system like MSN Messenger, many users would have used similar systems in the past. So the issue involves their messaging behavior, and trying to find out how much they know about messaging in general. Your familiarity with the program and the developers of the program will help you identify what issues to try to resolve in your interviews. Consider how the programs indicated in Table 5.3 raise different issues for user analysis.

How to Plan an Interview

An interview *plan* may seem like a waste of time rather than just getting down to the job of learning about your users; and a sketchy plan will, in fact, waste your time. But

TABLE 5.3   Kinds of Issues Raised in User Interviews

Program	Users	Issues
Microsoft Word (word processing)	All levels, very general	What tasks do most users do most regularly?
		What other word processing programs have users used?
RunLog (jogging records)	Novice	What motivates use?
		How do coaches and runners work together?
Network program (sets up email and file sharing)	All levels	What kinds of security do users need?
		How do users store and transfer information?
		What email behaviors do users have?
		What technical problems face users?
Pat's Food Shopper (organizes grocery store visits)	Novice	What shopping patterns does the user have?
Color Time (elementary school coloring book maker)	Experienced (teacher)	What goals do teachers have for elementary art students?
		What kinds of illustrations do teachers use for math, history, and spelling pages?
Phone Book Plus (manages client addresses/phone numbers, automatic dialing)	Experienced	What equipment does the user have to handle phone calls?
		How well does the user understand complex communication protocols?

investing time in an interview strategy can pay off in forestalling repeat visits or unproductive interviews. To plan for productive user interviews, follow these general steps:

1. Do preliminary research into the user's job and the programs already in use. Use some of the resources for researching occupations suggested earlier in this chapter.
2. Review the software program and identify the issues. If you can, review the **_program specifications_** for descriptions of users.
3. Establish the scope of your interviews (how many, with whom, etc.). This helps you set clear limits on your time. It takes at least 30 minutes to get good info from any interview, and no interview should last more than an hour.
4. Make a list of interview questions. This is your most important planning tool. Don't go into an interview without it.
5. Get permission (from bosses, from users, from security). Show respect for privacy and security.
6. Set up an interview schedule (dates, times, places). Do this for your own sanity and to project an ethos of organization and professionalism.
7. Plan a follow-up (thank you letters, reviews, testing). Remember that following a usability-based process you will want to involve users in subsequent writing stages.

When you interact with your users try to assess the verbal style that characterizes their work culture. Verbal style refers to the method of communicating based essentially on whether people take a direct or indirect approach. For instance, you may be able simply to ask the question, "What menu choices do you use to put the graphics into your reports?" For someone with a direct style that would sound reasonable because they would be used to articulating their trade secrets for you. On the other hand, for a person with a predominantly indirect style you might want to phrase the question differently. For instance, you might approach it indirectly, in the way Bill did in the following exchange.

> *Bill (writer): Mr. Y, your work on this project is coming along very well. Your reports have a lot of graphics in them and you do not seem to spend too much time finding the data you need.*
>
> *Mr. Y (user): Thank you. I'm probably wasting time using the program to search for data and my reports still are only the best I can do in the time provided.*
>
> *Bill: You must take pride in the way you're able to incorporate graphics into your reports. You use many functions and steps with the program in order to achieve such professional results.*
>
> *Mr. Y: Thank you. I only use a few of the operations, starting with . . . [here he describes the functions he uses.]*

In the second exchange the user, Mr. Y, who is used to an indirect style, feels that he is being respected by Bill's appraisal of his work. Had Bill used the direct approach, say as he might have with a computer professional in his own, US American culture, Mr. Y might not have told him exactly how he performed his work because he might have mistakenly thought Bill was being critical of his choice of program functions.[3]

How to Observe

Instead of (or in addition to) interviewing your potential users, you may wish to *observe* them. Observation consists of **shadowing:** spending time at the workplace watching users perform their jobs, recording the sequences of tasks, and also taking in the atmosphere of the workplace. In doing so you have a chance to capture some of the user's reasoning behind software use: reasoning that otherwise you would miss by just asking about activities. Certainly you will do some observing during all interviews, keeping your eyes open for telling details of the user's surroundings to add to your overall picture of the user. But a structured observation of users requires that you put some thought into the process ahead of time, that you understand some of the limitations of observation as a research technique, and that you follow some guidelines for maintaining your objectivity.

Above all, you should try to avoid the following two main types of distortion:

- **Getting too involved.** You can easily distort what a user does by allowing your presence to affect the sequence of the task. If you let the user know what you want to happen ahead of time you risk the chance of it happening whether or not it is "natural," because users generally want (even subconsciously) to satisfy the expectations of the person who has taken an interest in their work.

- **Not getting involved enough.** By not asking questions of your user, you risk focusing on the wrong details. You may take notes on a user's methods gathering of electronic information to prepare for a meeting, when in fact this particular meeting happens only rarely. You should also make sure you cross-check your information.

How to Write a Questionnaire

Like an interview or observation, the questionnaire enables you to get information from a variety of users. It increases your chances of getting a unique and very valuable piece of information, but more important, it can show a reliable pattern of use that you can't get from one user.

MAKE IT OPEN-ENDED. Some kinds of questionnaires, such as those you provide the user to help evaluate a manual or a help system, carefully guide the user's responses so you get specific information about the documents' design. These are called *product insert questionnaires* however, for user analysis, *open-ended questionnaires* (that ask essay-type questions instead of yes/no or multiple choice questions) usually work best. Figure 5.7 shows an example of a questionnaire with both open-ended questions and pre-selected questions.

PROVIDE CLEAR INSTRUCTIONS ON HOW TO FILL OUT THE QUESTIONS. Leave enough room for users to provide complete answers, and remember that it's rare for a user to write beyond the limits set by the space allotted.

PHRASE QUESTIONS AFFIRMATIVELY. Avoid the negativity of questions like "You wouldn't want to read a Reference Guide, would you?" or "What don't you like about tutorials?" Better phrasing would be "Have you used reference guides in the past?" or "Describe your experiences (if any) using tutorials."

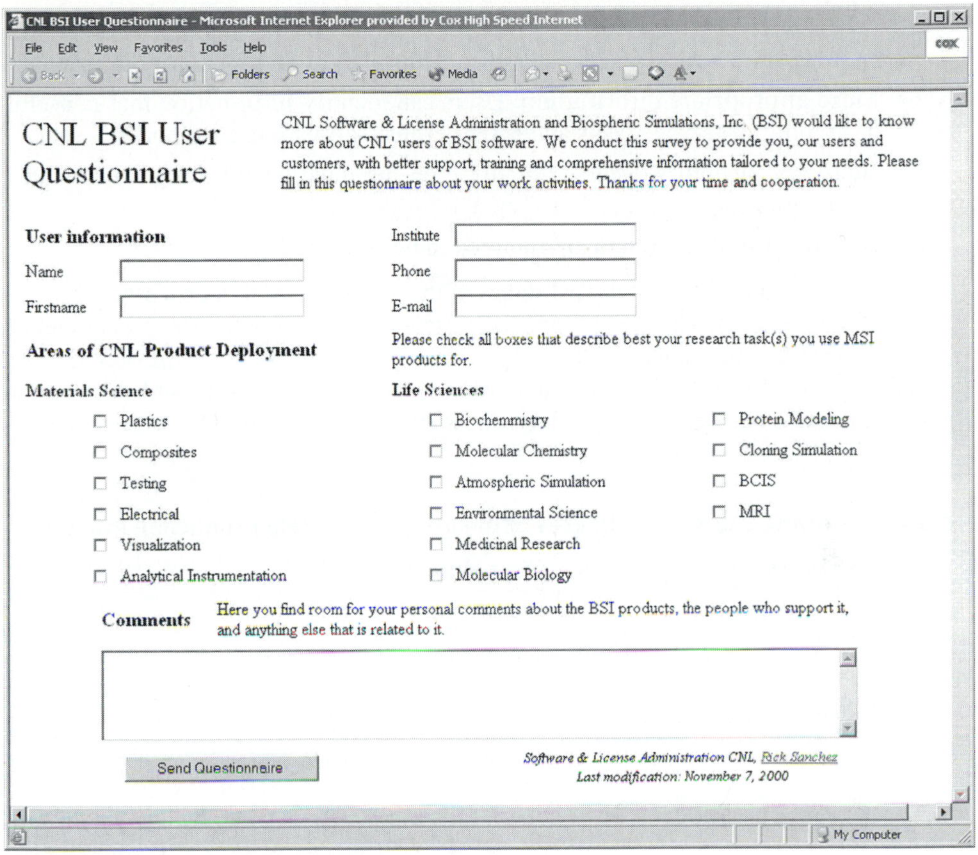

FIGURE 5.7   An Example of a User Questionnaire
This questionnaire helps documenters and systems administrators identify workplace actions and uses of the system.

STIMULATE RESPONSES BY INCLUDING SAMPLE PASSAGES. You can use mocked-up examples from manuals to test users' responses to certain design ideas you may have. You may, for example, want to show the user a sample layout of a tool description for a drawing program, or ask for their responses to a variety of ways of describing menus.

6   Involve Users in All Phases of the Project

A full user analysis should not only study users in the early phases, but should involve users in the entire documentation writing, reviewing, and testing process. So you should see the activities described here in the larger context of user involvement in your entire project. Subsequent chapters in this book rely on your establishing good relationships with your users early in the project. For example, where do you think you will find good user reviewers when you're sending your draft out (Chapter 7, "Getting Useful Reviews")? Similarly, your testing will require real persons from the user population to help ensure productive results (Chapter 8, "Conducting Usability Tests").

Involving the users in the process can have the following benefits:

- **Increased accuracy.** Users can indicate mistakes or lack of clarity in manuals.
- **More appropriate information.** Users can identify information that is useful for describing workplace tasks and activities that software use supports.
- **Increased usability.** Users can advise on the most useful design techniques for information.
- **Improved politics.** Users are flattered to be consulted.

Table 5.4 examines the standard stages of the document design process and ways you can involve users. The idea of involving users in the documentation process relates to two other key ideas: usability (making sure of the readability of the documentation and its suitability to the task) and collaborative writing (users and documentation specialists collaborating to produce the documentation product).

TABLE 5.4   Involving Users in the Phases of the Manual and Help Production Process

Production Phase	Ways to Involve the User
1. Start the Project	Identify users.
	Get permission for interviews.
2. Perform the User Analysis	Interview, observe, survey potential users.
	Provide examples/sample data.
3. Design the Documents	Check document objectives for suitability to users' needs.
	Get user feedback/reality check on design prototypes.
4. Write the Project Plan	Check outlines for appropriate task organization.
	Check list of hypertext links for online structures.
	Check outlines for completeness of information.
5. Write the Alpha Draft	Test prototype designs with users.
6. Conduct Reviews and Tests	Get user reviews for suitability, accuracy, completeness.
	Use users in usability testing of users guides and tutorials.
7. Revise and Edit	Reconfirm changes with users.
	Review graphics/tables/figures for accuracy.
8. Write a Final Draft	Confirm vocabulary decisions with users.
	Elicit user input on terminology used for indexing.
	Review definitions with users.
9. Conduct a Field Evaluation	Survey actual users for reactions, needed improvements.

Cultivate Relationships with Users

When you make your initial contact with users, plan to include them in as much of the documentation work as you can. Their perspective can inform your work and give you a reality check on the many design decisions you have to make as the project moves forward. If you have a relationship with a user you have a better chance to situate his or her actions in the workplace. Situated action (action particular to an individual or company) yields more useful knowledge than generalized action (such as you would get from some of the occupational research sources mentioned earlier in this chapter.) You will even find that what you learn about users on one project will be useful on subsequent projects, as you expand your knowledge of user categories of activities.

As a writer, you often take on the role of *user advocate*. Make the most of the relationships you form with your users as a useful counterbalance to the relationships you form with members of your development team: programmers, managers, testing specialists, and so on. This relationship can help you maintain a clear focus on your professional role.

Your relationship will also thrive if you exhibit a knowledge of the cultural characteristics of your users.[4] Those cultural characteristics can manifest themselves in ways that may or may not coincide with the characteristics of your culture. To develop an appropriate relationship with your users you should try to be mindful of the following four areas having to do with professional dealings in terms of software use.

DEGREE OF PERSONALNESS. "Personalness" means the sharing of personal, friendly information in a professional setting. For some cultures, like the Greeks, for instance, workplace relationships where software use takes place, are permeated with some degree of personal or friendly behavior. Among U. S. Americans often the friendly behavior takes a back seat as the person "gets down to business" with the interview. Among some cultures, such as the Japanese, friendliness parallels status, so if you're interviewing a person of higher professional status you can not expect to do much interaction on a personal level.

DEGREE OF FORMALITY. Some situations, like the teacher-student relationship, carry a greater degree of formality in some cultures. Among academics, for instance, it's customary to call someone by their title rather than their first name. Similarly, teachers who confess ignorance on a subject lose respect. So, depending on how much you know about the persons you're interviewing, you may or may not want to use first names.

DEGREE OF HIERARCHY. Often when interviewing about software use you will find yourself confronted with the boss or supervisor in an organization. Some organizations and national cultures take this hierarchical role very seriously and expect subordinates to do more listening than speaking, more following than questioning. Thus, if you ask a question of a superior from an Asian culture, he or she might tell you to look up the answer somewhere or might even avoid answering the question. The North American person might take this as evasiveness or lack of knowledge; the Asian person might see the questioner as disrespectful.

DEGREE OF DEVIATION ALLOWED. Deviation from cultural norms means that one can take a relaxed attitude to the constraints that those norms impose on your behavior. For instance, in Thailand it is possible to take a casual attitude toward deadlines and it might be acceptable for your interviewees not to show up on time. Whether or not you can relax cultural expectations depends on whether the culture is a "tight" or a "loose" structure. In North America, for instance, the culture is looser so violations of cultural norms are often tolerated for the sake of individualism. In Japan, on the other hand, the culture is tighter, and so violations of norms are not tolerated for the sake of the collective good.[5]

So in your dealings with users and your attempts to find out how they use your software in their work you face a real challenge: you have to examine as much of their work, their culture, and their activities as possible in order to see how your software fits in to their productivity. On the other hand, you need to pay close attention to how you approach them in interviews, focus groups, and with questionnaires so that your relationship with them doesn't falter because of violated cultural norms.

Do a Focus Group

To study users and to lock them into your plan throughout your project, consider conducting a focus group. Focus groups consist of groups of seven to ten participants who either have experience with previous releases of your documentation or whom you have identified as potential users. Focus groups can provide a rich wealth of knowledge about workplace into which you plan to situate your software product. Focus groups typically meet for one to two hours and attempt to come up with a variety of opinions about a product and its use. While the moderator—the writer or a hired consultant—structures the group's work around specific questions or workplace problem solving pertaining to the use of software and the features of documentation that would support the use, the group aims primarily to spark new ideas rather than come to a consensus.

As with all research methods, focus groups require some planning. To plan a successful focus group, follow these steps:

1. Locate potential participants.
2. Develop and administer a telephone screening questionnaire.
3. Confirm invitations in writing.
4. Draft open-ended questions and follow up questions, then revise.
5. Plan any hands-on activities (demos, dry runs, walkthroughs of procedures).
6. Make reminder calls.
7. Pilot-test the questions with one or two group members.[6]

Typically, your focus group will meet in a conference room or room with separate tables for smaller groups. Depending on the topics you want to explore, you can work with the entire group or, at times, break into smaller groups and then return to a whole-group format. On the simplest level, you would provide each member of the group with written questions, then discuss them one by one. Table 5.5, "Sample Focus Group Questions," illustrates some questions you might use.

TABLE 5.5   Sample Focus Group Questions

- In general, where is the first place you turn to find answers to your questions?
- In general, what usually drives you to consult a manual?
- When you do consult a manual, how do you locate the information you're looking for?
- If you could design an ideal set of manuals or an ideal help system for [program x] what features would it contain?
- What things in your existing manuals do you wish the writers had done better?
- How would [program x] fit in with the other programs you use at work?
- [Show a prototype and explain the proposed manual set. Make sure to say that this is very preliminary.] Overall, do you think the proposed set will be easier to use? Why? Why not?
- What support at your workplace do you think can help you use this program?

One procedure that focus group moderators sometimes follow involves asking group members to do a group brainstorm of the ideal characteristics for the support of a given program. The support could include manuals, training, online support, phone support, consultants—the works. After the moderator has collected (without commenting on their validity or feasibility) a list of characteristics, he or she breaks the group into smaller groups to come up with some suggestions as to what the writers could do to provide this support. After that the groups can come back together as a whole, with each group sharing their ideas. The moderator then records these and a general discussion follows. This procedure has the advantage of involving all members in the design process. Users who show enthusiasm for a program might even find themselves motivated to contribute to the support of a program by forming a special user's group.

Focus groups can go a long way towards giving your users the sense of involvement in the process of document production. By encouraging them in this way, you extend the possibilities of user-centered design.

7   Identify Document Goals

All human activities have goals: They keep people on track and give them something to measure their performance by. Spoken or unspoken, they lay the foundation for workplace activities. You can communicate your documentation goals—the activities you want to support, the user performance you want to empower—to other writers, managers, and clients. Documentation goals consist of statements of purpose that articulate what you expect to do for the user. Of course, they should reflect the goals of all software documentation—to encourage users to learn features of the program and to put those features to work—but they should state how you hope to do that for this particular set of users and this particular program. In short, they should reflect, as much as possible, the highly situated nature of efficient software use.

You may not have the writing-goals habit, being instead the kind of person who dives in, stays flexible, and adjusts along the way. Fine, but writing effective manuals isn't the same as playing a basketball game. The clearer your objectives, the better the chance that you will achieve them. Consider, too, that goals drive many parts

TABLE 5.6   Typical Goals Statements for Documentation Projects

Document Type	Goal Statement
Section of a document	The "Setting Parameters" section of the BitCom User's Guide will provide procedures for novice users in education and business to perform tasks associated with using the DOS mode command and the online setup utility.
Booklet	The BitCom Getting Started booklet will provide tutorial support for novice users in education and business to perform routine logons and logoffs.
Chapter	Chapter 5 of the PhoneNET Talk User's Guide will provide reference support for novice users in general professions to use all twenty of the ANET commands available through the command interpreter.

of the documentation process. You use them in document design as a way to guide decisions about page layout and text design (see Chapter 11). You use them in testing at various stages of the writing and editing (see Chapter 9).

It may help if you examine some examples of goals statements and consider the elements of a good one. The examples in Table 5.6 helped writers of those documents keep their projects on track. Goals statements can get very involved and you may need to take them to a fine level of detail, especially if you have a very narrowly defined spectrum of users. But the illustrations in Table 5.6 will work as models for most goals statements. They contain the name of the document or section, refer to the user type—novice, experienced, expert—and the kind of support you intend to provide: tutorial, procedures, or reference.

8   Tie the User Analysis to Documentation Features

The checklist in this chapter will help you organize your user analysis so you focus on, and gather, the information that will allow you the greatest success in designing your documents. But all the analysis you do will not help unless you tie the user analysis to documentation features: aspects of the documents that tailor information to specific users. You cannot, at the stage of user analysis, state the details of your document design: the page layout, choice of type size, style, fonts, or stylistic choices. That level of design results from considering the many design tools available to the writer. You should base all your design decisions on the user task needs that you discover in your user analysis. As you interview, observe, or survey your users, you should try to see how you can tailor the documents to the user's needs. Ideas that you get during your analysis can help to guide your later, more detailed, document design.

In Table 5.7 you will find a preliminary summary of some of the document features that you can use to meet some user needs. The list contains lots of good ideas to use as guidelines, but it by no means contains all the ideas that exist, or that you will come up with on your own. Use the list to keep yourself pointed in the direction of meeting user needs.

TABLE 5.7   Match Document Features with the User Analysis Results

Analysis Results	Features to Meet User Needs
Possible Users of the Software	
What recognizable groups (teacher, student, system administrator, end user) will your users comprise? Describe each group briefly.	Documents/sections for separate groups
	Illustrations targeting different groups
	Brief scenarios included for different groups
Provide a brief scenario of use by representative professions to act as a model of how the program will serve different professional user groups.	Special glossaries for individual groups
	Group-specific organizational schemes for different groups
Primary Issues Raised by the Software	
What learning difficulties, motivational problems, technical difficulties will your users face?	Focused tutorials for technical difficulties
	Explanations of workplace applications
	Suggestions for contact with user's groups
	Encouragement of use of online support
The User's Tasks	
What workplace jobs will the user do with the program?	Organization of sections, chapters, manuals around user-defined tasks
What important sequences will the user follow that can help you organize your program tasks?	Suggested sequences for difficult tasks
	Examples of tasks in procedural elaborations
What tasks can you use as examples in your online and hard-copy manuals?	Special job performance aids for difficult tasks
The User's Informational Needs	
What important kinds and forms of information does the user need in order to use the program?	Lists of sources of information for input to the program
Whom does the user communicate with, and in what forms and media?	Suggestions for importing and exporting data to and from other programs
	Emphasis of reports and documents the program produces
The User's Work Motivation	
What internal and external motivations affect the user?	Illustrations of efficient, creative, productive use of the program
	Emphasize the value of program output
	Reward accurate response to help screens
	Encourage independent thinking and problem solving

Continued

TABLE 5.7   *Continued*

Analysis Results	Features to Meet User Needs
The User's Range of Computer Experience	
Describe the kinds of novice, experienced, and expert behavior exhibited by the user. Pay attention to the numbers and kinds of programs, degree of technical knowledge, user attitudes towards computing, and the learning behavior of the user.	Analogies to other programs Types of documentation: tutorial, user's guide, command summaries Media of documentation: online, hard copy Support for technical background Full index of terms for novice users
The User's Knowledge of the Subject Matter of the Program	
What kinds of vocabulary reinforcement and subject/matter background will the documentation have to provide?	Glossary to reinforce terminology Encyclopedic overview of key concepts Suggestions for further reading Illustrations of application to workplace tasks
The User's Workplace	
What other computer programs are used at the user's job site?	References to other programs Background information on user groups
What user's groups does the user belong to or have available?	Encouragement/guidelines for group learning and use
Describe the degree of organizational support for computer use in your user's organization.	Suggestions of organizational support Encouragement of networked/integrated use
The User's Learning Preferences	
Describe the users preferences for instructor, manual, or computer-based learning.	Choice of manual, instructor-based, or computer-based tutorial Creation of workplace scenarios for training Suggestions for group learning
The User's Usage Pattern	
Describe the usage patterns users will exhibit in using the software in terms of regular, intermittent, and casual usage.	Organization of sections, documents, chapters around sets of features Emphasis on online support for intermittent users Quick overview cards for casual users

Discussion

The user analysis constitutes a major element in the design of task-oriented documentation. The user analysis allows you to study the user's workplace context: the personal needs and the organizational goals in which the software, the manual, and the help system must function, all the while recognizing and accommodating the user's cultural differences from yours. The user analysis can help you in many ways. It allows you to determine what actions the user needs to perform with the software. In addition, it gives you the occasion to apply other documenter's tools, such as user types and user cases, to the design problem. It also gives you a first contact with your actual users, through interviews and other forms of research, and so should be based on your before-hand knowledge of their workplace and cultural norms. From this contact you may identify potential topics for testing and persons to serve as test subjects.

Your user analysis will provide you with examples to use in your tutorials, but the user analysis does not just apply to the *teaching* level of task support. It applies to all levels of task support—tutorial, guidance, and reference—and to the document forms you design to operate on those levels. The user analysis helps you organize and write your table of contents by providing a task-oriented sequence for procedures. It applies to "getting started" sections by implying what kind of computer system, printer, and other hardware the user has. In tutorial documentation, the user analysis provides the case or problem-solving activity that the lessons focus on. Examples generated by the user analysis also show up in procedures and in reference documents. The user analysis guides you in your index preparation by suggesting vocabulary to cross-reference. Additionally, the user analysis helps you design the look of your manuals. It informs your choice of page layout, type design, and organization. It helps you decide what kinds of graphics to use and for what purpose. Table 5.8 illustrates some ways that manuals and online help reflect your user analysis.

Thus, your user analysis, if done well, can unify your documentation set. If ignored, it can contribute to your writing the default manual: a collection of explanations of toolbars, menus, commands and other interface elements divorced from the context of meaningful work that results in anxiety and frustration.

What Does *Use* Mean?

What exactly is meant by *use* of the program? As we saw in Chapter 1, use of a software program consists of applying the operations inherent in the user interface to actions or user tasks. In other words, software use is always *situated* in a user's workplace and cultural context. Figure 5.8 shows the components of such use. As you can see, for specific users, software use is made up of broad activities, which themselves consist of more narrow actions. These actions reside at a level above the operations, which are represented on the computer screen by the menus, buttons, dialog boxes, commands, and other elements of the interface.

The default manual (discussed in Chapter 1) consists of descriptions of the interface elements and operations the user can perform with them. But presenting information just at that level ignores the situated nature of software use and can

TABLE 5.8   Some Ways Manuals Reflect User Analysis

Element	Reflects this Aspect of the User Analysis
Installation	Kinds of hardware/operating systems used
Getting Started	User's subject matter background knowledge
	User's computer experience
Tutorial	Examples for lessons
	Background knowledge of subject matter and computer experience for transfer learning
User's Guide	Examples
	User goal statements
	Organization of tasks
Reference	Vocabulary
	Sequences for *hypertext links* in online help
Job Performance Aids	Needs for specialized support
Index	Terms for cross referencing
	User questions and problems

often leave the user frustrated and anxious. The solution, then, is to examine the user in enough detail at the level of the situated workplace context and develop from that investigation a number of documentation themes, design ideas, organizational strategies and so on so that it reflects the user integrating the program instead of just the program. That examination, which is called the user analysis, means asking and answering a number of questions about the user. The remainder of this chapter examines the kinds of questions you can ask and explains how the answers you get apply to the design of manuals and help systems.

What You Want to Know about Users

In preparing a user analysis, you basically ask the following eight questions:

1. What tasks will the user perform with the program?
2. What are the user's informational needs?
   - What information does the user need?
   - How does the user communicate?
   - What work motivations affect the software user?
3. What are the user's work motivations?
4. What's the user's range of computer experience: novice, experienced, expert?

Program: *Degree Audit*		
Knowledge Resources	**Components of software work**	**Documentation topics and themes**
User: Administrator		
Situated workplace context	Activities	Running an undergraduate program in engineering, planning curriculum, informing advisors and students, managing faculty resources, managing computer resources
	Actions	Update the degree plan database, facilitate software use for advisors, monitor student degree plans for successful graduation
Program interface	Operations	Administrator login, add a degree plan, delete a degree plan, add a class, delete a class, change a class name, add a substitute class, delete a substitute class, add a user, delete a user, add a pre-requisite, delete a pre-requisite
User: Advisor		
Situated workplace context	Activities	Advising students on degree plans and curriculum
	Activities	Initial meetings with students, degree auditing of students with existing degree plans, reporting to the administrator about progress
Program interface	Operations	Advisor login, set up a degree plan, retrieve student plan, add information to a degree plan, delete information from a degree plan

FIGURE 5.8   An Example of Situated Software Use
This example shows how software use can be made to tie in closely with workplace activities and actions.

5. How much does the user know of the subject matter of the program (e.g. accounting, writing, designing)?

6. What's the user's workplace environment: the user community (e.g. organizational structures/other software users)?

7. What's the user's preferred learning preference (instructor/manual/online)?

8. What's the user's usage pattern (learning curve regular, casual, intermittent)?

You may not *gather* information about all of these questions for a specific project. But you will need to *know* all of this information—either from past experience or the interviews, observations, and surveys discussed above. The questions you ask in investigating your users in these areas, as Table 5.9 shows, have a significant effect on your project.

TABLE 5.9   Topic Areas of Your User Analysis, Related to Your Documents

Topic Area	Effect on Documentation
Tasks	Organization of procedures
Informational Needs	Choice of emphasis in job performance aids, choice of examples
Work Motivations	Choice of tone, choice of examples
Computer Experience	Tutorial design, vocabulary, organization of procedures
Subject Matter	Choice of background information, vocabulary, special reference, aids/templates, configuration preferences
User Community	References to organizational support, choice of examples, reference design, patterns of integration of computers and help
Learning Preference	Tutorial design, choice of examples, choice of teaching media
Usage Pattern	Structure, page layout, documentation set design

If you look ahead to Chapter 10, "Designing for Task Orientation," you will see that the decisions you make for design of your document use the information you gather in these eight areas. This discussion will focus on each of these areas in turn, explaining the rationale for each and giving examples of the kinds of information each provides and how you can use that information in your manual project.

Tasks the User Will Perform with the Program

Knowing about job roles gives you somewhere to start in describing user actions. User activities or tasks form the basis of your task-oriented information product. If you learn nothing else about your user, learn the usual activities associated with his or her type of work.

The activities in Table 5.10 show a number of important characteristics. For one thing, they are not cut and dried, not easily describable, because each one could be done differently each time it is performed. For instance, no two teachers will organize lessons in the same way, nor will every retail manager write a report in the same way. In addition, no one software program will fit with one task. For sure, some programs try to, such as "suites" of programs for teachers that supposedly support "all your teaching needs" but for the most part a teacher might use a word processor *and* a spreadsheet program for planning curriculum, and so on. A designer will consult a software database of designs and then move to drawing software to execute a new design. Also, notice that each task requires a certain amount of information, as in when the oil land broker evaluates properties, he or she needs data (title histories, survey reports) about those properties to start with. The industrial designer needs to know about the operators of a piece of equipment, which he

TABLE 5.10   Actions Associated with Work

Retail Manager

- write reports to district headquarters
- analyze departmental budgets
- determine effective promotional activities
- direct promotional activities

Oil Land Broker

- meet with land owners and clients
- evaluate properties
- prepare sales packages
- oversee financial transactions

Engineering Designer

- organize design teams
- evaluate design software and hardware
- manage design libraries

Teacher

- plan curriculum
- prepare teaching materials
- organize lessons
- evaluate fellow teachers

or she gets from preliminary reports. Workplace activities exist within a context of information. Finally, each workplace task or activity requires communication and interaction with other persons in the workplace. In some cases the other person is a client, or customer, and in other cases a fellow worker or supervisor.

In sum, when you're identifying the tasks your user performs look for these characteristics:

- tasks that can be done uniquely reflecting the individual and the specific workplace
- tasks that may require more than one software program
- tasks that require information from data resources
- tasks that require communication with other persons

In the examples above, the most important elements lie in the professional's need to communicate with others and his need to solve complex, open-ended problems. Most professionals—engineers, attorneys, and so on—will use a program well only if they see how it helps support their workplace actions. To the extent that it does not, you will find very little energy expended on using the documentation for these programs. Documentation that truly supports professionals should make up a part of the toolkit those professionals use to further their workplace activities. Table 5.11 shows some of the kinds of tasks you can look for in the user's workplace.[7]

Your research, interviews, focus groups, questionnaires and other tools for finding out about the user should focus on identifying a few key tasks, such as the ones

TABLE 5.11   Tasks that Require Information

- Planning work activities
- Allocation of resources
- Problem solving
- Evaluation of work processes
- Writing reports and memos

in Table 5.11 that you can use as cornerstones for your manuals and help. Once you have identified these tasks or actions, you can proceed to planning information to support them using a software program.

The User's Informational Needs

Once you have identified the users workplace tasks you are ready to take your user analysis to the next level: identifying information needs to support the tasks. You need a picture of the setting in which persons work.

Increasingly, today's employees take on a more important role in their workplace than that of mere workers. Because of economic trends, increased availability of information technology, and knowledge management strategies, employees manage their own work, make independent decisions, and set their own production goals. This phenomenon sees each employee as more independent than before while also a part of a production team, and as needing the appropriate resources. For this reason, job motivation (independence), communication (being part of a production team), and information (having the appropriate resources) have become major priorities of computer end users.

Research into how computer users apply their programs to their expanded roles in the workplace focuses on three areas:

- Information needs
- Communication needs
- Work/professional motivation

These areas have received considerable research attention because they make up variables that affect software use. Information about each of these three areas can help you make accurate design decisions about what levels of support and what forms of documents to produce.

What Information Does the User Need?

Users need to know what's going on in their sphere of influence. For example, the winery manager needs to see lists of grape suppliers and the quality of their crop in order to decide which suppliers to use for the next quarter. Look at the actions you have identified. What will it take for the user to carry those tasks out? You need to supply information to help situate the task or you need to recognize the kinds of information they need, sometimes suggesting to the user where to get it. Table 5.12 suggests some of the kinds of information that users need to perform common workplace tasks.

TABLE 5.12   Tasks and the Information They Require

Planning work

- records of past activities
- records of production or work goals
- understanding of work processes

Allocation of resources

- data about resource need
- history of resource use
- workplace process goal statement

Problem solving

- written project goals
- data about options and alternatives

Evaluation of work processes

- data about past history
- objectives for evaluation

Writing reports and memos

- research data
- information resources on the Internet
- company archives
- customer information

For each action in the previous section, you can identify a communicative or informative dimension. To analyze this aspect of software use, you need to know the user's main sources of information. For example, teachers attend conferences or regular in-service meetings to keep up with changes in curriculum. Engineers study production reports and gather data from measuring instruments. Professionals read a wide variety of reports produced by their organizations and they keep up with a wide variety of journals, trade magazines, trade papers, and newsletters to get essential information to do their jobs.

Where Does Information Come From?

Information, from whatever source, bombards today's user as it never has before. Consider the winery manager described above. Her information about suppliers comes from personal experience, trade magazine reports, information about research on grapes, and other raw materials. Increasingly, professionals get their information electronically. Sources of electronic information include the Internet: a vast and growing network of information sources. With such areas as the World Wide Web and FTP sites (sites for transferring files and programs) the professional can find information on just about every topic. Many companies include company information on their home pages, and the Internet gives users a way to connect to other professionals.

- Databases and data files kept on customers, vendors, production levels, sales, commissions, inventory, and many other topics.
- Commercial databases such as ProQuest and WestLaw that provide software and hardware reviews, industry news, government policy, regulatory changes, and other national and international business information.
- Organizational databases and networks created specifically for the organization, containing investment statistics, stock information, production guidelines, design specifications, and specialized libraries.

What you know about information needs can help you point software users to information that can help them in using your program. The reason for this is that the operations in and of themselves (opening files, using toolbars, filling in forms) are given meaning through the information and communication that surround them. When your user is operating the program you really can exert little control over the exact outcome, because the work the user is doing is inherently creative and unique. But if you can provide the right kinds of information for the user, tell the user what he or she needs to accomplish an operation, you stand a little bit better chance of having the operation turn out successfully.

How Does the User Communicate?

Whatever forms of communication your user prefers, you should have a clear idea of them. They represent the ways users keep up with each other so their instructions get followed, their recommendations and suggestions get made, and their work gets appreciated. You should know what forms of communication your user needs and how your software fits into the picture through its report generating and other output functions. This way you can help the user situate the program operations into the workflow. Think how much the winery manager appreciates support for her electronic faxing of reports to colleagues, or of printing multiple copies for training sessions. Table 5.13 presents some typical forms of workplace communication.

Becoming familiar with the kinds of documents your users employ can help in another way. It's a key to the user's language: Using the terminology familiar to the user can trigger, in a psychological sense, a recognition of something familiar. Thus, when you are making up lists of terms to include in your index or keyword search list (Chapter 14, "Indexing and Keyword Searching"), you should spend some time looking over sample documents for important terminology.

The User's Work Motivations

What software documenter would not like to have an answer to this question: What motivates efficient and effective task-oriented software use? Part of the answer lies in the issues discussed above: awareness of how the users' tasks shape their information and communication needs. However, the user's situated, information-rich activity has another dimension: it is done with a purpose or motivation. When it comes to what motivates users, follow this simple principle: What motivates users professionally will also motivate them to do well with software—to assemble interface elements and basic

TABLE 5.13   Typical Forms of Communication in the Workplace

Presentations

Meetings	Articles
Interviews	
Conversations	

Documents

Reports	Memos
Letters	Proposals
Agendas	

Papers

Teleconferencing	Conferences
Internet information	Web pages
Sources	

Messages

Local area	Email
Wide area	Newsletters
Commercial	Distribution lists

operations into meaningful sequences leading to an objective. In fact, if we can tap into a person's desire to achieve these professional objectives, we can channel that energy into the use of software. Documenters can't become experts in human motivation, but an awareness of some basic ideas can help you design for effective use.

In general, people respond to different kinds of motivation. We refer to these motivations as needs. What needs motivate specific individuals can vary considerably. But psychologists have nevertheless attempted to classify people's needs as a way of understanding human behavior. They show us that not only do persons have multiple needs, but that these needs fall into categories depending on their importance. Thus, we speak of basic needs such as "food" and "shelter." We also have higher-level needs such as recognition by our peers and self-esteem.

The motivation of people in jobs has received a great deal of research attention because of its importance in enhancing employee productivity. Researchers have identified some motivations, specific to the work environment, called *work motivations.* Work motivations come from two sources, as shown in Table 5.14. They include such things as achievement, the use of one's abilities, and one's desire for status. Motivation by these needs relates directly to the idea of task orientation; the concept of task orientation means that the person operates out of his or her individual job initiative. People like to think of themselves as contributing, as important, as goal oriented. Some companies consider this motivation as a resource.

The writer of task-oriented software documentation thus is faced with the following question: what part does software use play in satisfying these work motivations?

TABLE 5.14   **Sources of Work Motivation among High-Tech Workers**

Internal Motivations	Environmental Motivations
Achievement	Independence
Use of Abilities	Job Security
Autonomy	Physical Conditions
Responsibility	Compensation
Creativity	Variety
Status	Activity Level
Authority	Upper Management
Recognition	Technical Management
Advancement	Company Policies

TABLE 5.15   **Ways to Use Motivational Information in Manuals**

Motivation	Manual Element
Achievement, efficiency	Online help, tips, layered page layout
Social needs	Examples involving realistic scenarios
Production	Tips for shortcuts
Creativity	Tips for customizing steps
Speed of use	Online help, quick-reference card

How can I design manuals and online help that tap into the user's motivational construct to channel job energies into the use of software?

You may wonder how the writer connects user's motivations and the design problems inherent in manuals and online help. In fact, you can evoke motivations in the manual elements listed on the left-hand column of Table 5.15: the background information section, the examples throughout the book or help system, the demos, and so on. The list on the right gives an example of how you might write these elements when you know what motivations you want to focus on.

Range of Computer Experience: Novice, Experienced, Expert

Some users may bring a different level of computer expertise to each new program they use. Some will learn a program faster than others because they have a broader range of computer experience. And because they have more experience, they will learn the program in different ways, using more guidance and support documentation, less tutorial. The range of computer experience they have, coupled with their motivation to use a program, determines, in part, the level of task support they require.

Attitude Can Make a Difference

A number of researchers have studied how anxiety (feelings of stress, fear, apprehension about computers) affects users' performance. Honeyman and White, citing other sources, note that computer anxiety relates to the feeling that the computer controls the events, not the user.[8] Also, many persons may feel a general fear of technology, and this can frustrate them in learning a program. Bracey, in another study, confirms what many researchers have noticed, that the more experienced the user, the less likely the user will feel computer anxiety.[9] Honeyman and White support this observation, noting that exposure to a positive experience with a computer can reduce anxiety. Users who once feared the "intelligence" of the computer begin to see it as a tool to help them reach their professional goals.

The Ability to Transfer Learning Can Make a Difference

Experience with computers creates a pattern of knowledge, what psychologists call a cognitive schema, that helps users transfer their knowledge from past experiences to current experiences. The underlying differences between the novice computer user and the experienced one pervade the research about learning computer programs. As one researcher puts it, you can see "a world of difference between the person for whom use of a computer-based system is routine and the novice or occasional user."[10] The novice basically exhibits very different degrees of anxiety and receptivity to different interface and media types than does the experienced user. Users with very little computer experience tend to take more time to learn functions, so the pitch of their learning curve remains gradual when compared to that of users with a lot of computer experience. They can transfer less of their previous experience into the mastery of their new experience.

Let's look at a discussion of three common user types based on their range of expertise: the novice user, the experienced user, and the expert user.

Novice Computer User

What kinds of behaviors in this example characterize a person as a novice computer user? For one thing, novice users have used few computer programs. Often they may know one program well, such as a word processor or an email program, but in learning new programs they do not have the benefit of knowing a variety of interfaces or computer types. This lack of experience makes learning difficult. The same principle holds true for learning in other situations. Many cars have similar dashboards. Thus, a person who has operated one or two late-model cars has an easier time dealing with the dashboard of a rental car than a person who does not have this experience. Such an experienced person would know where to expect to find the light switch, horn, and so on.

Furthermore, novice users have little or no technical knowledge (beyond the identification of drives and displays, etc.) of the computers they use. In their minds, they do not have elaborate, highly differentiated models of how computers work. Thus, their learning takes more time. Adding a command to their repertoire often means that they have to build the mental model, or adjust an existing one, to accommodate the new information. In a sense, they have to learn and assimilate information at the same time, instead of just learning it. Because they feel uncertain of their computer skills, novice users generally resist invitations to "explore" the software. They like you to tell them what keys to press.

Novice users often do not see a clear relationship between the program and their work. For novice users, inability to see the benefits of a program impedes their willingness to spend time learning it. And because they don't see the value of using a program they often form negative attitude toward new software. Writer Donald Norman refers to one typical novice attitude called "false causality;" this occurs when the computer coincidentally does something immediately after a user action and the user thinks he or she caused the effect. "Many of the peculiar behaviors of people using computer systems or complex household appliances result from such false coincidences," asserts Norman.[11] When the result is not what the user expected, the effect on the user's attitude is negative.

Another characteristic we see in the novice user tells us something about difficulties novices have in learning new programs. Because the novice had used few programs, he or she lacks the experience of learning programs. Learning programs requires understanding principles that may be very abstract, and often users have to chart a learning course that suits them. Some use tutorials. Some dive in and "follow the bouncing cursor." Some rely on friends. Whatever the pattern, novice users don't have one. They distrust manuals, feeling overwhelmed by their usually technical nature and arid prose. Manuals appear to be truth arranged in a meaningless way.

Experienced Computer User

The experienced user typically employs a variety of software programs. Additionally, the experienced user uses more open-ended programs (like computer languages) than does the novice. It takes an advanced degree of understanding of computer processing rules and methods to construct programs from computer languages. For experienced users, program use ranges beyond just knowing one program well: they use utility programs (like the virus protector), analysis programs, and drawing programs, among others. Software programs, for the experienced user, work together as part of a system.

Experienced users posses some technical knowledge of computers. They know how to install special components (such as modems) on a personal computer. This technical knowledge gives the experienced user a broader base of understanding—a more highly differentiated mental picture of how things work. This understanding affects both attitude and learning behavior. You can count on experienced users to have less trouble understanding installation procedures for software programs that require hardware installation, such as scanners, tape backup systems, and modems.

Perhaps the major difference between experienced users and novices lies in attitude and the way they relate programs to their work environment. Experienced users have moved beyond seeing the computer as an awesome thinking machine, to seeing it as a tool in their work. You can see evidence for this in the way the experienced user integrates a number of programs to do a single task or perform a single job role. For example, engineers might use both word processing and draw programs to create drawings. Experienced users see themselves as the most important element of the computer/job/person configuration. The computer serves the purposes they decide on, and if one program won't work, they try another. They exercise critical judgment over what software can do for them.

Experienced users, finally, have a more patterned learning behavior than novices while remaining open to learning in new ways. Because of their familiarity with programs, experienced users have begun to form definite likes and dislikes as to how they want to learn new skills and new programs, so we can see more of a pattern in their learning. On the other hand, they are less skeptical of manuals and documentation than novices, because they have begun to see the value of manuals and help in opening up programs for them. They accept online help systems more readily.

Expert Computer User

Expert users not only work with many programs, they work with an extraordinarily rich selection of programs: word processors, language compilers, and programmer's utility programs. Expert users often work in the computer industry, which encompasses the fields of programming, information science and systems, network services, and hardware maintenance. Professionals in the computer industry work with an extraordinary large number of programs and systems. In fact, most of these could be called open-ended programs, having a broad applicability. In contrast to experienced users, the expert user's familiarity with programs spans different operating systems and computer types.

The expert computer users' expertise allows them to dismantle computers with confidence, and trouble-shoot highly complex hardware configuration problems. They use this knowledge to set up systems and create software and hardware products. This technical knowledge creates a willingness and ability to learn technical details, and a well-defined background in programming languages and circuit board design.

Unlike the novice and the experienced users, the expert user often sees software as an end in itself. While experts uses many programs as tools—programming languages, for example—they go a step beyond that level and see them as ends in themselves. Experts care about how a program works, how it's assembled, and what kind of interface it has. They know the limitations of much of the software they use, and from that awareness derives principles to apply to their own programs.

Computer professionals learn new programs easily. They will quickly get the general idea of how a program works and apply the program to a specific use. Like other professionals, expert users consult manuals and online help frequently. They like online help because it doesn't take them away from the computer screen and it's sometimes easier to get to than a manual. In many ways, experts have highly distinctive learning patterns because they get repeated so often.

Table 5.16 summarizes the descriptions of the three types of software users. It also includes some generalizations about the kinds of documentation each prefers. Use the table carefully to remind you of the differences among users, remembering that user types represent generalizations rather than fact, and that documentation preferences will, of course, depend on your user's individual characteristics and needs.

Extent of Knowledge of Subject Matter of the Program

How much users know about the subject matter of a program can determine their ability to see the relationship between the program and the work they do. Subject matter knowledge is sometimes referred to as "domain knowledge." It means knowledge pertaining to the user's profession. For instance, domain knowledge for

TABLE 5.16   Characteristics of User Types

Characteristic	Novice	Experienced	Expert
Number of Programs Used	Few	Low	Many kinds
Degree of Technical Knowledge	Low	Some	High
Attitude	Vague, illogical, negative	Computer as tool; open	Programs as programs, not tools; for their own sake
Learning Behavior	Undifferentiated, resistant	Patterned, open, flexible	Highly differentiated
Documentation Preferences	Tutorials, index and table of contents, visuals, guided tours	User guide, job aids, online help, "Getting Started"	Command and task reference; online help; user guide

word processors would be principles of writing, for tax software it would be an understanding of tax codes. A high degree of subject matter knowledge can affect the amount of background knowledge you need to supply in your program. Take, for example, the lesson the producers of desktop publishing software learned about the amount of background subject matter to supply with their software. When the new software became popular, many computer users discovered a tool to re-do their newsletters, brochures and other documents, using powerful page layout, type fonts, and graphic placement features. These users generally had little knowledge of page layout; the result was newsletters with a jumble of fonts and pages cluttered with unneeded clip art. Subsequently, some desktop publishing software producers published a separate booklet with their program that covered the basics of page layout vocabulary, how to design a page, and many examples of layouts and their uses.

In a similar way, if you write a manual that automates accounting procedures and your user lacks a background in accounting—does not know the function of a general ledger or the meaning of terms like *accounts receivable* and *accounts payable*— then you can expect a more limited learning and use of the program. Table 5.17 shows some of the various kinds of subject matter knowledge needed for software use. You can find a number of ways to support subject matter knowledge in your manual. Depending on your design, you can use subject matter knowledge in introductions, examples, background sections, and elsewhere. Figure 5.9 shows how you can support subject matter knowledge in a glossary.

The Workplace Environment: User Communities

You will often find that within organizations or within the groups who use software the group surrounding the use forms into some sort of official group. Known generally as a "community," the groups that form to support software rely on the

TABLE 5.17   Kinds of Subject-Matter Knowledge Needed for Software

Program Example	Kind of Subject-Matter Knowledge Needed
Microsoft Paint	Composition, color mixing, art terminology (e.g., *brush feather, polygon*)
DAK Accounting	Bookkeeping, accounting
Microsoft Word	Writing techniques, formatting techniques, use of fonts
AutoCad	Engineering design, terminology (e.g., *layers, ikk lines, projection*)
HomeInventory	Financial planning, investment strategies
Probe (electronic circuit analysis)	Electronic circuit design, circuit terminology (e.g., *capacitors, inductors, diodes*)
Excel (spreadsheet)	Statistical functions, mathematics

ability of Internet communication among users and their willingness to share that information. It doesn't matter whether you are writing manuals for a very large program like Linux or Microsoft Windows, or a smaller, specialized application, the ease of communication afforded by the Internet and World Wide Web makes the phenomenon of user groups available and valuable.

User communities rely on the social dimension of computing. As we have seen, no work with a piece of software occurs in isolation and just for itself, but is part of larger, organized systems of human activity. User groups are a manifestation of the larger activity structures that surround all software use through interconnected technology. In fact, user groups serve both documentation or publications departments and also marketing and development. The information that user groups generate through sharing information about industries where software is used is valuable across the organization in planning new products and planning new approaches to software support.

As part of your user analysis you should investigate two things: the user groups that your users already participate in and your user's willingness to join groups for mutual support.

Existing User Communities

You will often find user groups forming around the use of a particular piece of software. This is especially true of software programming languages and large, adaptable systems. With these types of systems, the source code of the language or system is distributed freely to allow users to adapt it to their own purposes. Rental Master is one such system, and the remarkable aspect of this program is the amount of user support that has grown up around it. As you can see from Figure 5.10 the software homepage contains links to a number of user groups, developers' sites, forums and other elements of the "community" of use.

HELP FORUMS. Help forums are specialized group sites where users and help support technicians can post problems or questions and get help either from other users or support staff.

APPENDIX E: NETWORK TERMINOLOGY

NOTE: Not all of these terms were used in the text of this guide; however, you may find them helpful as you get started on the Internet.

archie	a system for locating files that are stored on FTP servers
ASCII	(American Standard Code for Information Interchange) default file transfer mode
asynchronous	transmission by individual bytes, not related to specific timing on the transmitting end
backbone	high-speed connection within a network which connects shorter (usually slower) branch circuits
bandwidth	the difference, in Hertz (Hz), between the highest and lowest frequencies of a transmission channel; the greater the bandwidth the "faster" the line
baud	unit of measure of data transmission speed; usually bits/second; may differ from the number of data bits transmitted per second by the use of techniques that encode two or more bits on a single cycle (i.e., 1200bps and 2400bps modems actually transmit at 600 baud)
binary	refers to a condition that has two possible different values; a number system having a base of two (0 and 1)
bps	(bits per second) measure of the rate at which data is transmitted
bridge	a device that acts as a connector between similar local area networks
broadcast	a packet delivery system that delivers a copy of a given packet to all hosts attached to it
client	the user of a network service
coaxial	cable comprised of a central wire surrounded by dielectric insulator, all encased in a protective sheathing

FIGURE 5.9   A Glossary Supports the User's Need for Subject-Matter Knowledge
Glossaries like this one from the Texas Tech University's *Network and E-Mail Reference Guide* support the user's subject-matter knowledge and can help the user see the relevance to workplace tasks.

SPECIAL INTEREST GROUPS. Special interest groups, known as SIGs, are user groups that focus on specific topic areas. They can be more or less formal, but usually have a leader and some way to accumulate and disseminate resources.

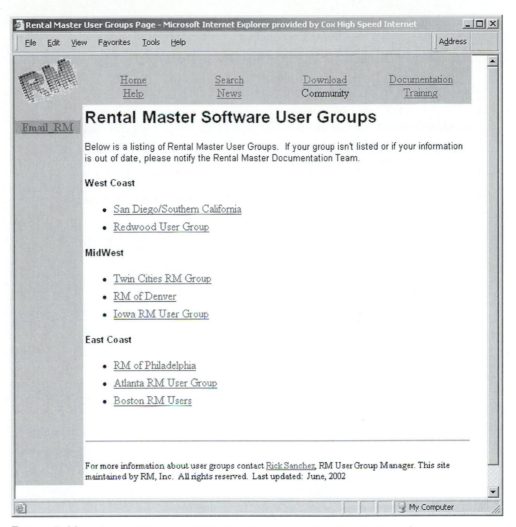

FIGURE 5.10   The Rental Master Web Site Shows an Active User Community
Rental Master Software actively cultivates and supports sharing of information and communication among its users.

NEWSGROUPS. Newsgroups are established online sites where you can post questions about a particular piece of software (or other topic) and get answers, sometimes from specialists, or sometimes from other persons who subscribe to the newsgroup. Newsgroups, such as the one shown in Figure 5.11, also contain update information posted by software companies.

USER GROUPS. User groups are subscription groups sponsored by software companies to allow for dissemination of information about the software to registered (and sometimes non-registered) users. According to the interBiz web page,

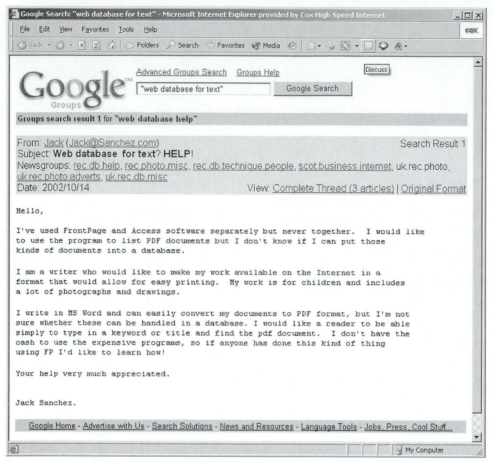

FIGURE 5.11   An Example of a Newsgroup Posting about Software
Newsgroups allow users to support one another and can provide valuable information about user information needs for documentation writers.

> *"User Groups help ensure that the interBiz family of products meet the evolving demands of the marketplace. User Groups elect their own officers and establish their own meeting schedules and agendas. Each group plays an active role in the product enhancement process, and is organized around a set of applications."[12]*

As this passage indicates, user groups at this company are determined by what kinds of programs users use, but beyond that, user groups play an integral part in the development of new products. Figure 5.12 illustrates an example of this kind of user group.

WEB RESOURCES. Web resources comprise a number of different information types, including those defined below. Offered as part of a company's web site these kinds of resources make up the extended web of support for any software. They refer to manuals and help and often contain the same information.

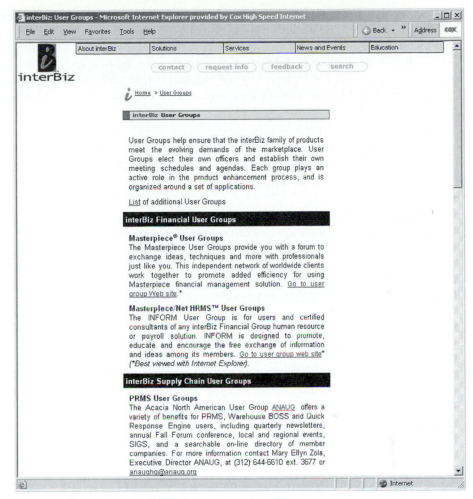

FIGURE 5.12   An Example of Users Groups Based on Application Type
The interBiz Banking web site identifies user groups according to their use of specific programs. This helps identify shared information needs among users.

- *Newsletters.* Online newsletters contain news about current releases, tips for software use, information about user groups and other support mechanisms and various other kinds of information about the software.

- *Third-Party Documentation.* Third-party documentation consists of documentation created by someone who doesn't work for the organization that created the software, or that was outsourced by the development company to create software. It is made up of books and online help, but can also be made up of highly focused documents for specialized users. HTML documentation or Java documentation often falls into this category. Everybody seems to write this.

- *FAQs.* FAQs are a mainstay of online support for software. Often they represent merely a re-organization of help topics according a re-statement of the topic in question form ("How do I Open a File?" "How do I set page borders?" and so on). But in their more useful form they actually respond to questions real users have asked that are not answered by the standard help topics. In this form they fill up a very valuable gap between descriptions of operations of the software and users' actual experience with the software.

- *Web Rings.* Web rings are software-driven connected groups of web sites, some devoted to user groups and some just sponsored by an individual or company. The thread of the ring is support for or use of software. Figure 5.13 shows an example of web rings.

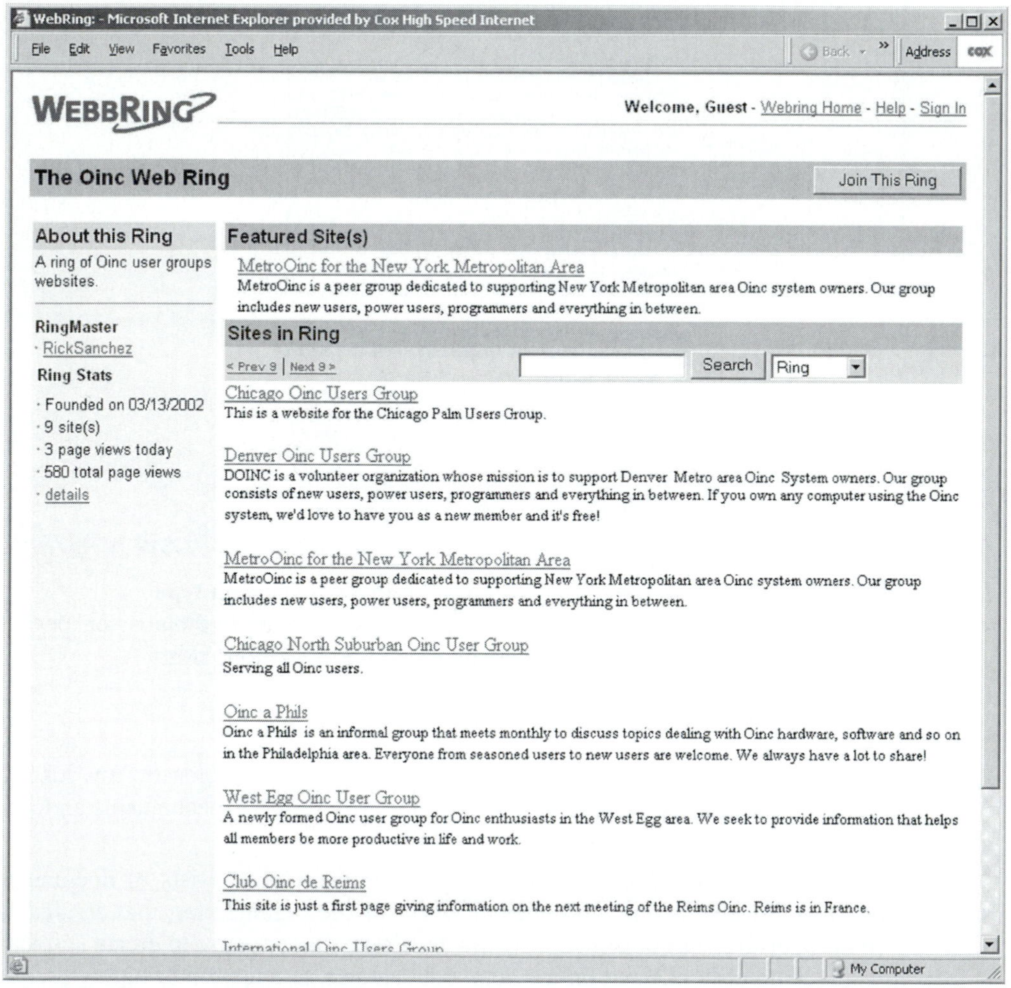

FIGURE 5.13   An Example of Web Rings
Web rings provide excellent sources of information about users of software programs and are often joined by software companies to promote their software.

- *List servers and mailing lists.* Mailing lists are based on email list server software and enable users to send information, questions, solutions, examples, queries, help calls and other dialogs to all members of the list. Thus a person could ask a question about software use and all members would have the opportunity to respond, and usually they do. The mailing list distributes the help function among the users and often is sponsored or moderated by a list manager who may be connected with the company or organization that created the software. Figure 5.14 shows an example of a list of lists.

Often you will be called upon to write specialized documentation for users of well known programs such as Microsoft Word or Access. The reason for this is because users in a specific environment need to use programs for highly specialized purposes

FIGURE 5.14   Examples of Mailing Lists Dedicated to Software Support
Mailing lists can provide you with interactive information about the use of a program and be a productive media for communicating with the user.

and the existing documentation is too broad and general to apply to their needs. Thus, you may write a guide to formatting in Word for users in your research and development group to help them create reports according to your company's format guidelines. On the other hand you may be called upon to document the use of a database created in Access that handles a membership list for your professional group. In both these cases, you will find yourself faced with the challenge of localizing existing documentation for this group.

Tip Your User to Help from Others

You should try to learn what user groups your user belongs to. If you're writing about a large program with established user groups, then you should pass along information about the group to your users. Often these groups have meetings and produce newsletters. If your software has other user categories, consider encouraging your users to consult them: Student users can consult with teacher users, data entry clerks can consult with analysis types. Suggest that these users ask others and, in your explanations and elaborations of procedures (Chapter 3, "Writing to Guide" can show you how to do this), acknowledge that the user doesn't work in a vacuum—that other persons use the program for related tasks.

Users' Learning Preferences

You may find that the organizational influences on the user also pattern another important element of user characteristics: the user's learning preferences. People who did not grow up with computers bear the imprint of their earliest successful learning, and they continue to want to learn in that way.

Learning preferences means the way your user likes to accumulate expertise in working with software. As many researchers have shown, no single method of training or presenting information for users works for all levels of expertise or job roles.[13] Fortunately, researchers have extensively examined learning preferences among software users over the past twenty years and we documenters can benefit from their discoveries. Knowing about the kinds of learning preferences users have and how to identify them can take you a long way toward designing useful print and online documentation for them. In particular, if you find the need to teach your user a subset of the overall tasks of a program—as discussed in Chapter 2, "Writing to Teach"—then what you learn in this phase of your user analysis will contribute greatly to the success of your information product. Software learning falls roughly into three categories: learning with an instructor, learning with a manual, and learning through a computer.

Learning from an Instructor

Learning from an instructor means you have a person who teaches users how to execute the features of the software and how to use them in a workplace.

SETTING. Often, instructor-based learning can occur in a classroom or training room or on the user's job site. Even if it occurs on the job site, more than one person can learn at a time.

SOURCE. Usually, organizations will sponsor instructor-based training, either by hiring consultants trained by a software company to come in and do classes, or using in-house training experts. The advantage of the in-house training expert lies in the fact that, in knowing a user's specific job demands and responsibilities, the instructor can tailor the instruction to very specific user.

VARIATIONS. You will find that users have experienced many variations of instructor-based training. Users may report having the attention of a single professional teacher, or having learned from a friend who knew the program. Some organizations even promote an "each one teach one" plan to encourage the spread of software learning. Other companies use the technique of putting a computer-automated employee in an adjacent cubicle to a non-computer automated employee and letting a natural sharing of expertise take place. Users like this method because of the immediacy of the learning. Their instructors can tailor their message to workplace goals.

MEDIA. Instructor-based training can occur in two forms: video or multimedia, and live. Video-taped instruction—or better, multimedia instruction—can be produced by in-house sources or by documentation specialists. Multimedia instruction has the advantage of using sight, sound, motion, and other effects to involve the user. The more common form of video learning is the videotape, often supplied by the software company, showing an instructor and screens. Often, users will watch while sitting at workstations (in their offices or in a training room), pausing the tape when they want. Live instruction, on the other hand, requires a set time, preparation by the instructor, and often overhead transparencies or workbooks. Usually, the instructor will demonstrate a procedure, then help the users perform the procedure on their workstations.

ADVANTAGES. The advantages of instructor-based learning include the ability to ask questions, and the close relationship of the instruction to one's workplace tasks. It also offers the security of having an expert at hand in case the learners get stuck or need extra help.

DISADVANTAGES. Instructor-based learning requires a degree of structure: someone has to set it up, plan it, and take time out of the usual workday to attend the classes. Often, too, users can feel intimidated by a class and an instructor, feeling that, as adults in a profession, they don't need an outsider to tell them how to do their job.

USER TYPES. Novices like instructor-led software learning because of the handholding it provides. Experienced users like this method for larger programs because it gets them started and they can take advantage of the instructor to help them realize workplace applications of a software program.

Learning with a Teaching Manual

Learning with a manual involves the user having some form of book, usually called a *tutorial,* containing lessons oriented toward learning and applying program features. The point with manual-based training is that the information on how to use and apply the program comes primarily through reading, not listening to an instructor.

SETTING. Manual or text-based learning often occurs on the job site, where the user has access to real-world tasks to perform. It requires some time taken away from the job. On the other hand, some companies have libraries of manuals and encourage users to check out manuals for new programs so they can study at home. Manual-based training can also occur in a training or class room, with or without an instructor or computer present.

SOURCE. Manuals usually are produced in-house by documentation departments, by the software companies themselves, as was the *Microsoft Windows User's Guide,* or by third-party commercial companies, such as Que books' *Windows for Dummies.* In some cases, manuals take the form of instructions written by a former employee about how to use a system.

VARIATIONS. Manual-based learning takes three basic forms: *tutorials* created to teach a specific set of lessons; *user's guides,* sets of procedures organized in various ways from alphabetical to task-sequenced; and *reference documentation,* lists and explanations of commands. The degree of structure and the degree of design with the clear intent to teach decreases from tutorials to reference documentation. Clearly, the greater the design to teach in these manuals, the better the chance that the user will have a successful learning experience.

MEDIA. The usual media for manuals is the book; however the booklet or quick overview can often do the job of teaching a specific task. Some manuals include workbooks for the users to calculate or record information.

ADVANTAGES. Many users like learning from manuals because they can do it alone, avoiding making any embarrassing errors in the presence of colleagues. Users also like the flexibility of the book: They can pace their own instruction and take the book to various settings, such as on a coffee break or overnight at home. Manuals also offer point-of-need support and teaching, which means that the user can consult the book when he or she has a problem. Too often, with more structured learning, such as in a classroom, the user will not think of a question until back at their desk confronted by workplace problems.

DISADVANTAGES. Often, manuals isolate the user from expert help; if a problem arises, the user can get stuck. Also, even though practically everyone has experience using books and print, some users may not have much experience learning highly abstract computer concepts using a manual. Studies have shown that manuals work best for users who have a history of using them.

USER TYPES. Experienced and expert users like manuals because they have had enough experience to make it over hurdles by themselves and they like the independence manuals provide.

Learning with a Computer

Learning with a computer differs from the two previous learning methods in that the information comes to the user primarily through the computer. Usually, such training takes the form of a computer-based tutorial on a CD, or a multimedia training package.

SETTING. Computer-based learning can occur on the job site or in a classroom. Often, users will have an instructor or monitor present, but the computer acts as the primary teacher, pacing the delivery of instructional information.

SOURCE. Computer-based training software often is packaged with a program. Commercial companies that do specialized training also create computer-based training programs for clients.

VARIATIONS. Computer-based training appears in many variations, as indicated in the list below.

- *Computer-Based Training Programs.* Lessons followed through choices on a menu
- *Manuals with Online Components.* Lessons that the user performs using the program on sample data
- *Online Help and Reference.* Procedures and descriptions of commands
- *Guided Explorations.* Hands-on exploration of the program while supporting recovery from errors
- *Trial and Error.* Users make menu choices on their own and try to figure out the program, taking cues from dialog boxes and relying on their overall knowledge of software

MEDIA. The primary media used in computer-based training is the computer itself, although some computer-based methods contain text components.

ADVANTAGES. The main advantage of the computer-based tutorial, from the user's standpoint, lies in the hands-on experience the computer can give. Also, users enjoy the sense of accomplishment they get from using a computer to learn *about* computers.

DISADVANTAGES. Computer-based instruction assumes some sophistication of users before they can succeed with it. They have to at least get the computer and the program started. And if the computer messes up, crashes, or can't find a needed information file, then the user can't proceed.

USER TYPES. Most users like this form of training because of its reliance on the computer. Novices tend to learn less from it.

Users have definite choices among kinds of learning, and often they base their preferences on past experiences. If they had success learning with a manual, then they will gravitate toward that method again, and so on. Often they will combine

FIGURE 5.15   An Example of Computer-Based Instruction Users of sites like the one in this example are familiar with learning about software in an online medium.

methods. As part of your user analysis, you should find out what experiences they have had with learning software and use it as a guide to design. Figure 5.15 shows an example of computer-based instruction.

Usage Patterns: Regular, Casual, Intermittent

The term *usage pattern* refers to the interaction of users with programs over time. Users don't learn and use all the features of a program the first time they use it. Instead, they learn a few, then more, and so on, accumulating those features that apply. We call this their usage pattern. You can easily conceive of a usage pattern by viewing it as a learning curve, which measures the number of features a user learns or uses regularly on the *x* axis and the time span of their use on the *y* axis. As shown in Figure 5.16, the use of a word processing program grows from the

FIGURE 5.16 Learning Curve Showing a Regular Usage Pattern
This graph plots the percent of features learned (*y* axis) over time (*x* axis).

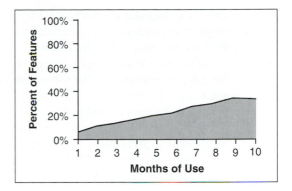

user knowing 5 percent of the features in month 1 to the user knowing 30 percent of the features after 10 months. (Actually, use would level off at about 10 percent—the estimated number of program functions that about 90 percent of the users employ.)

No ideal usage pattern exists, because of the uniqueness of each user's experience with a program. The learning curve provides us with a way to track how learning may or may not progress with a program. After users have been using a program for a while, their learning slows down because they tend to use only a subset of all the functions. They add to this store of knowledge less frequently than they did when they started. You can measure learning curves by presenting users with a list of program operations and asking them to identify the ones they perform more frequently than others. Done over time, this simple inquiry will tell you how their learning changes, or how they vary the number and kinds of tasks they perform.

What Causes Usage Patterns to Vary?

Usage patterns vary for a number of reasons. Users' motivation to learn new functions changes over time because the initial excitement and anticipation naturally decrease with familiarity. Also, users ordinarily undergo some sort of initial training that instills in them a basic competency. Without further training, their learning of new skills naturally levels off. But perhaps the main reason for variations in usage patterns lies in the nature of the features. Programs have subsets of basic and advanced features, resulting in users learning and using some on a more regular basis than others. A user's learning curve rises over time because some features get learned later. Your user analysis should include some investigation of the usage pattern you expect with your software. In general, you can expect three main types of usage patterns: regular usage, intermittent usage, and casual usage. The following paragraphs describe these patterns.

Regular Usage

Regular usage refers to the pattern a user would follow in using the program daily. Regular usage happens with people who use one or two programs as an integral part of their work. You find regular usage among secretaries and design engineers who use word processing and computer-assisted design software every day as the main way they do their jobs. Regular users become experts with one or two specific programs. Regular usage does not require frequent reference to manuals or help systems, and regular users do not make frequent errors. They become experts in the use of one program.

This pattern assumes **incremental learning,** meaning that a user learns some features before others, and that the learning of advanced features assumes knowledge of the basics. It refers to the pattern that a user would follow to get started. The most usual pattern goes something like the one illustrated in Table 5.18.

Knowing the usage pattern can help you design better manuals and help. Often, documents express the usual pattern in the *About This Manual* section in the front. The *About This Manual* section in Figure 5.17 shows you how to build information gathered about usage patterns into the documentation set, and how usage patterns can help you model the user's behavior. Clearly, the more accurate the model of the user's behavior, the more readily the user can identify with the usage pattern presented.

Intermittent Usage

Intermittent usage refers to usage by persons who know the software well enough to perform basic tasks but do not use the program as the primary software in their work. They use the software frequently and voluntarily, but at irregular intervals. Because of their intermittent pattern of usage, they make more mistakes than regular users. Making mistakes leads to frequent consultation of user's guides and online help systems. Intermittent usage occurs where software provides an important tool in a person's work, but their work does not require them to use the program all the time. Thus, their learning curve tends to plateau (indicating that they have learned a set of basic skills) at around 10 or 15 percent of the total functions of a program. In fact, where you find intermittent usage you will often find that users have learned and forgotten program features. For more information on how to design documents for intermittent usage, see Chapter 3, "Writing to Guide—Procedures."

Irregular behavior and frequent use of support documentation by intermittent users makes the intermittent usage pattern perhaps the most frequently encountered usage pattern among computer users, and the most likely pattern of usage to need support by online systems. Online help systems are particularly useful when you

TABLE 5.18   The Pattern of Regular Usage

Number of Features Learned	Activity	Documentation Needed
One or two	Taking a tour	*Guided Tour/Getting Started*
Several to many	Learning the basic functions	*Tutorial*
Some to few	Learning some advanced functions	*User's Guide* or *Online Help*
Few	Looking up advanced information/commands	*Online* or *Printed Reference*

About This Manual

The explanations, examples, and illustrations in this manual will help you take full advantage of Microsoft Word's many features. This manual is divided into three parts, "Learning Word," "Using Word," and "Word Reference."

Learning Word will teach you how to create, edit, format and print a short document. When you're finished, you'll understand how to use Word and you can build more complex documents to fit your needs.

Using Word is a guide to word processing with Microsoft Word. This part is organized into five general categories: Editing with Word, Formatting Your Work, Working with Complex Documents, Printing Documents, and Handling Documents. The hold heads in the margin show you where to find the step by step instructions for each task. "Using Word" also describes how you can use document windows to simplify complicated editing tasks, such as moving text between documents and how you can use the glossary window for inserting frequently used text into documents automatically.

Word Reference contains a directory of Word commands, a list of terms used in this manual and appendices that cover key sequences, preset page options, disk and memory management, the use of other Macintosh applications with Word, and daisywheel printers.

FIGURE 5.17   An *About This Manual* Section
This section from an early Microsoft Word *User's Guide* incorporates information about the regular usage pattern into the design of the document, providing a role for the user.

have an intermittent usage pattern. When users have forgotten a command, they can consult a help system to find answers (steps or explanations of functions) easily and quickly. Online systems work well where you find intermittent usage because the pattern reflects the behavior of experts as well as novices. Like that of expert usage, the intermittent usage pattern assumes a basic knowledge of the software and the online system so users possess the skills needed to use the online help effectively. However, like that for novice usage, the pattern assumes that users will progress more slowly, make more errors, and require support.

The existence of a production bias and of sub-optimal usage among intermittent users affects our view of users guides, indicating a main purpose of guidance documentation (Chapter 3, "Writing to Guide—Procedures"): to lead to a solution to a problem rather than teach skills. In fact, researchers have noted goal orientation and problem solving behavior as characteristic of users of procedural documentation. They have also mentioned that online systems should follow an organization around command types and user actions.[14]

Casual Usage

Casual usage refers to usage by persons with little or no formal training with the software system who need to use it immediately. In effect, they have no learning curve, because they never "learn" the system. They need to use it on the spot. Examples of this kind of usage occurs in library searches, searches for information in encyclopedias, electronic bulletin boards, and any variety of information systems. This kind of usage presents a problem to documenters because casual users rarely make use of *any* online documentation or hard copy, but prefer to peck at the keyboard or click with the mouse until the system coughs up what they need.

Casual users make up a distinct group of users among other users of software. You may have regular users who undergo training or use manuals, for whom you provide the usual kinds of support, patterned on incremental learning. But you may also need to provide some support for users who lack any prior familiarity with the system. Take a college library's catalog system, for example. The manuals for the librarian staff contain explanations of tasks; librarians can refer to these for details. But most students step up to the terminal and need to use it to find titles with no training at all. If your system has a group of users like this, you should carefully identify them and try to get a feel for the difficulties they will face.

Searchers of an encyclopedia would probably conceive of their problem as needing to locate a subject term; they would perform their search based on the alphabet. For these users, you need to point out not only that searches can use alphabetical sequences much like a printed encyclopedia, but that searches can also follow hypertext links. In the case of hypertext links, your task as documenter becomes more complicated because you have to present background information for the casual user and present it in terms that he or she can grasp immediately. One of the easiest ways to accommodate casual users is to provide user assistance that is embedded in the interface itself. Comments that take up unused space on the interface and that provide either pop-up definitions or short instructions can work wonders for users who are coming to your program for a one and only use session.

Glossary

casual user: a user who is a one-time-only user, who needs to operate a software program with no training or previous experience.

documentation plan: a document that lays out for the members of the documentation team the goals and plans for the documentation set. It consists of a summary of the user analysis, outlines of documents, and schedules for writing, testing, and editing.

expert user: a person who has learned both basic and advanced functions of a software program. A person who has a wide range of experience with computers and software.

incremental learning: a kind of learning where you build one lesson on another, and you cannot perform advanced skills until you have mastered the basics.

intermittent use: a pattern of software use characterized by users learning and then forgetting and having to re-learn a software operation.

learning preferences: methods of learning software that affect how users learn software. Preference include using a manual, learning in a classroom, using a tutorial, and using an online tutorial.

novice user: a user who may or may not have domain knowledge relevant to a program, but who is unfamiliar with the operations, menus, screens, and other elements of a program and has little idea of how the program applies to workplace actions.

program specification: documents written by a software developer that describe the functions, users, and user support (manuals and help) of a software program.

shadowing: a method of user analysis where you spend time in the person's workplace and follow him or her around, observing tasks and gathering information to give you a clear picture of how a particular software program will fit into the user's workplace activities.

tacit knowledge: knowledge that persons use in workplace tasks that is unspoken and unacknowledged, often used unconsciously. Tacit knowledge is the ability to do something without being able to explain why.

usability approach: the method of developing documentation that is based on user involvement in development and testing of manuals and help in workplace settings.

use case: an example of the application of software to workplace activities telling the workplace task, the user, and the sequence of actions incorporating the software into meaningful work.

work motivations: a group of psychological characteristics of people in their jobs. Work motivations meet needs that are internal (imposed by personality) and external (imposed by the job). Effective software and documentation taps into these needs, so showing the user how to integrate software into his or her work can be a way to encourage software and manual use.

☑ Checklist

After identifying your users, you should conduct research in the form of interviews, observations, and surveys to collect information usable in determining the documentation needed for your project. Use the following checklist as a way to remind yourself of the important elements.

User Analysis Checklist
Possible Users of the Program

❑ What recognizable groups (teacher/student, system administrator/end user) will your users comprise? Describe each group briefly.
❑ Provide a brief use case by representative professions to provide a role model of how the program will serve different professional user groups.

Primary Issues Raised by the Program

❑ What learning difficulties, motivational problems, and technical difficulties will your users face?

The User's Tasks

❑ What actions will the user perform with the program?
❑ What important activity sequences will the user follow that can help you organize your program operations or features?
❑ What actions can you use as examples in your manuals or help?

The User's Informational Needs

❑ What important kinds and forms of information does the user need in order to put the program to work?
❑ With whom does the user communicate, and in what forms and media?

The User's Work Motivation

❑ What work motivations affect the user?

The User's Range of Computer Experience

❑ Describe the kinds of novice, experienced, and expert behavior exhibited by the user. Pay attention to the numbers and kinds of programs, degree of technical knowledge, user attitudes toward computing, and the user's learning behavior.

The User's Knowledge of the Subject Matter of the Program

❑ What kinds of vocabulary reinforcement and subject/matter background will the documentation have to provide?

The User's Workplace Environment

❑ What user groups does the user belong to or have available?
❑ Have you identified: help forums, special interest groups, newsgroups, user groups and web resources available to your users?

The User's Learning Preferences

❑ Describe the user's preferences for instructor, manual, or computer-based learning.

Usage Pattern

❑ Describe the patterns users will exhibit: regular, intermittent, and casual usage.

Document Goals

❑ What goals in the area of meeting user needs do you envision for the documentation set you plan to write for the users you have described above?

Suggested Documentation Features

❑ Describe the documentation features that your user analysis suggests. Pay particular attention to areas of organization, emphases for motivation, separate sections, or other elements you want to provide to meet special needs and characteristics you discovered.

Practice/Problem Solving

1. Use the Checklist to Analyze Professionals for Their Computer Usage

Francis works as a graphic artist for a publishing and printing firm. She gets designs from clients, or comes up with her own, for book covers, spot art, technical art, drawings, photographs, and other visual elements of books, pamphlets, and posters.

Shun Yi works for the Hong Kong office of Global Accounting, Inc. She analyzes accounts for trends in client needs and writes reports to the Vice President for service management. She communicates with clients through the Global Accounting Knowledge portal which solicits suggestions of service from international customers.

Ray does reception, clerking, and other office work in a multiple-practice doctor's office. He welcomes patients, checks their appointments, and prints a daily routing sheet for each one. He sets up new appointments and processes checks and other payments.

Use the user analysis checklist to analyze the three professionals described in the scenarios above. Describe ways in which you think software could address their task needs. If you know what software programs they use, you might want to address some of the informational deficiencies of the documentation they use—where they may wish a manual would help them, but does not.

2. Practice Writing about Professional Roles You Know Little About

Pick an "obscure" professional role from the following list (or find an equally arcane profession on your own).

Hazardous waste/nuclear waste management-environmental compliance specialist

Business process reengineering, project management

Research scientist in meteorology (at a nationally funded R & D lab)

Medical Health Service Manager for the Health Board of Kenya

Quality-Assurance Specialist (hybrid microcircuit manufacturing)

Retail Farm Radio Marketing Account Executive

Find resources about this profession in your local job placement service, by contacting or visiting the company (calling the personnel director), or from a private or public employment agency. After studying the materials you find, answer the questions on the user analysis checklist about a person working in this profession. You may have to use your common sense to fill in any information gaps.

3. Plan an Interview

Select a program you know fairly well, such as a popular word processor or an Internet browser, and construct a plan for interviewing a user of a new version of that program. Follow the guidelines earlier in the chapter for identifying issues, questions, and methods of gathering information from the user. Write the questions and set up a schedule for the interview and a follow-up interview for that particular program and user.

4. Do a Survey of Software Learning Preferences

Identify a group of users to which you have access: office workers, service representatives, students. Survey them for the kinds of learning preferences outlined briefly in Chapter 2. In particular, focus on learning from an instructor, with a manual, or with a computer. Which method does this group employ primarily? Do they use a combination? How will knowing about their learning patterns help improve your documentation products? Write a brief memo about your findings.

CHAPTER 6

Planning and Writing Your Documents

This chapter explains the documentation process as a series of nine phases; a production task list for all nine phases is included. The chapter also examines some special considerations for developing online help systems. Examples show project plans for a documentation project, and the chapter gives guidelines for developing a documentation plan, including plans for designing the documentation set and managing the project. Each phase requires user involvement.

How to Read This Chapter

Both project-oriented users and beginners should read and study the Guidelines section in this chapter. It covers the basic process involved in planning and writing.

- If you've working on a project, you should read the entire chapter because it will become your guidebook. You will refer to many other chapters in reading this one, because you have to make decisions—planning decisions—based on what's in them, even though you may not have read them entirely yet.

- If you're reading to understand, then you should look in this chapter for ways to articulate document designs. Begin by reading about documentation processes in the Discussion section.

Example

Figure 6.1 shows plans for an online documentation project. The project required that the writing group gather and assemble style guidelines for company publications and put them online for the employees to access while writing. The plan shows how the members of the development team arranged the elements in the documentation process into a coordinated effort.

ONLINE NETWORK REFERENCE SYSTEM (ONRS) Tentative Schedule		
W = Writer	G = Graphics/Interface Designer	
E = Editor	T = Tester	
M = Manager	C = Client Contact	
Week	**Tasks**	**People Involved**
1st	Study existing documents	ALL
	Visit user site	assigned by M, C
	Learn HyperCard	ALL (coordinated by W)
	Research	ALL, C
2nd	Meeting to compare information	ALL
	Prepare survey for users	W, E, M
	Deliver surveys to users	assigned by M
	Outline information meeting	ALL
	Design document/interface	W, E, M, G
3rd	Initial writing of document	W, E
	Get back survey from users	M
	Revise written document	W, E
4th	Edit document	W, E
	Put document online	G
	Troubleshoot online document	G
	Choose test subjects	T
	Schedule tests	T, M
5th	Test document	T
	Review test results meeting	ALL
	Revise document	W, E, G
	Final writing and editing	W, E
6th	Select evaluation group	T, E, M
	Evaluation	T
	Study evaluation results	ALL
	Write up evaluation results	M
	Prepare maintenance guide	W, E, G

FIGURE 6.1   Task List and Schedule for Developing a Documentation Project

Guidelines

The key to producing quality documentation is to follow a process.[1] Processes vary, however, depending on the circumstances of how software is developed: some companies follow programming processes that differ depending on the characteristics of the company. Processes also differ depending on the degree of user involvement in

FIGURE 6.2   Guidelines for Planning a Documentation Project

1. Start the project.

2. Perform the user analysis.

3. Design the documents.

4. Plan the documentation project.

5. Write the alpha draft.

6. Conduct reviews and tests.

7. Revise and edit.

8. Write the final draft.

9. Conduct a field evaluation.

design. The documentation process described in this chapter follows nine phases, each building on the previous one, and each implying testing procedures and ways for documentation managers and writers to check their progress. The steps of the task-oriented documentation process involve building all the documentation products around workplace tasks you specify in the user analysis and user scenarios. Below, you will find descriptions of each of the stages of the process, as shown in Figure 6.2.

1 Start the Project

You may think of writing a software manual as simply sitting down with a program, learning how it works, and then describing what happened in a standard manual or help system format. Such an approach might have worked in the past, but given the complexities of the modern workplace it would probably result in a low-quality default manual instead of something truly useful. In the real world the start of a project is the chance for you to get to know the computer software you will write about, but one of the first things you should learn is that a good computer manual or help system is more based on the user than on the software. So one way to think about software is to see it as the raw material that you will adapt to the user; part, but not all, of the equation.

A lot of documentation gets done by the lone writer in a company or by a contract writer working on his or her own. But with larger companies like Adobe Systems, Inc., or Microsoft Corporation, the writing, like most other efforts in business today, gets done by teams. Even the lone contract writer is often brought on as a part of an existing team. Because team and group work represents the primary method of organizing software development work today, it is important for you to understand the basics of this kind of work.

Most software and software documentation is created by two kinds of teams: development teams and writing teams. The essential difference between them is that on development teams you work as the writer or editor with other professionals who do the programming, planning, and marketing of the product. Writing teams, on the other hand, consist primarily of people who write, edit, or test documents. A development team is sometimes called a cross-functional team[2], whereas on a writing team the members all focus on the writing or publications function.

You may find yourself on one of two kinds of teams: the development team—in charge of the development of the software and the documentation, or the writing team—in charge of the documentation only. The team is the starting place for your involvement in a manual or help project.

The Development Team

A *development team* develops the entire product: software and documentation. It assembles members usually from varied professional backgrounds and varied skills. The development team tends to define the team roles more distinctly, along clear-cut lines because of the backgrounds of its heterogeneous members. For this reason development team presents a greater danger of fragmenting than does the writing team.

You may find yourself on a team with different kinds of professionals than those described here, but for the most part, when you're on a development team you can expect some mix of the following persons.

PRODUCT DEVELOPER. The product developer designs the software program and often sets up the project team. In software development, the product developer often knows the subject matter (medicine, accounting, science education, etc.) well and has done research into the area.

PROJECT MANAGER. The project manager organizes the project, assigns tasks, and keeps the project on track. Often this person has a background in business management or organizational strategies.

MARKET/SYSTEMS ANALYST. The systems analyst studies the user situation and models the activities that the program will automate. The market/systems analyst often has a background in psychology, marketing, or information science. This person's main interest is in making the software product a commercial success.

TECHNICAL SPECIALIST/PROGRAMMER. The programmer knows the operating system well, and the language used for programming. This person actually writes and tests the program. He or she will have a background in programming or computer science. Most projects employ a number of programmers on any given project.

DOCUMENTATION SPECIALIST. The documentation specialist handles all the writing of the manuals and help for the project. This person has a background in technical communication (or in English) and has considerable editing skills. Often projects employ more than one writer or editor. With large projects there will be a lead writer, editor, and other writers and usability tester.

The Writing Team

Unlike the development team, the writing team develops just the documents. Working in the publications department of a company, the writing team may develop documents for more than one project at a time. On a writing team, especially one with a history of working on various projects, you will find fewer clear distinctions among the roles. Writing teams tend to show more cohesiveness because of the similarities of their backgrounds and the fact that they usually report to the same supervisor. Members of a writing team have to deal with developers, programmers, and others involved in the whole project, but these programmers and developers serve as subject matter experts and not members of the immediate work group.

MANAGER. The manager takes charge of the overall project and keeps it on track by creating and maintaining a schedule; meeting with the developer, client, or project supervisor; assigning tasks; tracking progress; and handling meetings. The manager usually writes and maintains the documentation plan.

LEAD WRITER. The writer conducts the user analysis and writes up the program task list and the drafts of documents and *topics*. While all the members of the team do some writing, this person develops the research information into actual drafts. The lead writer may or may not have other writers on the project, but he or she takes the initiative in assigning writing projects.

WRITER. This person is in charge of creating content for a project, or in re-writing pre-existing content according to the documentation plan. The writer is often responsible for interviewing subject-matter experts or researching information from written archives and other sources.

EDITOR. The editor edits the documents produced by the writer and also sets up the standards for consistency among the documents. This person creates a style guide for the team members to follow and edits the documents to make sure they do. The editor also takes charge of production duties for the team.

GRAPHICS DESIGNER. The graphics designer handles the technical aspects of screen captures and creating the illustration and artwork for the team. This person creates rough and final versions of all drawings and illustrations and tests them with users.

TESTER. The tester maintains quality in the documentation set by designing and conducting tests. This person contacts test subjects and communicates with them, sets up the test area, produces the test materials, and does the follow-up testing. This person may also, along with the manager, coordinate all the reviewing of the manuals and communicate the review information to the writer and editor.

Working on teams presents a number of challenges to your work habits and for many persons it's not something that comes naturally. You have to communicate much more, and you have to respect leaders and followers. In fact, much of what it takes to make it on a team involves psychology and the ability to communicate across national and corporate cultures.

Preliminary Research

At the start of a development and documentation project you will often have some preliminary research to do. When a software project starts someone—it could be the developer or the market analyst—will have examined the users and the industry and determined the need for the software. The justification for the software gets written down in what are known as project documents. Depending on the company or organization developing the software, you will find variations of the following types of documents.

PROJECT PLANS. Project plans identify the long and short term goals and justifications for the software itself. They articulate the problem in the workplace that the software intends to solve, such as an administrator's lack of organized knowledge in a certain area (need for a knowledge portal-type program) or a worker's need for a specialized tool (need for a business application-type program). Project plans identify the types of systems already in place, and they identify the users and the kinds of user documentation needed for the program. Project plans include preliminary schedules and task assignments.

PROGRAM SPECIFICATIONS. Program specifications describe the nuts and bolts of the program that will be created to meet the goals articulated in the project plans. These documents identify the programming language or development environment for the program, and give specific instructions to the programmers on how to structure the program and what operations or functionalities to create as they write the actual instructions.

MARKET ANALYSIS. Market analysis documents report on the market specialist's research into the sales and market potential of the software system. Market analysis documents reflect focus groups, interviews, surveys and other kinds of inquiries, very similar to those you will perform as part of the writer's user analysis. However, as Hackos points out, the analysis of the marketing specialist can not substitute for that done by writers. Marketing specialists focus their inquiries on the software buyer not the software user, and so they study what motivates people to buy rather than what motivates them to use or apply the software to their work activities.[3]

INFORMATION PLAN. The information plan identifies the kinds of documents needed by program users. This document is written by the developer or lead writer and indicates what manuals or help will accompany the program and how these documents will get produced. It often identifies the primary users of the program and contains the preliminary user analysis. It will reflect use cases, studies done of users showing how they will interact with the software. This document is very valuable to writers as it focuses on the user. For writers, the information plan resembles the design plan and documentation plan described later in this chapter. The distinction I make later between these two documents is that the design plan focuses on what the documentation will contain (content) and what it will look like (design); the documentation plan focuses on how it will be produced (management, scheduling, and so on) and is similar to the management plan.

MANAGEMENT PLAN. The management plan details the resources, people, and schedule for completing the project. Written by the project manager, the management plan represents a day-by-day strategy for completing the project. It identifies the computers and software systems and networks with which the program and documentation will be produced. It identifies work roles for all involved, and details duties and responsibilities. It sets project milestones, due dates, meetings, deadlines, edits, reviews, and all the other events that have to take place for a software system to be produced.

Overview of the Online Help Development Process

This discussion of documentation planning now turns to choosing online media and the special considerations for developing help systems. An online help system is a document that is delivered as software using the stand-alone WinHelp format or an HTML format. Online help is also delivered in portable document format (PDF), a format that uses a page layout but allows for hypertext linking. In terms of content, online help contains information about program operations and the interface of the program, arranged and presented within the context of the user's complex workplace activities. In terms of technology, however, online help systems offer the designer and user a number of features that allow you to design clever and useful ways to present information. Figure 6.3

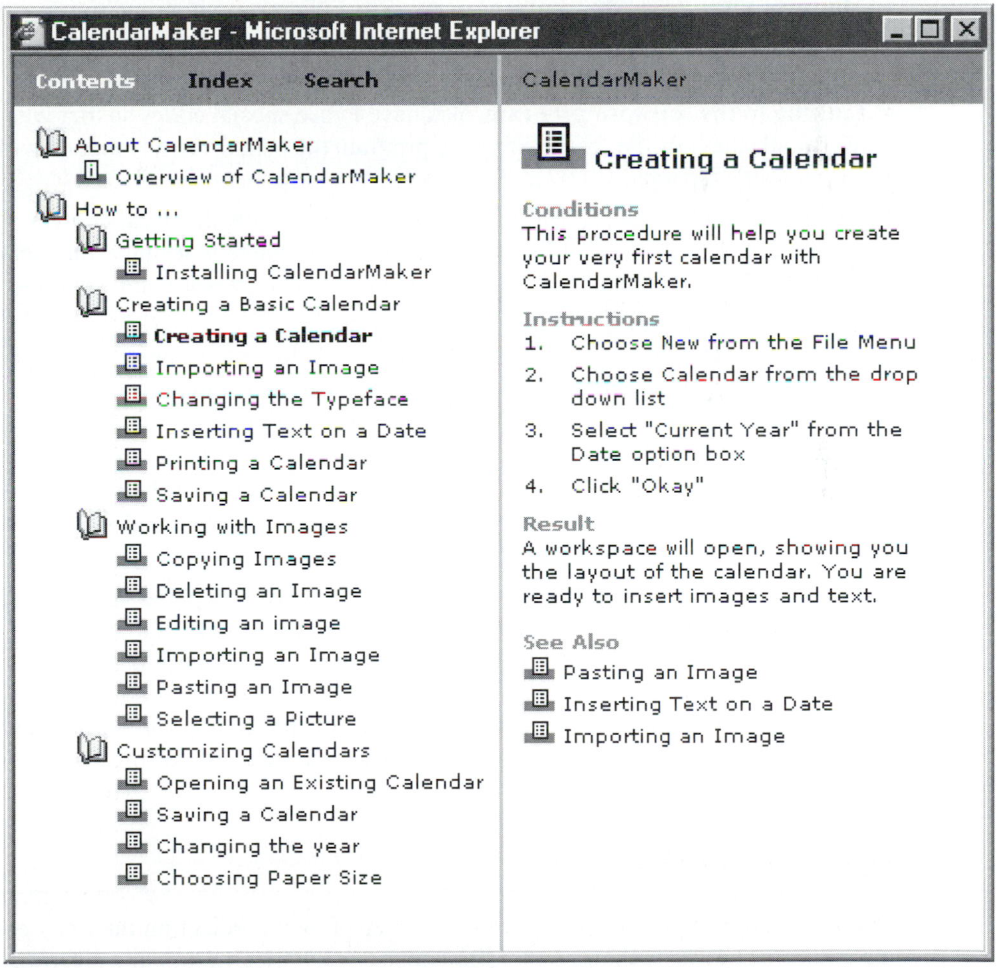

FIGURE 6.3   An Example of an Online Help System
This online help system organizes program operations according to categories meaningful to users.

shows an example of an online help system for the CalendarMaker program. Notice how the topics, operations of the program, are arranged according to user activities.

The processes that you need to follow in arranging the workload of an online documentation project basically follow the nine-step list you will find below. However, designing and managing an online help project includes some extra steps because of the technical aspects of help. Because of the tools and sometimes the personnel required to produce online help, you need to plan extra time for some project phases.

The stages of developing an online help system mirror that of developing a user's guide or tutorial except for these key differences:

- **User analysis.** The user analysis needs to focus on workplace activities of the user, but you also need to have some sense of the user's familiarity with help systems and technical knowledge.

- **Mastering the authoring environment.** Depending on which authoring system you use for a project you will need to schedule an amount of time to learn the software and overcome the inevitable delays that occur once you get started.

- **Linking to the software program.** You have to use special codes so that when the users calls the help file from within the program the help system responds with the appropriate help topic.

- **Testing the help system.** You have to test the links as well as the content. Troubleshooting becomes more difficult. Testing also requires working with users at a computer. You can't send the book but you have to send the software, get it installed and overcome other technical hurdles.

- **Testing in different user environments.** Online help systems need to work across computer platforms (Windows, Macintosh, Linux, Unix, Internet, Intranet) and to accommodate these different environments you need to try out the help system in each one. This testing means that you need installations in these environments and the ability to test and record how well your software works in each of them, often accounting for changes in operating system versions. This element can drastically slow down the development of an online help system.

Creating a document seems relatively quick and easy compared to learning the authoring software you need and the complication of creating files that the help software can handle consistently. Testing, in particular, takes much more time because you have to see to it that all the links work.

The following brief overview of the process of developing online help can get you started and show ways you will need to adjust your plans for producing in this type of media.

Starting a Help Project

One of the challenges you face in starting a help project is selecting the right authoring environment. An authoring environment is a software program that allows you to identify and write topics for your program, manage your project, and create help files in the format you need for your program. Below is a list of some, not all, of the most well-known authoring environments for online help.

COMPANY PRODUCT WEB SITE FEATURES

- ComponentOne, *Doc-To-Help,* www.componentone.com
- Sevensteps, Inc., *Sevensteps,* www.sevensteps.com
- Microsoft, Inc., *Microsoft HTML Help1.3 SDK,* www.microsoft.com
- eHelp, Inc., *RoboHELP,* www.ehelp.com
- AuthorIt Software Corporation, *AuthorIt,* www.author-it.com

Most of these authoring systems write help files that have similar characteristics: a three-pane format showing contents, an index, and a search page. Individual programs will differ slightly in how they look, but they all use essentially the same overall format. However, in selecting an authoring environment for your project (assuming that cost is not a factor) you should look at some of the following features.

Single-Source Capabilities

Single sourcing with help files essentially means that you write your help topics once and then can use them in different formats easily. Some authoring programs keep all the topics as word documents or as *HTML* files and allow you to assemble various published formats using the same files. The advantage with single source authoring is that you don't have to update information in various locations but can just update it once and then re-publish the files.

Authoring Features

Authors and designers of help systems need to know what features the authoring program offers (Figure 6.4). Usually you can count on an array of features including the following: pop-up notes, marginal notes, glossary creation, keywords, see also links, image editing, indexing capabilities, jumps (from one topic to another). Some programs make the writing of help using these features easier than others. For example, with HTML Help from Microsoft you don't get to see what the HTML file looks like until you view it in your browser, whereas other authoring programs show you the finished product as you edit. Also, some authoring programs provide capabilities for advanced features like using *dynamic HTML* codes for special effects or building browse sequences into the help system. Some also provide capabilities for showing *multi-media* files and allow you to include *script files* easily.

Your consideration of the authoring environment in terms of authoring features should be based on the kinds of capabilities you want to provide to your users. When you're considering one program over another it's a good idea to make a list of what features help the writers and what features help the users. The list of features that help the users will help you identify the ways you want to provide information and the kinds of information your reader needs.

Management Features

Management features include the ability to plan and keep track of projects. Some authoring environments offer more or less help in this area. Some allow you to identify what

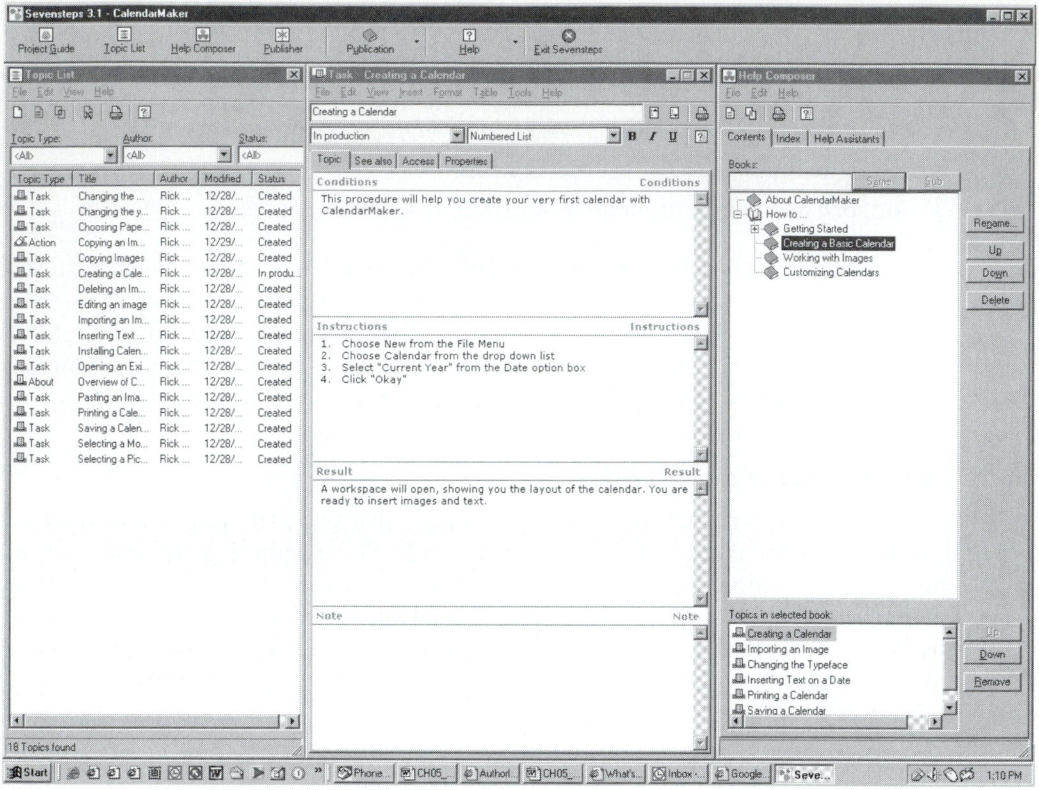

FIGURE 6.4   An Example of a Help Authoring Environment
The panes in this help writing environment allow you to organize help authoring and manage the connections among help topics.

stage of completion a topic is in, and who is responsible for it. Some programs offer forms to help you plan and others help with scheduling, showing, as in Figure 6.5, the dates for project events. You should also look for the kinds of reports that these authoring systems allow you to produce. Reports include information on topic types, numbers of topics, progress, overviews of systems, and so on. These reports help you organize large projects.

Types of Help Formats Supported

Not all authoring systems write to all formats available or desired by your project. Some write for all, some just for one or two. Below is a list of common help formats.

- **Word Documents for print or *PDF*.** Consists of documents that look just like printed pages and often are intended to be printed when necessary.
- **Windows Help.** This format represents the industry standard format until about 1997 when most help began migrating to the Microsoft-supported HTML format (Figure 6.6).

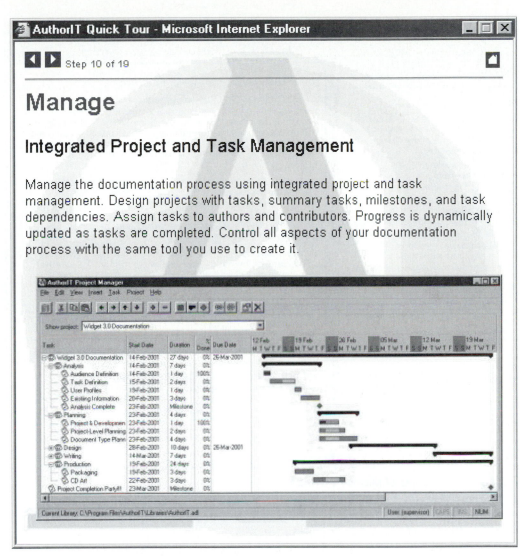

FIGURE 6.5   An example of Management Support with an Online Help Authoring Program
Notice how this task management system schedules the phases of the help development process.

- **HTML.** This format uses HTML pages and is intended to be viewed with browser programs.
- **XHTML.** This format is useful when working in a single-sourcing environment because it allows topics to include special codes useful in assembling documents.
- **Microsoft HTML Help.** This popular format is supported by Microsoft, which produces a compiler program that takes all the files in an HTML format and creates a single .chm (for "compiled html") format file for easy transporting to network locations.

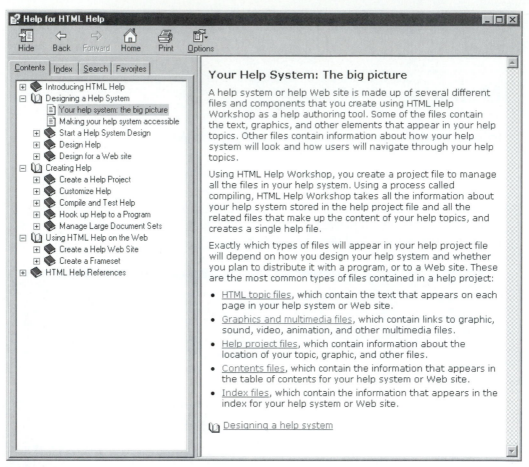

FIGURE 6.6   Features of Microsoft HTML Help
HTML Help allows you to document web applications easily without having to manage large
numbers of files. As you can see from this example the help file contains the three main parts:
navigation tools, contents, and topic panes.

- **Java Help.** This format of help is used to support applications written using Java
  as the programming language.

2   Perform the User Analysis

During this stage you research a number of elements pertinent to effective software use,
mainly focusing on the workplace activities of your users that involve software. Your
analysis of users means interviewing and observing them in order to determine their user
type, learning preferences, and other elements described in Chapter 5, "Analyzing Your
Users." The user analysis results in lists of activities for each of the primary users of the
program, which in turn allow you to identify which of the program operations the user
would need. The user analysis is the key to the table of contents of your manuals and help.

The activities in the user analysis should allow you to group the program operations for your table of contents. A table of contents based on the menu structure of the program will result in a version of the *default manual:* a manual that simply records the functions of all the menu items, commands, and dialog boxes, but doesn't structure the information according to user tasks. Figure 6.7 illustrates this sort of table of contents. As you can see, it adequately describes the functions of the program but fails to reflect anything about the use of the program for meaningful workplace

Installation
Using the File Menu
Create a Calendar
Open an Existing Calendar
Save an Untitled Calendar
Save a Titled Calendar Under Another Name
Using the Edit Menu
Edit an Image
Select a Picture Provided by CalendarMaker
Select a Picture From Another Document
Select a Month
Inserting Text on a Date
Using the Format Menu
Change the Day of Week Typeface
Change the Date Typeface
Change the Text Typeface
Change the Typeface
Using the Print Menu
Print a Calendar
Using the Image Menu
Use the Clipboard
Copy Images
Import an Image into CalendarMaker
Paste Images into the Calendar
Delete Calendar Images
Using the Advanced Menu
Change the Year
Choose the Paper Size

FIGURE 6.7   An Example of a Default Manual Table of Contents
The *default* manual merely describes the interface and can be written without a user analysis.

activities. Instead, this table of contents reflects the structure of the program, something that may only be of passing interest to the users. This method of organizing the information also does not suggest that the user learns or moves from beginning activities to advanced activities, in other words, it doesn't suggest learning.

In Figure 6.8, on the other hand, the operations of the program have been grouped according to how they would be used by actual users. As you can see, the table of contents focuses on the kinds of activities users actually do with the program: they create basic calendars, do work with images to get them right for each month,

---

Getting Started
Installing the Software
Creating a Basic Calendar
Creating a Calendar
Importing an Image into CalendarMaker
Changing the Typeface
Inserting Text on a Date
Printing a Calendar
Saving an Untitled Calendar
Working with Images
Editing an Image
Using the Clipboard
Selecting a Picture Provided by CalendarMaker
Selecting a Picture From Another Document
Copying Images
Pasting Images into the Calendar
Deleting Calendar Images
Customizing Calendars
Opening an Existing Calendar
Saving a Titled Calendar Under Another Name
Choosing the Paper Size
Changing the Year
Selecting a Month
Changing the Day of Week Typeface
Changing the Date Typeface
Changing the Text Typeface

FIGURE 6.8   An Example of a Task-Oriented Manual
Table of Contents
The *task-oriented manual* arranges the program interface information in meaningful ways based on a user analysis.

and they customize existing calendars, or otherwise make calendars that have complicated text and image features. The table of contents for the task-oriented manual also suggests a growth in learning of program features, something that the user would naturally expect as he or she became more adept with the program.

Perform the User Analysis for a Help Project

The user analysis for a help project needs to focus on workplace activities of the user, but you also need to have some sense of the user's familiarity with help systems and technical knowledge. The Sevensteps authoring environment includes a number of forms that help the project manager handle the planning of a help project. Figure 6.9 shows one such form. Notice that one of the behaviors asked about in the form is the user's experience with help systems. This kind of computer experience is very useful because it can tell you about the users willingness to use your system instead of a manual and it can tell you what kinds of features the user expects in a help system.

3 Design the Documents

During the design phase you apply the three types of document forms—tutorial, procedures, and reference—to your user's needs. The design phase also means that you outline the documents and decide about their layout: pages, text style, size, font, and language, as discussed in Chapter 11, "Laying Out Pages and Screens." At the design stage you write the titles of your documents and finalize what they will look like.

For online help, you need to decide on the types of products you will produce (help, training, guided tour, templates, messages, tips, etc.), make a detailed list of help topics, and determine their layout. During your work in this stage of the development process you should consult the chapters in "Section 3: Document Design Tools" as necessary to gain the right background information for making design decisions.

At the design phase of your project you set out the content and look of your documents but decisions at this phase are important for planning, and may change over time. This change results from your trying out your designs with actual users and reviewers. As you will see below, some development methodologies actually require you to sketch out a prototype of draft of a document and then test it repeatedly until you feel it meets your goal of meeting the users' information needs. For this reason you should look ahead to the chapters in Section 3 to get a sense of what options you have and how you want your documents to look, and then use this knowledge as a starting point as you plan your documentation project.

Design the Help System

Chapter 10, "Designing for Task Orientation," covers the design process for online help systems. At the planning stage you need to know what topics you want to include and how they relate to one another. You need to begin to build a list of keywords and glossary terms that you want to include in your help system (for more on preparing a list of keywords, see Chapter14, "Designing Indexes and Searches"). Finally, you

Seven•steps

Form for defining user groups

Name Project

CalendarMaker 1.0

History

Version	Date	Author	Remarks
1.0	10/03	Rick Sanchez	Preliminary analysis

Characteristic	User group 1	User group 2	User group 3
How do these users call themselves? (name of the group)	Teachers	Office Managers	Project leaders
What is the overall goal of these peoples' jobs?	to manage education	to manage an office staff	to manage project members
What is their background in education?	college	college	college
How often will they use the business application?	once per semester	on-going, daily updates	beginning a project
How long will they work with the business application?	3 hours per semester	30 minutes per week	3 hours during a project
What is this user's experience with help systems	uses Word help, likes to print help files	uses Help frequently	too busy to use print manuals
Is use of the new application mandatory?	no	no	no
What is the users' level of experience in the use of applications in a Windows environment?	some, does grades on a spreadsheet	lots, works with Office 2000	Uses Microsoft Office and Project
Will the new application implicate a whole new way of working or do processes remain unchanged?	teachers will abandon hand-made calendars	same processes	same processes

FIGURE 6.9   A Form for Analyzing Help Users

This analytical form allows you to compare characteristics across user groups.

should have a table of contents topics that indicates the types of topics you have in the document: commands, step-by-step, etc.

4 **Plan the Documentation Project**

If you are managing a project you need to know how to write a project plan for your documents because the writers and others involved will need it as a guidepost for their activities. They will look to it for deadlines, to clarify their responsibilities, to identify knowledge resources they have, and so on. On the other hand, if you're a writer or editor and a *member* of a team, you need to know about the plan from the manager's point of view so you can understand and respect the kind of work the manager put into it and the way it plays a key role in directing your activities. The description below focuses on the management aspects of the documentation project plan rather than the design aspects. The design of the documents, as treated below, gets articulated in the ***design plan,*** which you write after you have completed your user analysis. The reason for this is that the design plan needs to be based on a thorough user analysis, rather than the preliminary one done by the developers at the very start of the project.

As you can see, a well-organized and quality document is not the result of a haphazard approach to writing, but a carefully planned project that makes the most of your users, the writers and programmers on your team, your hardware and software resources, all within an allotted time frame. The key document in organizing a documentation project is the ***documentation plan.*** The documentation plan allows you to specify the manuals and online help you identified during the user analysis, and add information about the entire project. In writing the project plan, you must describe the management aspects of your work: schedules of drafts and tests, people and hardware resources, and time/page estimates. The ***project plan*** culminates your research and design work on a project. You should review the project plan with all the people involved before going on to the next stage in order to ensure appropriate document and project guidelines.

Remember, of course, that you can't always get the management of a documentation project down ahead of time, so that all you and your team members have to do is follow along. Writing manuals and help is an extremely complicated process, and until you have experience following the process you should expect to make important decisions as the project unfolds. You will find a detailed outline of the documentation project plan later in this chapter.

List Project Events

Documentation projects involve a series of events, sometimes called "project tasks" that allow you to manage the effort it takes to write a software manual or help system.

As you look over the suggested tasks in the list in Table 6.1, realize that not all of them will apply to your project. Use those that do, and invent ones you foresee that don't appear here. After you have your list together, share it with the other people in the project and get their approval and input.

TABLE 6.1   Project Events

Start the Project	Obtain working copies of the program
	Meet with the development or writing team
	Review development documents
Perform the User Analysis	Identify and make contact with potential users
	Design user interview materials, questionnaires, and other forms
	Perform user interviews, observations, focus groups
	Administer user surveys or questionnaires
	List user activities
	Write user scenarios
Design the Documents	Identify the content of the documents
	Identify suitable writing and graphics software for the project
	Review design options
	Create mockups and descriptions of all documents
	Set up a style sheet for the documents
	Write the document **design plan**
Write the Project Plan	Set documentation goals
	Identify and assemble the writing team
	Identify and negotiate project resources
	Arrange for design testing and reviewing
	Set up a project schedule
	Circulate **project plan** to team members
Write the Alpha Draft	Carry out writing tasks identified in the project plan
	Test prototype designs and drafts with users
	Maintain the style sheets for the project
	Maintain project records and archives
Conduct Reviews and Tests	Get user reviews for suitability, accuracy, and completeness
	Arrange for technical reviews and review meetings
	Contact users for usability tests
	Make arrangements for use of usability lab
	Conduct usability tests
	Write a usability report
	Maintain project records and archives
Revise and Edit	Incorporate review information into drafts
	Reconfirm changes with users
	Schedule document edits
	Schedule review meetings
	Assemble and manage all the graphics
	Review graphics and tables for accuracy
	Maintain project records and archives
Write a Final Draft	Confirm vocabulary decisions with users
	Maintain the file system for drafts
	Write the index if necessary
	Maintain project records and archives
Conduct a Field Evaluation	Survey actual users for reactions, needed improvements
	Contact customer support for records of user questions
	Write an evaluation report on the project
	Maintain project records and archives

What Goes in the Project Plan?

The project plan section should detail the tasks and schedule you intend to follow to complete your project. The introduction should give an overview to the section and should introduce the main phases of document production. In addition, it should mention the final delivery date (just in case the reader misses it in the executive summary or the schedule section). The project plan consists of the following three main parts:

- Schedule of events for completion of your project
- Plans for using resources
- Time/page estimates

Each of these should follow a specific heading. You should introduce these subsections as you do all sections, with a brief, even one- or- two-line overview.

Decide What to Schedule

When scheduling a documentation project, you need to schedule the overall phases of document development and specific events, such as meetings and deadlines, that occur within the overall phases.

Aside from identifying the overall phases, your schedule should include dates for the following six kinds of events:

- Meetings
- Deadlines for drafts
- Project report due dates
- Test completion
- Review deadlines
- Edits

You should introduce your schedule, telling its purpose. Then you should create a table or tables to present the names of the events, known as project milestones, and their completion dates as illustrated in Figure 6.10. Your reader will examine your schedule for its completeness and its good sense. Don't, for example, expect a schedule of two days for a document review to get accepted—they take longer than that. In Figure 6.10 you will find an example of part of a work schedule.

MEETINGS. Meetings usually fall into two types: regularly scheduled ones to go over progress and problems with development or writing team members, and special meetings that occur at project milestones. Both types of meetings usually take about an hour. Usually meetings coincide with completion of drafts, tests, and reviews and are used to bring the writing or development team back together after some *milestone* has been reached.

Special meetings involve other personnel associated with your project such as sponsors and users. In scheduling special meetings, consider both the people who attend (and their possible schedule conflicts) and the purpose of the meeting. Usually the manager or supervisor will attend most meetings; Table 6.2 indicates attendees whose input to the documentation process comprises the main reason for the meeting.

Event	Start Date	End Date	Personnel
complete user analysis	4/15	5/1	manager, writer, user group
progress report		5/15 (due)	manager
write draft	6/21	7/7	writer
do user walkthrough (print) (meeting)	7/12	7/12	user group, writer, editor
do user walkthrough (online) (meeting)	7/14	7/14	user group, writer, editor, test coordinator
user walkthrough report		7/16	manager
technical review (meeting)	7/15	7/15	programmer
editing	8/9	8/25	writer, editor
. . .	. . .	. . .	. . .

FIGURE 6.10   An Example of a Schedule for a Project

TABLE 6.2   Kinds of Meetings Associated with the Documentation Process

Type	Attendees	Purpose
Review meetings	All persons who reviewed a draft	Go over the review information and resolve conflicts
User walkthroughs	Selected users	Go over the documentation design plans for suitability, ease of use, and usability
Technical walkthroughs	Programmers	Go over the program task list for inaccuracies, omissions, logic

In addition, all meetings should aim to increase communication generally and foster a team spirit. The "purpose" indicated in Table 6.2 tells what specific new information or chore the meeting focuses on.

SET DEADLINES FOR DRAFTS. Set the dates for your alpha draft and any other drafts for all documents. (Note: The day you will "complete" them is usually counted as the date when you will put them in the recipient's hands or upload them to the project web site.)

SET PROJECT REPORT DUE DATES. Any development project requires some reports so others involved in the process can keep track of it. You, as a writer, benefit from this because those involved can see your work and approve it along the way. Report types (see Table 6.3) include the following: evaluation reports, progress reports, oral briefings, and test reports.

TABLE 6.3   Kinds of Reports Associated with the Documentation Process

Report	Writer	Reader	Description
Evaluation	Team leader	Supervisor/ sponsor	Describes and evaluates work of team members
Progress	Writer/ team leader	Manager/ client	Describes work to date and plans for completion
Test	Writer/tester	Manager/writer	Describes the results of tests and recommends revisions
Oral briefing	Writer/team leader	Manager/ development team	Present new tasks, review progress, discuss team business

SET USABILITY TESTS COMPLETION DATES. Usability tests fall into three categories: predictive tests, developmental tests, and evaluative tests. Predictive tests cover your design plans and documentation plans. They occur early in the process. Developmental tests cover elements of your document as you work on them. These occur while you write and the information feeds back into your continuing work. Evaluative tests occur after you have completed the document. Information from evaluative tests gets used in planning your next project. You can learn more about tests in Chapter 8, "Conducting Usability Tests".

ANTICIPATE REVIEW SCHEDULES. Reviews need special attention for scheduling because they take more time than other events. You have to really plan ahead for these. You may have to set up times for the following interested persons: managers, users, technical personnel, attorneys, or clients/sponsors. Indicate on your schedule when you will put the documents in the reviewers' hands and when they will put the documents back in your hands. For more information on reviews see Chapter 7, "Getting Useful Reviews".

If your program has more than user group (say both novice and experienced; or managers and technicians), you will probably want to get reviews from all types of users. The reader of your documentation plan will expect you to have anticipated this and scheduled it.

SCHEDULE EDITS. Edits focus specifically on your text. Usually you would edit after the final draft. If you arrange to have your work edited by someone else (a good idea), you should schedule this event carefully. Depending on the level of edit that your work requires, an edit may require more or less time. For more information on edits see Chapter 9, "Editing and Fine Tuning".

Estimate How Long Project Tasks Will Take

No two projects take the same time, and depending on how many persons you have working on a specific system, you can shorten or lengthen the overall time. If you're writing online help you will need to schedule extra time for coding, compiling, and testing (see Chapter 8, "Conducting Usability Tests"), and for involvement of programmers and extra reviewers (see Chapter 7, "Getting Useful Reviews"). Figure 6.11 illustrates the times that the development phases take in a typical project.

User's Guide	
Phase	%
1. Start the project	8
2. Perform the user analysis	15
3. Write the project plan	15
4. Design the documents	8
5. Write the alpha draft	15
6. Conduct reviews and tests	15
7. Revise and edit	8
8. Write the final draft	8
9. Conduct a field evaluation	8
Total	100

FIGURE 6.11   Time Estimates for Project Phases: User's Guides and Help
These phases are just estimates, but you can use them to block out the main activities of your documentation project. With a help system the testing phase can get drawn out because of the need to test the software functions built into the system.

Phases in an online help project take a different proportion of the total time needed in a project because of the technical requirements in creating a help system. As you will see below, help systems require you to work with programmers and manage the technical aspects of linking the help system with the software program. For this reason a greater proportion of the total development time gets allocated to the later parts of the schedule.

Of course, when you schedule the phases of the document development process, they don't always come out in a neat, linear form. In fact, they always overlap one another and get spread out at various times over the total duration of the project. Figure 6.12 depicts the overlapping of phases during a typical project.

PLAN HUMAN AND MATERIAL RESOURCES. No project can move forward without human and material resources; you should plan yours carefully. When you present descriptions of what resources you will use to complete the project, consider that the sponsor or client who reads your documentation plan needs convincing that you have carefully planned and that the resources you specify will actually allow you to meet your documentation objectives.

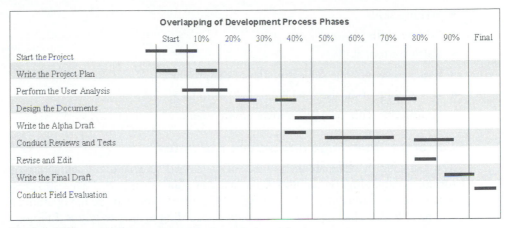

FIGURE 6.12   Overlapping Development Process Phases
Rarely do you complete one phase of development before starting another. This figure shows areas that often overlap.

- **Human resources.** Of course, your work makes up the main resource of a small project. Convince your reader that you realize this and have examined your human resources carefully, and that you know who you will have to involve in the project. Consider the efforts of the sponsors, users, testers, reviewers, and editors. They value their time, and their time adds to the value of your product. Plan wisely for their time on your project.

- **Material resources (computers and equipment).** Think through what machines you will use in a way that will benefit you and your reader. Assure yourself that the technology you will need will work and that you know how to use it. Consider and discuss what computers, printer(s), software, and other equipment (scanners, servers, copiers) you will use. Explain your choices in a way that lets your reader know you have a valid plan.

Assign People to Tasks

If you're the manager of your documentation team then you determine the roles team members will play: writer, editor, tester, graphics designer, review coordinator. Most of the tasks in Table 6.1 will fall into these categories. After you have identified the tasks associated with your project you then need to assign people to them.

The Discussion section of this chapter details the kinds of people involved in writing and development projects, but as a matter of planning you should try to assign people to tasks as early as possible. Assigning tasks means deciding who will take responsibility for making sure a task gets completed. This person becomes accountable for that task. Others may help the person with the task but usually one person reports on its completion.

As a manager you should know the talents and experience of your team members. When assigning tasks, keep in mind the following characteristics:

- Writing skills
- Editing skills
- Software tool skills
- Experience with the subject matter of the program
- Knowledge of the user and the user's workplace
- Familiarity with the development environment.

Calculate Time/Page and Screen Estimates

You can roughly calculate the number of pages and screens in a document by using formulas depending on the user activities you plan to support. Page estimates start with your table of contents. In your table of contents you can see the operations you will have to describe for your user, grouped according to user activities. In Figure 6.8, for example, you can see that the action of "Creating a Basic Calendar" is going to require the user to be able to perform six operations. Table 6.4 shows how you would estimate the number of pages it would take to create the documentation for the "Creating a Basic Calendar" action.

You can roughly estimate the number of pages in your final documentation set by calculating the number of overall program operations in the software multiplied by the average page length for each task. Most program operations (that is to say, functions like "save a file" or "open a new document") take between 1/2 and 1 page to write. Add in extra pages for front matter (introductions, overviews, license agreements, etc.), and count cover pages and all other pages.

Multiply the grand total of all pages that you will produce by the industry standard of 2.9 hours per page to arrive at an overall estimate of the number of hours required to produce your documentation set. This number includes development time, user analysis time, reviews, testing and all the many activities it takes you to write a "page" (or "screen" in the case of online help). In the example above,

TABLE 6.4   Page Estimation Table for "Creating a Basic Calendar" Activity

Operations Needed	Page length Long = 1 Med. = .5 Short = .25	Screen Shot Included = .5 Spot graphic = .25 Not needed = 0	Total
Activity: Creating a Basic Calendar			
Creating a Calendar	1	.5	1.5
Importing an Image	1	.5	1.5
Changing the Typeface	.5	.5	1
Inserting Text on a Date	.5	.25	.75
Printing a Calendar	.1	.5	1.5
Saving an Untitled Calendar	.5	0	.5
		Total pages	6.75

a paper manual to cover this one user activity would take approximately 19.75 hours (6.75 pages multiplied by 2.9 hours per page).

Of course this limited example doesn't take into consideration a number of variables that could cause the total page count to be different. These differences follow the lines below.

- **Type of documentation.** Depending on whether you're writing tutorial, procedures, or reference you will need to add information. The 2.9 magic number mentioned earlier is for a standard procedure written on a seven by nine inch page (the standard format for documents.) Tutorials take roughly 50% more pages to cover the same material because you have to include more explanations, overviews, and summaries and other information to fulfill the teaching function. Conversely, reference documentation takes roughly 50% less space (or even much less) because all you have to do is list the bare bones information and let the user figure out the rest.

- **Availability of information.** If you have to travel many miles to interview a subject matter expert just to figure out how to work the operations or to understand the user activity, then it will take you longer than 2.9 hours per page to write the tasks.

- **Experience of writers.** Writers with experience can crank out procedures much faster than beginners simply because they have learned the basics of how to write steps, and know what to include with each one. Writing steps is not a divine gift and so it takes time to perfect the skill. Chapter 3 on writing procedures can help you with this skill, as can Chapter 12 on language.

- **Reliability of the program.** If your program is being tested and revised while you're writing there's every chance that you will have to revise your procedures, which can cause that 2.9 hours per page figure to stretch to something like 5 hours per page.

Page and screen estimation is a fine art that follows some very predictable patterns, so you can learn it as you go along. You can easily find more advice on estimating lengths at the web site www.writingsoftwaredocumenation.com. There you will also find page and screen estimation spreadsheets that can help with your page and screen estimation tasks.

Write the Help Project Plan

The project plan for an online help system includes elements that you don't have to think about with a print project. You may, of course, undergo a project that uses both print and online documentation, in which case your documentation plan will reflect both products.

Most help development teams are cross functional in that they include a non-writer, someone to assume the technical role of linking the help system to the program. The example in Figure 6.13 shows how the people and tasks are handled in the Sevensteps authoring environment. The program is designed to work with four persons: a project leader (manager of the project), an author (who writes the actual topics), a domain expert (who knows the industry in which the users of the software work), and a developer (a person who handles the technical aspects of the project; linking the help system to the program.) The program comes with an estimator of

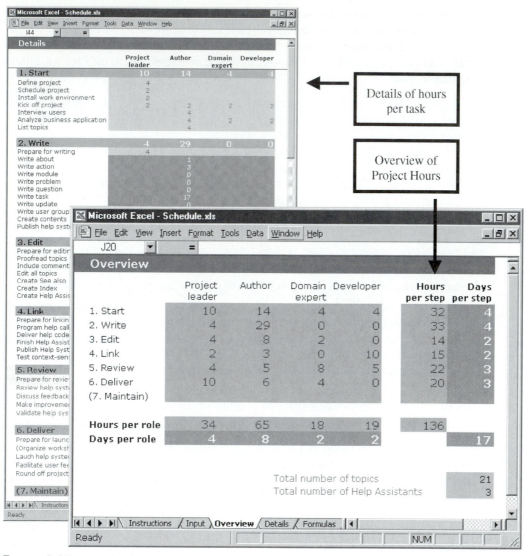

FIGURE 6.13   A Spreadsheet for Estimating Online Project Tasks
This spreadsheet calculates the days and hours needed to create an online help system based on
the numbers of topics and the personnel involved. Note the extensive task list in the left-hand
screen, reflecting the complexity of an online help project.

projects that uses formulas to calculate how many hours each person working on the
project needs to put in in order to create a given number of tasks. Figure 6.13 shows
how the estimator works. In the overview panel the spreadsheet lists the seven stages
of the project in the first column, and shows the four persons involved in the project
across the top.

In the spreadsheet in Figure 6.13, you can see that for a project that has 17 topics (the CalendarMaker project) the project leader will put in a lot of hours (10) at the beginning of the project, doing the planning work. In contrast, the author will put in approximately 29 hours during the writing stage of the process, as this person or persons is in charge of writing the topics. During the linking stage (step 4) the developer will put in 10 hours because it is during this phase that the special codes typed into each of the help system topics get listed in the program so that when the user clicks on the help file in the program the program knows which help topic to display.

Reviewing the Documentation Plan

Your documentation plan should undergo thorough testing and review by managers, clients, and users. Your coordinating editor, if you have one, should also examine your document plan. The design plan requires a much more extensive review than your user analysis received. The design plan review should cover the actual design of the document (pages, text, language, and other features) and the project plan (allocation of computer and human resources and scheduling of drafts and other production activities).

The documentation plan gives you the opportunity to hold a user *walkthrough* (going over the important design elements of your document) and a technical walkthrough (going over the accuracy and completeness of the program task list).

5 Write the Alpha Draft

The alpha draft represents your first complete document, including all the front matter, text, graphics, appendixes, indexes and associated documentation set materials. As a written document, the alpha draft is tested, reviewed, and edited—all according to the specifications laid out in the documentation plan. It's a good idea to make up an alpha draft checklist, showing all the elements you need in the complete document. Writing also requires that you do preliminary, often informal testing on the material you're writing. It's a good idea to show your procedures to fellow writers and users to get feedback as you go.

Writing also means that you follow the style guidelines for the project. It may turn out that you actually write and certainly test those rules as you write. Your job as a writer is to work with the program interface and your knowledge of the user to produce descriptions, procedures, overviews, explanations, and all the content of the document. That content shares your wisdom and experience with the program in such a way that your users can follow easily and productively.

Draft the Help Topics

Writing topics is a matter of either writing new content using the authoring environment of a help system, or importing content from files that you will also use in a print document. Writing help topics requires that you not only supply content for each topic, but supply information about the topic that the system uses to relate it to other topics. In Figure 6.3, for example, you can see in the left pane a list of existing topics. Above this list the author can identify the "task type" (about, action, module, problem, question, task, update, group), the "author" (by selecting from a list of authors in the drop-down menu), and also tell the "status" of the topic (created, in production, draft, complete, content checked, finished.) The environment also allows the author to indicate what "book" or workplace category the topic is to reside in, as well as specify the "see-also" links to and from other topics (on the "see also" tab at the top of the center pane), and the index terms that relate to this topic.

So you can see that the writing of topics involves creating content, but also creating important links and interconnected relationships among topics and recording them using the help authoring system. This is the reason you need to identify these elements in the planning stage and communicate them to all the writers on the team.

Once the topics are written a special program called a *help compiler* is needed to read the topic files and create a third file, the actual help document. Help compiler programs are included with help authoring systems. The compiler reads the topic files, follows the instructions it contains as to how to set up the help system, and creates the system. In some systems this process is called "publishing." Often you will encounter errors in your topic files such as unresolved cross-references or missing information. You will have to correct these errors and compile the help system again. Once the compiler creates this third document with no errors, you can read and test your system.

6 | Conduct Reviews and Tests

Because your alpha draft contains all the elements of your product, you can send it out for review by clients, executives, and managers, as well as users. At the same time, you can design usability tests using the original documentation objectives, to test for elements such as accuracy, task orientation, and so on. Information from reviews and tests provides feedback for the next draft of the set. For information on these topics, consult Chapter 7, "Getting Useful Reviews," and Chapter 8, "Conducting Usability Tests."

Conducting Reviews and Tests of Online Help

Testing the system occurs after the compiler has successfully created a version of the help system that you can circulate among users and clients. The problems with testing a help system relate not to the technical aspects of the system, but to the information in it. Part of the task consists of clicking on all the links and pop-ups in order to make sure they do what they are intended to do. Testing, both of print manuals and help systems, is covered in greater detail in Chapter 8, and reviewing is covered in Chapter 7. With a help system you usually go through these phases after you have completed a whole version of the system rather than reviewing parts at a time, because the elements (topics and so on) are so interconnected and you can't see how one part works in isolation from the others. Also the testing has to coincide with the release of a version of the program so that you can test the context-sensitive links.

For planning purposes you need to make sure to schedule the following tests:

- technical elements
- content accuracy

7 Revise and Edit

While the reviews and tests provide feedback from external sources (managers, users, clients, and so on), revising and editing also allow you to submit your work to an editor or edit your own document, applying an editor's skills in reorganizing and checking for accuracy on many levels. Consult Chapter 9, "Editing and Fine Tuning" for information not only in what to look for in editing, but in ways to structure the editing process to make it efficient.

Revising and Editing Online Help

Revising and editing of a help system means that you go back into the authoring environment and record and respond to all the feedback obtained during the reviewing and testing phase. This procedure, where you edit at the level grammar, punctuation, and mechanics is one of the last stages where you can add value and improvements to the system. Editing of online help systems is covered in Chapter 9.

8 Write a Final Draft

The final draft revision contains information gathered from the activities in the two previous stages. If you do them thoroughly, you will find that your document improves greatly at this stage. Incorporating feedback into your document will result in a camera-ready copy that you can hand to the printer or, in the case of help systems and other online support, deliver ready for distribution with the program.

Writing the Final Draft of a Help System

Writing the final draft of a help system mainly consists of preparing the help file for distribution with the finished program. Scheduling of the help system at this stage should coincide with development plans.

9 Conduct a Field Evaluation

After the user has installed and operated the program, the last stage of the development process happens: the field evaluation. This special kind of test enables you to gauge how well your manual met the task needs of the intended user. Information from this evaluation usually ends up in an evaluation report and provides input for your next project.

Because of the differences between the print and online media, these phases will grow and shrink in proportion to one another. But when you develop complex electronic documents, help systems included, you face technological and managerial challenges that you can weather if you anticipate and plan well for them. The following section focuses on the impact that developing a help component in your documentation set can have on your schedules.

Chapter 8, "Conducting Usability Tests," gives advice on testing that you can follow for evaluating the success of your project.

Conduct a Field Evaluation of a Help System

The field evaluation entails gathering feedback from users after the program and help system have been delivered. You can plan for this in a number of ways:

USE FEEDBACK LINKS IN THE HTML SYSTEM. This method gathers feedback information from users of an HTML-based system and stores it in a database on the server where the application or help system resides.

USE OF EMAIL LINKS. This method is easier to use and only requires that you set up an email address (example: feedback@calendarmaker.com) that invites users to respond to any problems they have with specific tasks.

Alternately you can set up an electronic survey on a web site that asks specific questions about the help system and stores the information in the survey database.

Discussion

In writing software documentation in a professional setting you usually will find two main kinds of projects: stand-alone projects and development projects. The stand-alone project is where you are assigned or contracted to write documentation for a software application that has already been written or is being revised. Often this happens when the program has simply been developed without plans for a manual or help system (or with a rudimentary help system written by programmers), or a well known program such as Microsoft Word or Excel requires in-house documents for specific readers within an organization. For example, a writer might be assigned to write a manual to help museum department heads format their yearly reports using Microsoft Word. Such documents would be highly focused on user activities because the domain of workplace knowledge is limited to a specific company and users.

The development project is the other very common kind of manual and help project. Development projects occur in organizations that create software as their main products, such as Computer Associates, Inc., or BMC Software, Inc. These companies study markets for products and have standard processes in place in their software and information development organizations for creating software. Technical communicators work for these organizations either as regular employees or as contractors and consultants on a project-by-project basis.

Differences between stand-alone projects fall into three areas: team structures, work processes, and kinds of development documents required.

Team Structures

In stand-alone projects writers have the entire program at their disposal before they start. Writers can then learn the program, become familiar with its features with no input into the interface. The writers in this case usually form writing teams with programmers, technical experts, clients, sponsors, and user representatives as secondary members. The writing team thus has the advantage of sharing information and supporting one another in writing tasks. On the other hand, development projects differ from stand-alone projects in that the writers are involved in the project from the design stages onward and so they have more input into the usability and interface of the project. As writers and experts in user analysis for software, writers can suggest screen layout, help and document formats, and usability testing procedures to insure that the product is designed to meet users' needs and not merely as a collection of nifty features. The team may be composed of writers but the team itself interacts with programmers and developers as part of a larger development effort.

Kinds of Development Documents

With stand-alone project the writers are in charge and so can create the kinds of project documents (plans, designs, reports, etc.) specifically pertaining to the writing. On development projects, writers have access to program development documents such as program specifications, actual printouts of programming instructions or code, marketing analyses, user models and cases, and project proposals. These documents form the textual resources of the project and can be very valuable to writers on the team.

Processes

In a stand-alone project the writers essentially follow the writing processes described in this chapter in the Guidelines section above. This is for writing manuals or help. When individual writers or a writing team is part of a larger development project the writing often has to parallel the development process that the larger project follows. In software development today there are three main development methodologies in place; the waterfall method, the rapid-development method, and the object modeling method.

THE WATERFALL METHOD. The waterfall method is so called because it follows a strategy of identifying the user and program specifications at the beginning of a project, and then following the plan through a series of stages until the final product and manual are produced. Based on the philosophy that it is possible to understand all the parts of a program, documentation, and user at the beginning, this method uses product specifications as the guidelines for team members. Writers and programmers have to follow the same requirements set at the beginning. The process goes through phases of gathering

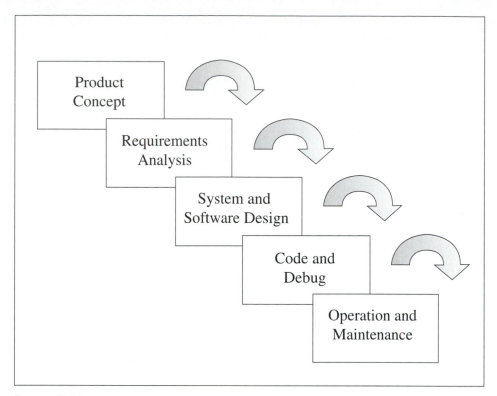

FIGURE 6.14   The Waterfall Method
The waterfall method of software development depends on understanding the users'
requirements at the beginning of the project and then carefully designing a system to
meet them. Ideally, each phase is done before the next begins.

analytical information about users and computer environments, writing programs and
manuals and help, testing all the systems together, and producing the final product all
at the same time. The requirements may change as the process develops, but usually
this change is minimal. Figure 6.14 shows an example of the phases of the waterfall
method. This method is more "traditional" in that it has been around since the early
years of software development. The main problem with this method is that it is very
difficult to say at the beginning of a one-year project exactly what the user will need
and so the program and documents tend to diverge from the original requirements during development as programmers and writers naturally respond to user needs.

THE RAPID-DEVELOPMENT METHOD. The rapid-development method is used when the
development team wants to bring products to market faster than is possible with the
waterfall method. Also, this method subscribes to the philosophy that the users' software and information needs should be incorporated into the development process in
order to make sure that what comes out at the production end of the cycle actually
meets user needs. The rapid-development method uses two things to assure that user

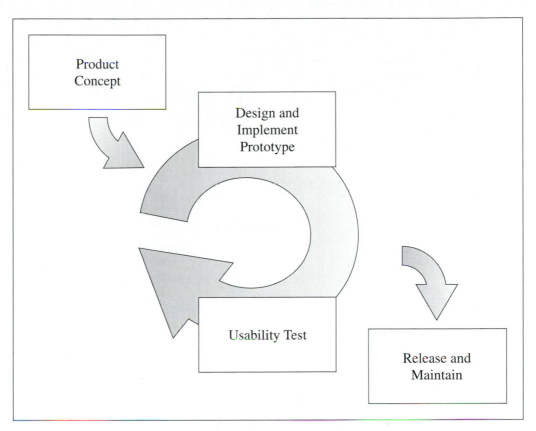

FIGURE 6.15   The Rapid-Development Method
The rapid-development method uses a process of usability testing and prototyping to test
designs out during development. Once the right design is found the program is complete.
User requirements are refined as the product unfolds.

needs are met: prototypes and usability testing. As Figure 6.15 shows, the process
requires that the team put together a prototype of the product (program and documents) and then test them with users for suitability, quality, applicability, and so on.
The team then re-designs the prototype and tests again, following the process until
program and document usability goals are achieved.

The rapid-development method is very popular for developing web-based applications because the technology and user needs change so fast. Gone are the days
where users used just one or two programs in easily definable environments. To
ensure quality in programs and documents, developers have had to shift to this
method, which can capture user actions and needs for software and information support directly, helping to "build in" quality and applicability.

THE OBJECT-MODELING METHOD. The object-modeling development method is sort of
a hybrid of the waterfall method and the rapid-development method (Figure 6.16). It
uses the sequential development model of the waterfall method in that it specifies

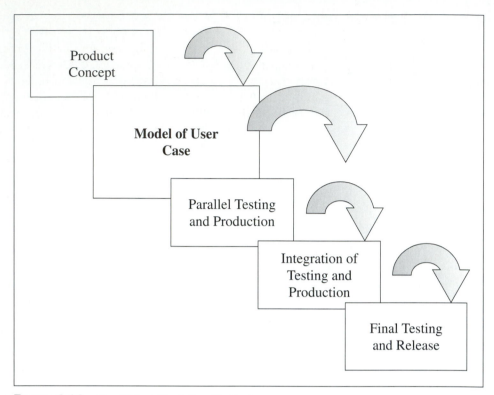

FIGURE 6.16   The Object-Modeling Method
The object-modeling method uses a complex system of symbols and descriptions to create an abstract design for the system that can be turned into software and documentation with a high degree of consistency.

what the program will do at the beginning of the process. And it uses the usability oriented approach of the rapid-development method by identifying user needs and building them into the program.

The object-modeling method, as it is sometimes called, identifies user needs through a development tool known as a "use case" model. The user case model is a plan that tells exactly how the user will employ the program in the workplace. For instance, a use case might describe a transaction carried out by a clerk in a video rental store. The clerk would get the customer information, check for outstanding videos, record new rentals, record coupons, record the cash transaction, and update the database. For programming purposes, each of the events get recorded as a programming function (called an "object") which is described as using information from other objects, occurring in sequence with other objects, and so on. Programmers can then write the program for these "objects" and assemble them in to useful programs. As long as the use case is valid and realistic, the program will fit the user's needs. Writers in a project using this development method can use the user cases as actions and build their documentation around the same objects as the program. Figure 6.16 shows an example of the phases in the object-modeling method.

The Documentation Plan

When you begin documentation planning you should set for yourself a number of goals, among them efficiency and logic in the whole process. Often you will have to meet with other persons involved in a software development effort (clients, executives, managers, programmers, and so on). Lay your ideas out for them as to how you will go about producing a user-friendly manual on time, using your staff to its fullest—and at a minimum cost. You may achieve these goals through the sheer power of your personality, your charismatic leadership, your gift for higher math, and your financial savvy. If you're like me, on the other hand, you need some help. I think of management plans as what we have instead of genius. For me, a solid, time-honored documentation management plan—reasonably thrifty and sensible—has to do.

Why Write a Plan?

If clients, managers, and programmers accept task orientation as a basic principle, you have half the battle won. They will understand that defining menu items does not necessarily constitute useful "how-to documentation." They will understand the distinction between teaching the system and teaching how to use the system. They will more than likely understand and accept the logic behind your plans for audience analysis and document testing. To them, your plans will make good management sense. Your plans should use personnel, computer resources, and budgets efficiently—maintaining close records of performance and a flexible and pleasant work environment. As a final element—probably most important—your plans should involve users in crucial development phases.

Strategies to Make a Documentation Plan Persuasive

A persuasive documentation plan will cause your reader—a client, a sponsor, a writing team, a development team—to believe that the project will fulfill its purposes and that they should invest their time and resources in it. The suggestions below can help you polish and refine this persuasive purpose.

- **Use an executive summary.** Give the bottom line up-front to help the users decide in favor of your project.
- **Have a goal orientation.** Set out objectives the reader (client, developer, programmer, executive) can identify with.
- **Do the math.** Go over budget figures carefully so detractors can't argue over them.
- **Show a team orientation.** Emphasize the contributions you will make to the overall value of the project.

Strategies to Make a Documentation Plan Easy to Follow

You should keep in mind that these plans—the design plan and the project plan—all make up significant progress toward a successful manual. Write them with enough care and completeness that another writer could use them to produce the manual set as you envision it. In fact, that scenario often happens in the software industry: A consultant will design a manual or help system and then other writers will follow the design—like builders follow an architect's blueprints—to produce the final product.

Below you will find a selection of strategies that you can employ in writing the design plan so as to make it easy for another writer to complete.

- **Standardize your terminology.** Keep the same names and titles for separate sections and documents. This way you don't confuse the writer, who sees projects in terms of filenames and titles.

- **Include sample pages.** Clarity in exactly what you want the pages to look like can help any writer who has to follow your plans. To this end, sample pages work better than brain-numbing lists of fonts and point sizes. Also, if you know the production tools (graphics programs, spreadsheets, book-writing programs like FrameMaker), you can specify the styles in the conventions of those production tools.

- **Don't stint on detail in the outlines.** Put as much well-considered detail into your outlines as you can so the logic of the document sections appears clearly to the writer. That will help the writer in deciding what to include in the sections you specify.

Two Parts of a Documentation Plan

Overall, the documentation plan does two things: first, it describes the manuals and help and second, it describes the documentation project. In the first part, the *design plan,* (sometimes referred to as the "content specifications") you tell what your manuals will contain (content) and what they will look like (forms, layout, language, graphics, etc.). In the second part, the *project plan,* tell *how* you will produce your manuals: the schedule, resources, and time/page estimates. The documentation plan may also contain appendixes about the users and the program. Thus, the project plan represents a culmination of your research work so far on the documentation project.

What Goes in the Design Plan?

In the design plan, you specify what the actual documentation products will contain and what they will look like. The design plan contains three main parts: the description of the users, the description of the goals of the manuals, and the description of their content (outlines and layout).

DESCRIBE THE USERS. In the users section of your design plan, summarize the results of your research and analysis efforts. Specifically, describe your user or users in terms that indicate the kinds of tasks they perform and their level of computer expertise. A typical description would first name the user, in terms of job title, followed by a discussion of the user's job related responsibilities.

Avoid describing the users as "anybody" unless you plan to write a default manual. The default manual doesn't reflect any user analysis because of its primary focus on the interface of the system. A user analysis yields characteristics that help you design manuals and help focused on actual workplace usage.

SET OUT THE DOCUMENTATION OBJECTIVES. Having discussed the users of your manuals, you should say what documents your project will produce. If you have separate documents, some longer, some shorter, your should list the titles of each. If you have a longer, continuously bound document containing sections for different users (such as a tutorial, and a user's guide, etc.), you should treat each section separately, listing its title. Table 6.5 describes some frequently used titles. Most of the titles would include the program title, as in *AccountMaster User's Guide,* or *NetHog Reference Guide.*

Objectives for task-oriented documentation usually contain three things: the user of the manual, the activities the user performs, and the corresponding document. Study the objectives statements below.

- The manual will allow financial managers and consultants to apply the program to the following five workplace activities: how to create financial plans, analyze investment portfolios, write tax and other financial reports, use advanced spreadsheets for analysis, and upload data to web-based portfolios.
- The manual will introduce the user to the management functions and the uses of the program, and will cross-reference advanced functions with the reference guide.

TABLE 6.5   Sample Titles for Software Manuals and Help Systems

Titles	Description	Support Level
User's Guide	Contains procedures for most program functions. Most titles with Guide in them contain mostly procedures, as in *Installation Guide.*	Guidance
Manual	Contains various sections including user's guide, installation section, reference section, and so forth. A good title for the all-inclusive manual.	All levels
Tutorial	A special book or booklet containing lessons that cover the basic and/or advanced features of a program.	Teaching
User's Manual	Like *Manual,* this title indicates an all-inclusive manual for a specific user. Alternatives include: *System Operator's Manual, Teacher's Manual, Administrator's Manual.*	All levels
Help	The title for most online help systems. The help system usually contains an index and jumps to online tutorials and procedural documentation.	All levels
Quick Reference Card	This title indicates a brief overview of the commands, menus, tools, or other interface objects in the program. Usually printed on card stock.	Support
Pocket Reference	This title refers to a small brochure containing essential program and background information.	Support
Getting Started	This title indicates that the document will contain an overview, or perhaps a walk-through of some basic program features, using examples.	Guidance
Reference Guide	This title often gets used for manuals that include all three levels of documentation support. Technically it should just include reference or support-level documentation.	Support

- The user will use the manual to basic program skills in the classroom, and also use it later as a reference.

- The help system will provide support for experienced manufacturing design users who need recall support at the point of need.

- The document will provide step-by-step instructions in using the program for file management, reporting, and writing form letters for sales associates, account supervisors, and other professionals in product support.

PROVIDE OUTLINES OF INDIVIDUAL DOCUMENTS. You should include outlines for each individual document you plan to produce (manuals and help). Each outline includes the name of the document and the estimated number of pages. The outline should include complete section titles and other divisions of information in the document, down to the level of the individual task. An outline starts with the title page and ends with the index, and lists all sections and headings in between. Outlines look like tables of contents.

For online documents you need to describe the organization of topics. Like tasks and other content elements of hard-copy documents, online documents require organization according to a number of schemes—almost all the same ones described in Chapter 2, "Writing to Teach," Chapter 3, "Writing to Guide," and Chapter 4, "Writing to Support."

LAY OUT INDIVIDUAL DOCUMENTS. Describe the layout for each of your individual documents with reference to thumbnail sketches or sample pages that you include in the appendices. The layout should contain enough detail so that if someone were to read your plan he or she could complete the documentation set exactly as you planned it and it would meet your documentation objectives.

How much detail should you include? When in doubt, spell it out. Consult Chapter 13, "Using Graphics Effectively," which contains the most examples of sample formats of any chapter. Find a format that you think would appeal to your user and circulate sample pages for initial review. To specify the layout for your documents include the following information for each document:

- Page size
- Column specifications (for all page types)
- Table specifications (for all table types)
- Body text style, size, font
- Style specifications for:
  - Section and other headings
  - Task names
  - Steps
- Cuing patterns
- Notational conventions
- Binding and boxing specifications
- Any special formatting or page layout instructions

Reviewing the Documentation Plan

Your documentation plan should undergo thorough review by managers, clients, and users. The design plan requires a much more extensive review than your user analysis received. The documentation plan gives you the opportunity to hold a user *walkthrough* (going over the important design elements of your document), a technical walkthrough (going over the accuracy and completeness of the content), and a project team meeting.

An Outline for a Documentation Plan

This chapter has presented many elements that could go into a documentation plan. Given the details of your project you may include all of them, invent some of your own, or only use a portion of them. The following outline will give you a start in putting together your documentation plan.

THE EXECUTIVE SUMMARY

- Purpose statement
- Brief description of the software application
- List of primary users and user categories
- Overall goals of the documentation set: what kind of work will you support
- List of the main print and online components of the documentation set
- List of the deadlines for the alpha and final draft (and any meetings, review deadlines, test deadlines, date of final delivery)

THE DESIGN PLAN

- Purpose statement
- The Documentation users. Describe user groups (one brief paragraph each) telling the following:
    1. User type
    2. Professional role
    3. Primary work activities using the software
    4. Information needs
    5. Computer experience
    6. Subject matter knowledge
    7. Workplace environment
    8. Learning preferences
    9. Usage patterns
- Document types. List each document giving the following:
    1. Title
    2. Kind of document (tutorial, guidance, reference)

3. Media (print/online)
4. Outline
   - Title
   - Sections from cover/title page to index and all sections in between
   - Estimated page length for each section
5. Layout
   - Page size
   - Column sizes
   - Table sizes and styles
   - Body text size, style, font
   - Styles for sections, headings, steps
   - Cuing patterns and notational conventions
   - Binding and box specifications
   - Special format or layout

THE PROJECT PLAN INTRODUCTION AND PURPOSE STATEMENT SCHEDULE
A table or graph showing each of the following.

- Drafts: types and delivery dates
- Reports: types and delivery dates
  Report types may include: progress, evaluation, team briefings, walkthroughs, test reports.
- Tests: types and dates
  Types may include: predictive, developmental, or evaluative (see Chapter 8 for details on these types of reports.)
- Reviews: types and dates
  Types may include: managerial, user, technical, legal, sponsor/client (see Chapter 7 for details on these types of reports.)
- Edits: types and dates

RESOURCES
- Personnel resources: consider the efforts of all involved. Personnel include: sponsors, users, testers, reviewers, and editors.
- Computers and other equipment: think through the machines and software you'll need. Equipment resources include: computers, printers, scanners, copiers, paper, toner, and software.
- Software: word processing, desktop publishing, help authoring, book assembling, browsers, and any software needed for testing in user work environments.
- Time/Page Estimates. A table or graph showing the following:
  - Total number of tasks and topics in each document
  - Estimated average number of hours per task (your formula, usually 2.9)

- Grand total of hours for the project
- Total cost for the project

APPENDIXES

Put the following elements in the appendix (if needed).

- Pertaining to the user:
  1. The transcript or report of the user analysis
  2. Task/user matrixes
  3. User scenarios
- Pertaining to the documents:
  1. Thumbnail sketches of pages and screens
  2. Sample pages
  3. Sample layout elements: tables, lists, etc.

Glossary

authoring environment: a computer program that allows you to create a help system. Example: Sevensteps, Authorit, eHelp, Forehelp, DocToHelp.

help compiler: a type of program that creates help systems out of text files. Help compilers read files and transform them into help programs that users can run on their computers. The most common help compiler programs are WinHelp and WinHTML, both from Microsoft.

documentation plan: a written document describing a manual or help system for a software program containing specifications for the content and layout of the documents *(design plan)* and a management plan for creating it.

default manual: a manual created by transcribing the user interface of a software program into a series of procedures and definitions without need of user analysis.

design plan: a document that describes the design of content and the layout of pages of a manual or help system.

development team: a team consisting of writers working with software developers, programmers, usability specialists and other persons who are producing both a software and an information product.

dynamic HTML: a special version of HTML that allows for special effects in presenting browser pages. The special effects include fancy cursors and moving text, among others.

HTML: stands for hypertext markup language and indicates a system of special codes inserted into text that give instructions to browser programs (such as Internet Explorer) to enable the browser to display the text correctly. Web pages are created using HTML codes in text files.

milestone: events in a project that indicate you have completed a phase and need to have a meeting. Often a milestone requires a written product, such as a report, and may entail a change in personnel (e.g., the project gets handed off to the editors).

multi-media: refers to a way of presenting information to a user, reader, or viewer that uses more than one method. This presentation method usually consists of video pictures, sound, text, and animation. Usually multimedia gets distributed on CD Rom disks because of the amount of space needed.

PDF: stands for portable document format. This format shows pages online as they appear in print, allowing for online annotation and hypertext links.

program specifications: long- and short-term plans for a software program detailing the requirements for programmers and enabling them to build the program.

project plan: a document that tells the persons, resources, schedule, and time estimates for managing a manual or help project. The *project plan* and the *design plan* make up the development plan.

script files: files containing codes that enable special effects using the JavaScript programming language that are used to enhance the capabilities of web pages.

single source: a method of organizing the content of manuals and help so that all the content is created once but used in a variety of formats and media.

topic: in help systems, the basic unit of information. The topic can contain a step-by-step procedure, a definition, an explanation, or other useful information.

writing team: a team consisting primarily of writers, publications managers, and editors dedicated to creating an information product (manual or help system).

walkthrough: a kind of meeting where you step through a document or part of a document for the purpose of examining its design elements.

☑ Checklist

Use the following checklist as a way to evaluate the efficiency of your task descriptions and task list. Depending on the level of detail you choose for your list, some of these items may not apply.

Documentation Plan Checklist
The Overall Plan

❑ Does the documentation plan contain elements to ensure its persuasiveness? (executive summary, goal orientation, persuasive figures, team orientation)

The Design Plan

❑ Adequate user descriptions
❑ Objectives for the documentation set
❑ Descriptions of print and online documentation types
❑ Outlines of individual documents (tasks and topics grouped according to user actions)
Specifies layout clearly for the following elements:

❑ page size	❑ sections and headings	❑ notational conventions
❑ columns	❑ task and topic names	❑ binding and packaging
❑ tables	❑ steps	❑ special layout
❑ body text style, size, font	❑ cuing patterns	

The Project Plan

Schedule includes all events occurring during the development process.

☐ Meetings   ☐ Test completion dates
☐ Deadlines for drafts   ☐ Review deadlines
☐ Project report due dates   ☐ Editing deadlines

Schedule includes the list of resources.

☐ Personnel on the team   ☐ Testing facilities and meeting rooms
☐ Budgets   ☐ Software for writing and authoring
☐ Computers, servers, and other   ☐ Time/page estimates
   equipment

Practice/Problem Solving

1. Reverse Engineering a Project

Reverse engineering means you take a product—a manual or help system—and examine the way the authors might have constructed it. You can do this with a manual for a program you work with, such as Microsoft Word or your virus protection program. What user activities and actions do you see implicit in the design? What organizational or managerial difficulties would have presented themselves to the team, and how might the manager have overcome them?

Your first step in this practice exercise will be to select a manual to "reverse-engineer." You don't have to choose an accounting program manual as in the example. Any manual will do. Review the information about scheduling and project planning in this book, study the manual you have selected, and come up with reasonable answers to the following questions:

- How long is the document?
- How long did it take to create it from start to finish?
- How much did the manual cost to produce?
- What user needs drove the design of the document?
- How many people worked on the document?
- What testing or reviewing problems did the document offer?
- What kinds of equipment did the writing team use?
- What kinds of management reports, if any, did the writing team have to produce?
- What kinds of meetings did the team hold?

2. Adapt Production Tasks to a Specific Solution

Study the list of tasks in Table 6.1 (page 190). These tasks cover the design and production of a document set from scratch. However, many documents don't start this way, but from existing manuals that have become outdated because of revisions in the program or a need to be re-designed for greater task orientation.

Find a default manual that you think could undergo this kind of revision. Maybe you would turn a default manual into a task-oriented manual. Or you might want to create a tutorial section that the previous document didn't contain. Perhaps you think the revised document should contain more graphics illustrating user actions. Write out one or two goals you would have for your revision.

Then, try to imagine the production tasks that would be required for your imaginary revision project. You would follow the same nine phases as in Table 6.1, but you might add or change specific tasks within each phase. For example, under "Perform the User Analysis" you might want to add the task "Review the user evaluations of the previous release of the document."

Go over the list of tasks in Table 6.1, selecting those that would be appropriate to the revision project outlined above. Consider what tasks the revision project would entail, and add them to the list. Write a brief paragraph explaining and justifying the changes you made to the project task list.

CHAPTER 7

Getting Useful Reviews

To review documentation, you send it out to get the reactions of other people who are involved in the project. This chapter covers review procedures designed to provide information relating to the task orientation of your documents as well as to ensure their technical accuracy and conformance with company policy. It shows a page from a reviewed manual and an example of a review form you would use as a cover sheet for review copies. It details steps for planning and conducting reviews by users, managers, programmers, subject-matter professionals, and clients. It focuses on the review walk-through as a way to gain specific information about the task issues your users face and to assess how well your document meets the needs of task-oriented employees.

How to Read This Chapter

All readers should study the examples and follow the guidelines for managing reviews. These guidelines can help ensure product success throughout the documentation process. Ongoing review keeps you on track and keeps managers, clients, and users informed and on board.

- If you're working on a project, you should read the entire chapter. You may want to skim it in the early stages of a project, to get a sense of what's ahead, then read it in more detail later in the project when you have a first draft.

- If you're reading to understand, you should study the Discussion section first in order to get a broad understanding of the process and its issues, then read the Guidelines as necessary.

Example

The document review form in Figure 7.1 illustrates a typical transmittal letter for reviews used at the Documentation and Training Center of Excellence (DTCE) at Texas Instruments in Dallas, Texas. It represents one of a number of checklists and forms that managers at DTCE use to control their documentation process. The document review form in Figure 7.1 illustrates a number of factors that contribute to successful reviews.

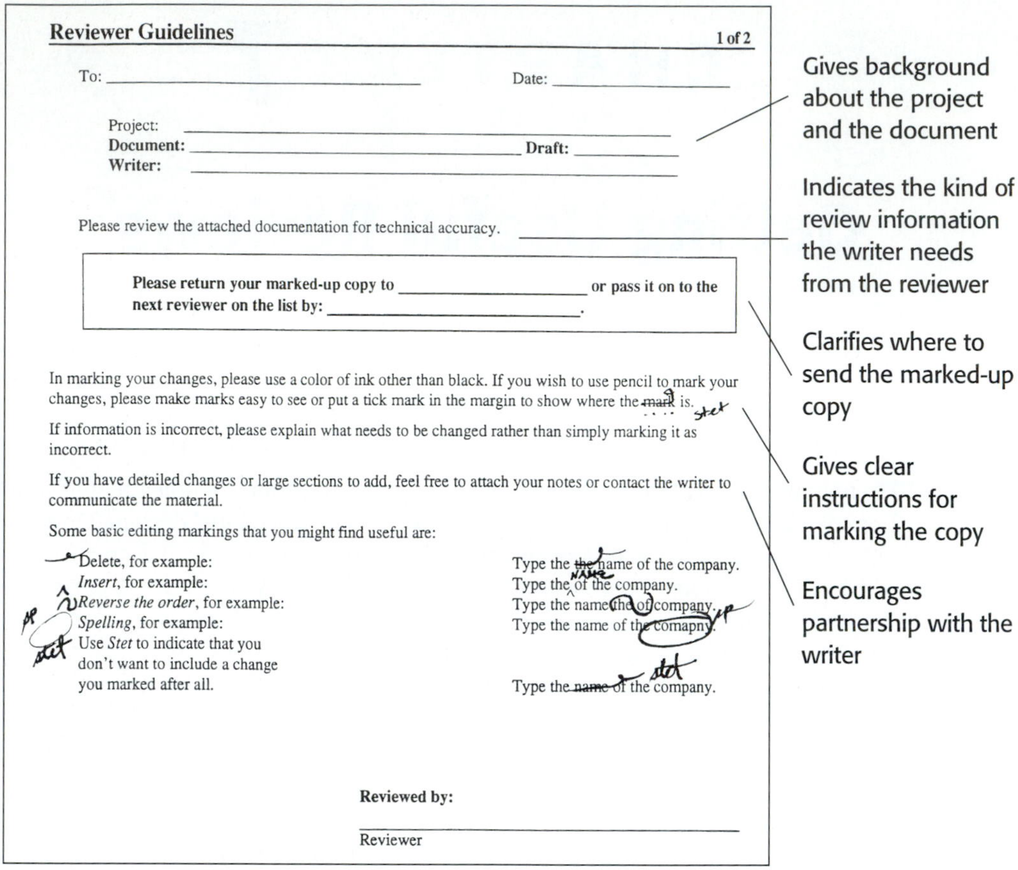

Reviewer Guidelines   1 of 2

To: _____   Date: _____

Project: _____
Document: _____ Draft: _____
Writer: _____

Please review the attached documentation for technical accuracy. _____

> Please return your marked-up copy to _____ or pass it on to the
> next reviewer on the list by: _____.

In marking your changes, please use a color of ink other than black. If you wish to use pencil to mark your changes, please make marks easy to see or put a tick mark in the margin to show where the mark is.

If information is incorrect, please explain what needs to be changed rather than simply marking it as incorrect.

If you have detailed changes or large sections to add, feel free to attach your notes or contact the writer to communicate the material.

Some basic editing markings that you might find useful are:

Delete, for example:
Insert, for example:
Reverse the order, for example:
Spelling, for example:
Use Stet to indicate that you
don't want to include a change
you marked after all.

Type the the name of the company.
Type the of the company.
Type the name the of company.
Type the name of the comapny.

Type the name of the company.

Reviewed by:

_____
Reviewer

Gives background about the project and the document

Indicates the kind of review information the writer needs from the reviewer

Clarifies where to send the marked-up copy

Gives clear instructions for marking the copy

Encourages partnership with the writer

FIGURE 7.1   The DCTE Review Form
This form from the Documentation and Training Center of Excellence at Texas Instruments, Inc., helps ensure accurate and consistent review information.

While technically you don't expect reviewers to edit a document the way a staff editor might, you realize that they will want to add, delete, and mark text. The form includes basic editing marks as a way of helping reviewers respond consistently. Also, a reviewer's list (shown later in the chapter) accompanies the form. Forms encourage accurate and specific information from reviewers. They indicate, through their professionalism, that the writer takes the reviewer's comments seriously as an added value to the documentation product.

Guidelines

Of all the activities associated with document production, reviewing may take more of your time and require more careful planning than any other. To make the most of your investment, approach the review stage with great care and with the attitude that the review can really increase the usability of your document. When you have to rely on users, managers, and technical experts to read your documents, you need to plan carefully how you will use their time. Following the steps shown in Figure 7.2 can help you get the most out of your review cycles.

1   ## Review the Document Objectives from the Documentation Plan

With reviews, as with all elements of document design for information-oriented work, you, like your users, should consider document objectives. Each kind of document you have (tutorial, user's guide, etc.) and each feature it has (elaborate background information, layered presentation of material, etc.) and its text and page design (columns, screen size, etc.) result from a specifically articulated purpose. Your development team should have arrived at the purpose for your documentation set during the planning stage, and you should easily find those objectives in the documentation plan.

Consider this example of document objectives and the questions they would pose to reviewers. The program Grabber captures screens and converts them into a variety of formats (JPG, GIF, TIF, BMP, PCX, etc.). The documentation plan for the Grabber User's Guide identified a number of objectives, listed in Table 7.1, with an explanation of how you would address them in reviews.

Document objectives also play a role as part of managerial review. Managers and supervisors review your documentation to make sure that it meets company policy and reflects well on the company. But they also have other responsibilities that relate to the documentation. Managers and supervisors need to review the existing documentation to verify that the features it implements carry out the objectives stated in the documentation plan. For these reasons, you should restate those objectives in the review materials for these persons.

2   ## Determine the Type of Review Needed

When you review, you send documents to other persons with cover sheets asking them to read or examine your materials and respond to issues in their area of expertise. Usually you do the following four types of reviews:

FIGURE 7.2   Guidelines for Managing Document Reviews

1. Review the document objectives from the documentation plan.

2. Determine the type of review needed.

3. Establish a review schedule.

4. Plan the reviews.

5. Write a cover letter with questions for reviewers.

6. Prepare feedback material for reviewers.

TABLE 7.1   Tie Document Objectives to Reviews

Examples of Objectives	Review Type	Questions Answered
Provide task-oriented examples of processed images	User	Do the examples reflect real-world tasks?
Encourage use of online help	User	Does the online help facility look usable or resemble others you have used?
Provide guidance to facilitate the easy translation of graphic file formats to desktop publishing programs	Technical	Do the translation procedures cover all supported formats?
Provide task-oriented guidance support for all program functions	User	Do the task names reflect workplace tasks?
Provide tutorial support for basic graphics conversion and basic screen capture	User	Do the tutorial lessons take up too much time for the normal user?
Provide extensive reference support for expert users in: 1)graphics file formats, 2) error messages, 3) frequently encountered image formats	Technical	Does the reference section cover all existing error messages and image formats?
Provide background information about graphic files and types	Subject Matter	Does the background section sufficiently explain graphics files?
Encourage efficient choices of file formats and conversion techniques	User	Do the examples of file formats match those you use in your work?
Support data transfer to programs users use most frequently	User	Does the manual need to cover data transfer to programs *not* mentioned in the manual?

- **User reviews**. Reviews by the actual intended users of the document.
- **Management reviews**. Reviews by managers and supervisors associated with your documentation project.
- **Technical reviews**. Reviews by programmers and developers of the software.
- **Client reviews**. Reviews by the people or department paying for the software and documentation.
- **Subject-matter expert reviews**. Reviews by experts in the professional field represented in the software.

The kinds of reviews you design will depend, in part, on the kinds of persons who have a stake in the documentation. These will include other members of the development team, or, in the case of technical writers doing independent contracting, programmers and clients. Each person on a team and persons sponsoring or paying for documentation represent a concern for the quality of the overall software product. Carefully planned reviews, like those described in Table 7.2, capitalize on the help you can get from these reviewers.

TABLE 7.2   Types of Reviews

Type of Review	Persons Involved	Concern
Managerial review	Managers, supervisors, team leaders	Staying in budget, meeting document objectives, quality control
User review	Users, operators, system supervisors	Ease of use, applicability to task needs of empowered workers
Technical review	Programmers	Technical accuracy of procedures, coverage of functions
Subject-matter review	Professionals in a certain field	Accuracy of background information
Editorial review	In-house editors, team editor, writer in editor's role	Meeting standards of grammar, organization, format
Sponsor review	Client, sponsor	Meeting diverse user needs, getting value for money

All reviews offer challenges and problems, because not all of the persons whom you need to do the reviewing may be at your job site, or they may lack familiarity with your project. In these cases you need to focus the reviewers by providing specific questions you want them to answer. But beyond that, the reviews present logistical problems that you must surmount. In Table 7.3 you will find some of the problems you might face with each type of review.

No two review situations present the same problems; however, you can bet that careful scheduling and planning will go a long way toward productivity in this stage of the documentation process. On the other hand, a haphazard approach can cost you your profit on a project.

3   Establish a Review Schedule

Scheduling a review means that you have to give your reviewers enough time to prepare their response, and you have to decide on the right circulation strategy.

TABLE 7.3   Problems Encountered with Reviews

Type of Review	Problem You Must Surmount
Managerial review	Making sure they review the document objectives, and see the document in the context of the whole software project
User review	Making sure reviewers represent the target users of the system
Technical review	Getting programmers to pay attention to a program they have stopped thinking about; getting them to care about something outside the technical area
Subject-matter review	Finding the experts in specialized fields willing to do the review
Editorial review	Making sure editors have a clear understanding of user needs
Sponsor review	Scheduling time with busy professionals

Make sure you give your reviewers enough time to fit the reading into their schedule of other projects. They may have to read all or parts of documents, and that takes time. A careful review takes about one hour per 15 to 20 pages. When reviewers need to write extensive comments, that rate slows drastically, to about five or eight pages per hour.

Circulation strategy refers to the way you circulate copies of your document to reviewers. You can choose one of two methods of circulating documents: sequential, where you send one copy to multiple reviewers, or simultaneous, where you send multiple copies to multiple reviewers.

Sequential Circulation

Sequential circulation entails making one copy of the document for each category of reviewer, then scheduling reviews for each reviewer in turn. Figure 7.3 shows an example of a cover sheet for a sequential review. Each reviewer reads the document and then either gives it back to you for you to give to the next person, or passes it to the next person for you.

List of Reviewers   2 of 2

When you have finished your review, please pass the attached document to the next person on the review list.

Reviewer: _____
Date to complete review: _____
Date passed on: _____

Reviewer: _____
Date to complete review: _____
Date passed on: _____

Reviewer: _____
Date to complete review: _____
Date passed on: _____

Reviewer: _____
Date to complete review: _____
Date passed on: _____

Reviewer: _____
Date to complete review: _____
Date passed on: _____

FIGURE 7.3   Example of a Reviewer's List for Sequential Circulation
This sequential circulation list from DTCE at Texas Instruments, Inc., a companion form to the reviewer's guidelines form in Figure 7.2, tells readers where to send the document after reviewing it.

ADVANTAGES OF SEQUENTIAL CIRCULATION

- **Low cost.** The primary advantage of circulating sequentially lies in the cost. You only have to make one copy of a document, which can represent a significant savings on larger documents. At 5 cents per page, a 200-page manual costs $10 per document: a lot for a small project.

- **Less hassle.** Sequential circulation may also save you by leveraging the reviewer's time and resources to circulate the document; they put it in the mail to the next person. Usually, with this method you attach a circulation list for each reviewer to check off his or her name in turn. This method works best when reviewers are near one another, such as in the same department.

- **Encourages team spirit.** Sequential circulation requires each reviewer to see himself or herself as part of a team, as indicated by the list of other reviewers in the project. It's up to each person to contact the next reviewer or send it along, which may cultivate communication among the reviewers that ultimately could benefit the project.

DISADVANTAGES OF SEQUENTIAL CIRCULATION

- **Spawns margin arguments.** Your document may spark arguments in the margins over certain features the reviewers encounter. While these arguments can enlighten you about the content, they can cause reviewers to ignore other important points.

- **Early reviewers affect later reviewers.** One person's review may have an effect on subsequent reviewers. Say the first reviewer trashes your manual for an irrelevant reason (or because he or she hates your boss). That may cause remaining readers to lose their objectivity and result in a waste of time for you.

- **Causes political problems.** It's often hard to say who gets the document first and who gets it later. The person last on the list may take offense at this, which may affect the outcome of the review.

- **Hard to control.** When you depend on the reviewers to circulate the document, you are relying on the weak link in the chain. If one person lets your document sit on the desk for an extra week it can stretch your schedule.

- **Takes extra time.** Circulating from one person to another means that some readers have to wait to get the document, and by the time it gets to them it may show signs of shop wear. Both of these facts can affect the reader's response.

Simultaneous Circulation

Simultaneous circulation entails making multiple copies of a document, one for each reviewer. Under this circulation method each reviewer gets a fresh copy, reads and records, and then gives it back to you. Online circulation is simultaneous in that reviewers can all download and review the document at once.

ADVANTAGES OF SIMULTANEOUS CIRCULATION

- **Fast.** With each person commenting on his or her own copy the process goes faster. You don't have to wait for the document to crawl from one office or cubicle to another or get lost in interoffice mail.

- **Good for geographically diverse reviewers.** This method works well for circulating documents to reviewers at far-flung sites who would have to use the mail to send the document on.

- **Fosters a one-on-one relationship.** This method makes reviewers feel special. Each person gets a fresh copy of the document and gives an individual reply that is unaffected by that of other users.

- **Easy to control.** With simultaneous circulation you can track each document separately, and easily spot a review that is late. You can intervene by investigating and making adjustments (reminder memos, offers to help, reassignment of the review to someone with more time, etc.) to keep the whole schedule on track.

- **Easy, when online.** Online documents can easily get circulated to more than one reviewer at a time because of the low cost of copying a document. Caution: Make sure the version you circulate doesn't get lost on someone's hard drive, only to crop up later and cause confusion with a later version.

DISADVANTAGES OF SIMULTANEOUS CIRCULATION

- **Expensive.** With print documents you have to have a budget for duplicating in order to create the copies to circulate. Plus, you have to pay for mailing and duplicating cover letters. Bound documents add even more to the cost.

- **Takes more of your time.** You may find yourself investing more time in duplicating and circulating documents simultaneously because you have to run the copy machine, make trips to the mail room, and pester your reviewers with reminders of deadlines.

- **Fosters redundant comments.** With simultaneous circulation you may find all the reviewers spending time on one problematic element of a document. Had a previous reviewer already dealt with a problem and provided a suggested solution, the reviewer would know to move on to something the previous reviewer missed.

- **Causes version mix-ups with online.** Unless you can control the access to your simultaneously circulated online help system, you may have trouble making sure that reviewers read the most current version, and that all previous versions (containing potentially embarrassing mistakes) get deleted from people's hard drives.

4 Plan the Reviews

Your reviews will go more smoothly and produce better results if you plan them carefully and write up the strategy and schedule in your documentation plan. A well-articulated review plan encourages you to think the process through by making it a goal-driven process. A plan provides background material to help reviewers respond more productively, but it also acts as a communication device for review meetings: it opens up the process to colleagues and builds team spirit.

Encouraging team spirit is particularly important when you are communicating with persons from different cultures. The Japanese, for instance, may not see your reviews as an opportunity to add to the process, but primarily as a way to reinforce the relationships inherent in the project.[1] In addition to expressing your awareness of the differences between your culture and the culture of your reviewer, you should take extra pains to explain your goals and objectives and to reinforce the personal relationship that exists and that underlies and gives validity to your asking for reviews. Without the acknowledgement of this relationship your reviewers may hesitate to say anything critical about your document.

While it's not entirely linear, the process of setting up and executing specific reviews follows the fairly standard procedure outlined in these guidelines. In your documentation plan you should identify the kinds of reviews needed and when they will occur during the process. Will they be reviews of design, planning, documents, or what? If you plan review meetings (called *walkthroughs*) you should schedule them and indicate who will be attending.

5 Write a Cover Letter with Questions for the Reviewers

The key to getting useful information back from your reviewers lies in directing their efforts. If you just hand them a manuscript and ask them to read it over, you may get back a careful, studious review based on their responsibilities and expertise, but more likely you will get back a shallow reading focusing on idiosyncratic, arcane grammar rules the readers remember from college. Such a review does you no good. This is where a carefully worded cover letter and review sheet come in.

To get relevant information out of your reviewers you should provide them with a cover sheet specifying the kinds of things you want them to examine and acquainting them with your project and your schedules. This letter has a huge job to do and should sound as professional as possible. The more you can convince your readers of the importance of the review and their contribution to it, the better their response. Compose your cover letter using the following guidelines:

INDICATE DOCUMENT OBJECTIVES AND BENEFITS TO THE REVIEWER. Tell the reviewers the objectives of the overall document; what specific kinds of work you want to support; how the document and program will get used. Use a statement like "The overall objectives of this document include. . . ." Also, point out the benefits of the reader's participation in the review process. Your reviewer may enjoy specific benefits, but you can always point out these general ones to reviewers:

- Reviewers benefit from increased quality: their work helps ensure quality documentation products. Reviewers need to know that their efforts add real value to your product.

- Reviewers benefit from increased communication among company members: their participation helps the company grow by departments sharing information. Reviewers need to know how the information they learn about your product can reflect on their own management concerns.

- Reviewers benefit from increased visibility as team players: employees like to create an image of themselves as cooperative and supportive. They need to know that you will alert their superiors as to their cooperation and the value they add to your document.

ASK FOR SPECIFIC ADVICE/COMMENTS. Tell the reviewer exactly what you want them to do: read the document and look at specific aspects of it pertaining to their expertise. This will forestall their focusing on editing details inappropriately. If you ask specific questions of your reviewers, you get usable answers. Examples include those indicated in Table 7.4.

PROVIDE THE NECESSARY BACKGROUND. Where necessary, provide the appropriate background for your users so that they can respond better. Background can include a number of items, such as the following:

- Background about the user: review the user's task demands, computer proficiency level and user type, amount of subject matter knowledge the user has, other programs the user knows
- Background about the project: remind the user of the version and draft numbers, what department(s)and what people have committed to the project, the writer, the editor, the other reviewers
- Background about the product: list the main features of the product, platforms it runs on, relevant revision history of the product, current versions, how the product relates to other company products
- Background about the subject matter of the program: include definitions of key terms, brief overviews of the topic areas where necessary (how project tracking works, why do we want to model escaping gas from a chemical plant, etc.)

TELL REVIEWERS HOW TO MARK OR COMMENT. Usually you only have to remind reviewers to make marks in the margins, and because you don't expect them to edit the document, you don't have to tell them how to make editing marks, such as insertion carets and delete marks. If you do, you might give a brief list of marks such as indicated in Figure 7.1. If you do a sequential circulation, you might want to ask the reviewers to use a certain color of ink to help you distinguish their comments from those of others. If they do their reviews online, ask them to include their initials with their comments so you will know who said what.

GIVE DATES AND PLACES FOR RETURN. You want your reviewers to get the document back to you on schedule, so tell them when you want or need their comments. Use a simple reminder like "To include your comments in the next draft I will need them back by. . . . " In some cases, with reviewers you know will put off reviewing or perhaps have difficulty meeting deadlines, you might offer to drop by their office to pick up the document. Tell them, "I'll be by next week to pick up the manual." This strategy helps them get ready in time and offers you a chance to discuss any troublesome areas that their review might have uncovered. Include the draft number and other details pertinent to the project.

THANK YOUR REVIEWERS. Thank your reviewers for the time they have put into reading your document. Mention, in an email or written memo, the contribution they have made to your document's success.

TABLE 7.4   Sample Questions for Different Types of Reviewers

Type of Review	Sample Questions
User	Does the table of contents reflect the order of tasks as you would perform them?
	Can you understand the example on page (n)?
	Does the binding and format make the document easy to use in your workplace?
	Do you think you will use the list of figures and list of tables, or do you think they could get left out?
	Take a look at the index: does it contain terms you would use in your daily work?
Managerial	In your view, does the document meet the objectives set out in the documentation plan?
	Does the document reflect the company and product favorably?
	Does the document meet all legal commitments required by our company?
	Does the binding and presentation of the document meet the standards for our other company publications?
Subject Matter	Does the document reflect the subject matter (e. g. accounting, statistics, geology) accurately?
	Does the glossary contain adequate and correct definitions of key terms? Can you add new ones?
	Do the examples used in the tutorial reflect realistic professional problems in the subject matter area?
Technical	Does the document reflect the functions of the program accurately?
	Does the *User's Guide* cover all the existing program functions?
	Have you added any new commands to the program since the release used to write this document?
	Have the installation steps changed since this version?
Sponsor	Does this document look like you thought it would look when you commissioned it?
	Do you think your managers and supervisors can use this document easily?
	Does the document present an accurate and acceptable representation of your company?

Figure 7.4 shows a review sheet incorporating some of the guidelines in the previous paragraphs, but adapted to a specific situation.

6   Prepare Feedback Materials for Reviewers

Reviewers need to know that they count as partners in your documentation process. To foster this, show them, in a letter or a note, that you read their comments and paid

REVIEW SHEET

Project	*A Simple Introduction to PowerPlay Models*
Reviewer	Rick Sanchez, BevCo, Inc.
Author	Thomas Barker
Program	PowerPlay 2.0
User (groups)	Retail product managers
Review deadline	April 21, 2003

Review instructions

These review sheets will make sure we agree on the design at this stage. It helps us both because it clarifies our assumptions. Important: the attached document represents rough drafts.

Background

Based on my notes of our last meeting, we agreed on the following:

- One installation task
- Three program tasks involving opening and exploring three of the six models (others suggested and mentioned but not explored)
- Two program tasks in graphic format, one program task in a' cross-tab format"
- Each program task will have different content: "How to drill down to the SKU level," "How to use the add categories dialog box," "How to drag from the dimension line to the legend box."

Questions

Do the following tasks cover your requirements?   __Yes   __No

1. "How to Install the Models"
2. "Open and Explore a 2-year value trend graph using the dimension bar."
3. "Open and explore a 2-year variance trend graph using the add-categories dialog box."
4. "Open and explore a year-to-date sub-category report by drilling down to the SKU level."

Comments and Suggestions Please comments on other elements of the procedures.

FIGURE 7.4   A Review Sheet Giving Specific Guidelines to the Reviewer
This cover sheet encourages the cooperation of reviewers.

attention to them by making the required changes in the document. You won't and can't change everything a reviewer finds objectionable in a document. But a professional job of review management will certainly yield much useful information. Try to give some credit for that to each person on your review team. Let each person know the effect of his or her work on your project.

Memos and thank-you letters work well for getting feedback to your reviewers. Refer specifically to items reviewers mentioned. You might not want to send them a whole new version of the document reflecting their comments, but you can photocopy the pages they marked significantly and attach them to revised pages, so the reviewers can see how you interpreted and acted on their comments. Remember that in most cases, as with user reviews or reviews by subject-matter professionals, you may not need their official approval of your changes, as you might with a managerial or sponsor review. But it doesn't hurt to share your revisions with them.

Discussion

Sometimes the terminology used in the documentation process can get confusing. All three functions of reviewing, testing, and editing represent the formal management procedures that constitute the quality-control activities of your documentation project. In all three functions, your work gets scrutinized in specific ways for specific reasons, with the goal of improving the product. Some organizations include editors as reviewers and so would include an "editorial review" in the above list of functions. In this book, editing is treated in Chapter 9, "Editing and Fine Tuning," and therefore is not included here, but you should realize that for many professional documenters the word "review" refers to editing as well as gaining comments from others on a draft. Reviews can be classified according to when they occur in the documentation process. Thus we have the "design review" that asks managers, users, programmers, and others to respond to the documentation plan as a statement of the design of the document and the plan for producing it.

Reviewing Differs from Testing

The fact that reviews require comments from the array of persons listed in Guideline 2 above, indicates one way they differ from testing (see Chapter 8, "Conducting Usability Tests"). Testing tends to concentrate on users and issues of accuracy and statistics. Reviews, on the other hand, develop information about conformance of a product to management schedules and company policy: "Is this the kind of document you ordered?" Additionally, reviews don't produce quantitative data about a document. Reviews don't provide you with statistics about how accurately a procedure describes a function or how long it takes users to perform the function. Reviews produce *comments,* reflections from users about issues such as the suitability of an example or whether or not a task sequence matches a user's expectations. A test might tell you whether the access elements of a document (the index, the table of contents, and so on) helped users find specific information in a certain period of time. A review would provide you with a number of users' reactions to the format of the index or table of contents, from which you would make decisions on possible design changes. Finally, reviews don't occur in laboratory settings under controlled conditions. Reviews occur in offices and workshops of reviewers and usually take the week or so it takes to look over a whole document.

Reviewing Differs from Editing

When you circulate drafts for reviewing, you can expect to get information different from that you would get from an editor. Editors bring their training in editing to bear on your document, often using professional procedures such as the levels and types of edit described in Chapter 9. Reviewers, on the other hand, bring their professional training as managers, subject-matter experts, and programmers to bear on your document. Usually you submit a document once to a single editor; you may submit the same document to a number of reviewers, expecting to get different information back from each. With reviewing, you get more specialized kinds of information: on accuracy, suitability of information, compliance with management policies. While you may get some details of editing during your review process, usually you get responses and reactions to questions of suitability of an element or the usability of a specific feature.

The Purpose of Reviews

Reviews can serve a number of purposes in your project:

- **Communication function.** Reviews help you communicate with people associated with your project. If you work on a development team, reviews will reinforce the communication already set up and working among the other writers, the software developers and programmers, and the editors. If you work more or less on your own, reviews provide you with a way to share your work with others, to let them know what you're doing and to get their contribution.

- **Management function.** Reviews help you manage your project. Reviewing allows you to touch base with everyone's schedule and helps you keep your own production on track.

- **Quality assurance function.** Reviews help you maintain the quality of your product. Reviews offer you the opportunity to benefit from the insights of others concerned with usable documentation. Careful planning and preparation for reviews will reward you in helping you produce a better document than you could have without the review.

As you approach the task of having your documents reviewed, remember the main purpose of the process: to benefit from the reactions of others, with the aim of improving your work. Thus, when you plan your reviews, approach the job with the right attitude: that what users say about your document can help you shape it to meet the two goals of ease of use and usability. Reviewers represent workplace perspectives: those of clients, of subject-matter professionals and especially of users. Use their knowledge and energies to help make the document better. Empower your reviewers by telling them your shared organizational and professional goals and letting them benefit from helping you achieve them.

Reviewing throughout the Documentation Process

Often we think of reviewing as occurring only during the later stages of a project, after you have finished a complete alpha draft. In fact, that alpha draft can have fewer troubles along its way to becoming a final draft if you have taken time to have it reviewed at earlier stages. Reviewing as a part of the documentation process requires that you have some kind of document to send out. The earlier the stage of the process the more explanation you will have to provide about the background of the program and the project. You can find out about scheduling reviews and the place this schedule plays in your documentation plan by consulting the topic "Anticipate Review Schedules" in Chapter 6, "Planning and Writing Your Documents." Usually you can count on getting reviews at four key stages of the development process: after conducting the user analysis, after determining the task list, after designing the documentation set, and after writing drafts.

PLANNING AND DESIGN STAGE. Your documentation plan should undergo thorough testing and review by managers, clients, and users. Your coordinating editor, if you have one, should also examine your document plan. The design plan requires a much more extensive review than you got for your user analysis. It should cover the actual design of the document (pages, text, language, and other features), and the project plan (allocation of computer and human resources and scheduling of drafts and other production activities).

USER ANALYSIS STAGE. If you have written up your audience analysis as a set of parameters, you should show them to your managers, clients, and users as a way of checking their validity and accuracy.

DEVELOPMENT AND WRITING STAGE. Early in your development process you should validate the accuracy of your task list. A technical review or technical walkthrough with programmers, or a technical walkthrough (meeting to validate accuracy) can assure that you will get lighter markups on later technical reviews.

DRAFT STAGE. As opposed to a formal, scheduled review of your alpha draft of a document, you may want to schedule and implement draft reviews of parts of a document that have reached some milestone of completion: such as a portion of a tutorial, or an early draft of a quick-start section of a user's guide. Make sure you indicate in your cover letter what stage your document is in. Reviewers may take issue with some things (say the lack of adequate running headers and footers) if they don't understand that these elements won't get put in until a later draft.

Reviewers as Partners

Good reviewing has to do with attitude: Reviewers can provide real help to your project. Imagine reviewing not as a chore, but as an opportunity to learn and get feedback and subsequently improve the task-orientation of your document. With the well-coordinated help of a series of reviews from users, managers, programmers, subject-matter professionals, and sponsors, you can't help but turn out a very high-quality, usable document. The first step in developing that attitude toward reviews comes with encouraging a sense of partnership among your reviewers. The following list outlines some techniques for fostering this kind of productive partnership.

- **Tell them the benefits of participation.** Make sure your reviewers understand that they not only benefit you by reading your documents, but they also benefit themselves. A mutual benefit results in a win-win situation.

- **Don't abuse the privilege.** Avoid going back again and again to a favorite reviewer because of his or her thoroughness. Make sure you get his or her permission before sending the document. Handle the whole process professionally because the offended or abused reviewer may have a long memory.

- **Show them revisions.** Sharing revisions with reviewers can help them see themselves as co-writers or co-designers. If you have managed to get good feedback from reviewers, let them know how you used it.

- **Hold review meetings or walkthroughs.** To encourage partnership with reviewers, call them together in a meeting and let them meet the other partners in the reviewing team. Plan the meeting well in order to foster a cooperative team spirit. Go over the document, focusing on areas where you have questions about reviewers' responses or areas where you got conflicting feedback. This method works especially well for detecting technical errors early in the documentation process. You'll find guidelines for holding a user walkthrough below.

- **Keep contact over time.** Keep in touch with good reviewers by keeping a file of information about the work they have done for you. This file can help you avoid over-using reviewers and help you establish long-term contacts.

- **Return the favor.** Reviewing means often that you ask someone to read something without compensating them, as a favor to you. It's good business practice to return these courtesies in meaningful ways, like reviewing their documents or participating in activities that they sponsor within your organization, such as voluntary management meetings or focus groups.

- **Thank them in print.** We have all seen passages at the prefaces of books where authors thank those who contributed to their manual. When you have the opportunity, list the names of those who reviewed for you. It doesn't take much room, and it demonstrates that you recognize the document reflects the best thinking of many thoughtful people.

Negotiate Changes Diplomatically

Very often you will find yourself confronted with a conflict between reviewers and no easy way to satisfy both. For example, despite your document style guidelines, your reviewers may disagree over hyphenation in the word "online," about the suitability of your examples, over the size of your page numbers or your style of numbering. In these situations you risk diminishing the productivity of your reviews if the conflict spirals.

More often than not differences stem from participants coming at the review process from different corporate or national cultures that see the review process, and the whole process of software documentation from different perspectives. Rather than assuming that you can negotiate in a way that seems familiar to you, consider that you might have to resort to some of the strategies suggested by Elaine Winters in *Cultural Issues in Business Communication.*

BE FIRM ONLY WHEN NECESSARY. The more you respect another culture's tendency to leave things undecided, the better chance you have of discovering a solution in the future. Try "Let's leave this issue of content open while we gather more information to help us all."

REFER TO THE RELATIONSHIP RATHER THAN TO THE PERSON. It's always a good idea to reinforce relationships when dealing with cultural conflict. "We agree on the importance of our partnership in this development project, so let's strive for a mutual solution."

ACKNOWLEDGE CULTURAL DIFFERENCES AND GIVE THEM VALUE. If the issue is one of the way people react to one another, let your participants know that. Engineers come from a "high-context" culture where what other engineers think makes a lot of difference, whereas writers come from a (relatively) low-context culture where often writers respect one another's individuality and consistency in documents and don't rely on arcane chestnuts of grammar. Sometimes a compromise on one issue can smooth negotiations on others.[2]

Do a User Walkthrough

A user walkthrough is a one-hour meeting in which you present the document from end to end and focus questions on the key issues of usability. Of all the reviews you do, this kind of user review will contribute most to helping you meet the task needs. Technical and management reviews focus mainly on accuracy and conformance with company publications and other policy. The kinds of information illustrated in the following list can only come from actual users.

- Does the document reflect recognizable workplace activities?
- Does the background information support users' professional knowledge?
- In what order would the user expect to see the sections of the document?
- Are any important tasks omitted or treated inadequately?
- What tasks does the document not focus on that users would find important?
- Does the document accurately reflect workplace goals?
- Does the tone of the document suit users' workplace discourse?
- Does the document support efficient information processing?

A Walkthrough Integrates the Documentation Team

To get this kind of information in a brief and very effective way you assemble a group of your users for the purpose of going over the document from front to back, asking questions like those listed above, recording responses, and making changes on the spot. Writers who have used walkthroughs find that they give their users a sense of participation in the writing process, but mostly they constitute an efficient way to obtain a large amount of information quickly. They report that, after a technical walkthrough, their conventional reviews turned up many fewer errors in the documents.[3]

As in the case of a technical walkthrough, a user walkthrough can keep users focused on the important issues of ease of use and task orientation. Also, a technical walkthrough allows you to get a lot of information in short amount of time about how well your document meets user needs. And because individually-circulated reviews often take so much time and present such scheduling challenges, you should consider the user walkthrough as an efficient way to bypass the hassles of circulation and still increase the task-orientation of your documents.

How to Set Up a User Walkthrough

To set up and conduct a user walkthrough requires some planning. Overall, you can follow this process:

1. Decide on the issues you want to examine. You want to make sure you have the correct task orientation in your manual or help system. Look over your documentation plan and identify questionable areas, especially user activity categories.

2. Choose the attendees. Select your meeting attendees carefully so they represent your actual users. Contact the users you discovered during the user analysis. You may also want to invite development sponsors or clients, who often have insights into workplace applications. Often these people represent your users' executives or managers, who have a concern for the user's workplace objectives. When you have distinct user groups, such as a teacher group and a student group, you may want to hold separate meetings.

3. Prepare a meeting agenda based on these; prepare specific passages or document sections to present for user approval. Write out questions or make up a questionnaire for users to complete during the walkthrough.

4. Make copies and provide files for all attendees. During the walkthrough, you want all your attendees to have their own copy of the document, and preferably to have looked it over ahead of time. However, if you have clearly focused questions and representative sections to examine, you can probably conduct the meeting without requiring attendees to read ahead of time.

5. Run the walkthrough. The writer leads the walkthrough. Begin by announcing the agenda, which will keep the group focused on the issues you need to try to resolve. Make it clear that you welcome editorial comments (on spelling and punctuation) on later drafts and that attendees will have a chance to comment on that later. Go through the document section by section or part by part, focusing on

key examples, page designs, cuing patterns, and user scenarios that you want your users to comment on. Record comments on a master copy of the document and remind attendees that they will receive a copy of it later. Allow attendees to discuss and try to resolve issues that come up.

6. Do a follow-up review. After the meeting, send copies with the suggested changes to users after the meeting. Again, remind them of the key issues you want to resolve in your document, and solicit comments.

The user walkthrough takes extra effort, but offers a number of advantages for writers and users. It allows users to have a say in the development of the documentation, and it allows you to gain valuable insights into the workplace usability of your document. Carefully planned and conducted, user walkthroughs can result in fewer negative comments on subsequent reviews.

Review Form

The form shown in Figure 7.5 is a useful form you can use as a cover sheet. It includes all the elements you need to remember and use to remind your reviewers of their roles. You can use it in conjunction with the sequential circulation page to keep track of these reviewers.

Glossary

cover letter: a letter of transmittal that accompanies a draft for review that explains the goals of the review and instructions for reviewing.

design review: a review that focuses on planning the design specifications for the document. It covers preliminary layout and overall document set design.

document review: a review that focuses on a draft of the finished product and whether or not it met the design goals set for it.

document review form: a transmittal document that you attach to the front of review copies, providing guidelines for the reviewer to follow.

sequential circulation: sending documents for review by giving it first to one reviewer, then another, then another through a sequence until all reviewers have read the same document.

simultaneous circulation: sending documents for review by giving each reviewer his or her own copy and having all the reviews done at once.

walkthrough: a meeting of writers and reviewers at which the writer goes over the document from front to back to get and record reviewers' reactions.

☑ Checklist

Review Planning Checklist
Use the following checklist to help you plan for your reviews.

REVIEW SHEET

Project	
Reviewer	
Author	
Program	
User (groups)	
Review deadline	

Review instructions

Background

Questions

Comments and Suggestions Please comments on other elements of the procedures.

FIGURE 7.5 A Sample Review Sheet
Use this form as a cover letter for circulating drafts of print and online documents.

Document Objectives

❏ List the objectives from the documentation plan that you need to validate through your review.
❏ Review product specifications from the client or sponsor for objectives.
❏ List management or policy objectives your document must meet.

Stage of the Document Process

❏ User analysis review: Review user descriptions and scenarios for use in tutorials and as user role models.
❏ Technical walkthrough: Review the accuracy of the procedures and resolve any questions about the functionality of the program.
❏ Design review: Review or walkthrough design principles for the documents.
❏ Draft review: Review complete drafts of documents as they are completed.

Type of Review Needed

❏ Managerial review (managers, supervisors, team leaders).
❏ User review (users, operators, system supervisors).
❏ Technical review (programmers, developers).
❏ Subject matter review (professionals from representative fields).
❏ Editorial review (staff editors, editorial department, team editor).
❏ Sponsor review (client, sponsor).
❏ Identify any special issues to address or problems encountered with any of the above reviews.

Cover Letter with Questions for the Reviewers

❏ Indicate document objectives and benefits to the reviewer (increased quality, increased communication, increased visibility as team players).
❏ Ask for specific advice and comments.
❏ Tell reviewers how to mark or comment.
❏ Give dates and places for return.
❏ Thank your reviewers.

Set Up a Review Schedule

❏ Sequential circulation (one copy: each reviewer passes the document to the next person or you pass the same document to each person in turn).
❏ Simultaneous circulation (multiple copies: each reviewer returns the document to you).

Feedback Materials for Reviewers

❏ Clean copies of documents showing revisions with specific pages indicated (smaller projects)
❏ Copies of pages containing revisions stapled to copies of pages with reviewer's marks (larger projects)

User Walkthrough Planning Checklist

Use the following checklist to help you plan for your user walkthroughs.

❑ Decide on the issues you want to examine.

❑ Choose the attendees.

❑ Write an agenda for the meeting.

❑ Run the walkthrough.

❑ Do a follow-up review.

Practice/Problem Solving

1. Associate Types of Feedback

As you know, testing, reviewing, and editing address different aspects of a document's quality as a tool designed for access to information. For practice in managing this kind of information, consider a manual that you know well, such as one you did for a previous project or one you might currently have on the drawing board.

Put yourself in the manager's position. Undoubtedly the nature of the software and the software's users—needs, experience, and so on—will result in some quality issues. For instance, adapting the software to different user groups, or helping users understand with arcane and complicated subject matter.

Consider this example: a program that manages a stamp collection called "StampMaster." This program logs, tracks, and calculates the value of stamp collections. It also contains reference information about stamps and stamp collecting for beginners.

Here's a list of problems and issues that the program raises:

- How much background information about stamps should the program give the user?

- How many examples should the sample database contain?

- Does the user's guide comply with the company image of a trendy, specialty producer?

- Should the manual contain a list of Web sites pertaining to stamp collecting?

- Should the documents contain methods of using the program for unintended purposes (such as for a coin collection, or butterfly collection)?

- Is the manual technically accurate?

- What terminology should the document use in referring to the technical aspects of stamp collecting?

Consider the differences between reviewing, testing, and editing. Which of these problems pertain to your testing efforts, your reviewing efforts, and your editing efforts? Which method of assessment would help you solve which problems?

2. Prepare a Review Cover Sheet

Imagine that you have been asked to rewrite a manual or help system for a software program. (You can pick any program and document you like). To accomplish this rewrite you might want to start by doing a review of the present document to see how it meets users' needs. Who are the users or user groups for the software? What issues does the document raise? Compose what you think would be a suitable cover letter for the review.

3. Conduct a Review

Using the review cover letter you prepared for exercise 3 (or one that you prepare for another manual of your choice), conduct a review of the document. Use either sequential or simultaneous circulation. When you get your review materials back, write a brief summary report including recommendations for revising the manual.

CHAPTER **8**

Conducting Usability Tests

Testing documentation consists of procedures for gaining empirical data about the usability of documentation products. This chapter covers three basic types of tests: tests for task performance (procedures), tests for skill and understanding (tutorials), and tests for access to information (references). It recommends a ten-step test plan covering the main tasks you will need to perform when conducting usability tests for both small and large projects. The guidelines and discussion focus on the importance of user testing and emphasize three types of field testing to ensure task orientation. Additionally, the chapter discusses ways of interpreting user test data.

How to Read This Chapter

- If you're working on a project, you can adapt the procedure test form in Figure 8.1 to your immediate needs. If you plan to conduct extensive testing, you should read the whole chapter. The Guidelines, in particular, can help you plan extensive tests.

- If you're reading to understand, you should study the procedure test form in Figure 8.1, and then refer back to it as you read the Discussion section. This will provide the needed overview of the process so that you can use the Guidelines effectively.

Although this chapter discusses a variety of test forms, the one in Figure 8.1 covers many of the points you want to look for when evaluating procedures (or other step-by-step documentation.) As this chapter suggests, you should adapt this form, and others, to the specific needs of your documentation project. I would like to thank Mr. Mac Katzin, author and technical communicator, for the inspiration for this form.

Guidelines

Your tests will go more smoothly and produce better results if you plan them carefully. While not entirely linear, the process of setting up and executing specific tests follows a fairly standard procedure. Figure 8.2 lists guidelines for testing documentation.

Procedure Test Form

Part 1: Information about the procedure and document being tested

Document title:

Procedure name:	Section #:	Page #:

Part 2: Information about the evaluator

Your name (or evaluator number):

Exact start time (hour/minute/second):	Exact end time (hour/minute/second):

Thank you for helping to find out how well this procedure works. Follow the steps as carefully and as far as you can, and then record your experience. Take your time. When you're ready, fill out Part 3.

Part 3: Evaluation of the procedure

Progress

❑ got to the end	❑ got to step #:

Steps

❑ OK	❑ out of order	❑ inconsistent	❑ incomplete

Indicate which steps need correcting:

Graphics/screens

❑ OK	❑ showed the wrong thing	❑ too small	❑ confusing

Indicate which graphics/screens need correcting:

Explanations

❑ OK	❑ incorrect	❑ not clear	❑ not relevant	❑ not enough detail

Indicate which explanations need re-writing:

Explanations

❑ OK	❑ incorrect	❑ not clear	❑ not relevant	❑ not enough detail

Indicate which explanations need re-writing:

Terms

❑ OK	❑ not clear	❑ too technical	❑ too simple	❑ other

Indicate which terms you had problems with:

Comments that might clarify your experience/reaction to doing the procedure:

FIGURE 8.1   A Procedure Test Form

FIGURE 8.2   Guidelines for Testing
Documentation
You can use a form like this to test
procedures. It helps the elevator focus
on the document rather than his or her
performance.

1. Decide what to test.
2. Select the test points.
3. Choose the type of test.
4. Set performance and learning
   objectives.
5. Select testers and evaluators.
6. Prepare the test materials.
7. Set up the test environment.
8. Record information accurately.
9. Interpret the data.
10. Incorporate the feedback.

1 Decide When to Test

You can test at any time during the nine stages of the documentation development process, even though stage six (review and test) focuses on testing (see Chapter 6, "Planning and Writing Your Documents"). Usually you test after you have a draft finished, can see areas that need testing, and can still make changes. But, as Table 8.1 shows, you can test at roughly three stages: during design, during writing or development, and after the document set goes to the customer.

Testing and Document Goals

In testing your documents, you'll have to find what parts you want to test. The sad fact that you can't test all your procedures leads you to this conclusion. You don't have time, and not all of your procedures, tutorials, and references need testing. As

TABLE 8.1   Decide When to Test

Development Phase	Kind of Test	Description
Phase 3: Design	Predictive	Done at the design stage to test the suitability of design specifications and production goals. High degree of flexibility in making changes based on results.
Phase 5: Writing/Drafting	Remedial	Done while drafting or writing. Corrections made immediately and re-tested. Moderate degree of flexibility in making changes based on results.
Phase 9: Field Evaluation	Evaluative	Done after finishing and shipping a product. No chance to change the existing product. Changes have to wait for the next release.

you have seen in the previous guideline, the objectives you set for your manuals and online help can guide you to test points. Your document objectives state how a manual or help system applies the program to the user's work environment. Focusing on objectives can help you maintain a view of the program in the user's workplace and give parameters for your test efforts. Also, the document features that you identified during the user analysis probably need testing. In general, whatever you try out as new, as innovative, or as different—which should make up a lot of your efforts—should get tested.

What drives you to design different formats or incorporate different strategies? Often you have to support more than one user group of a program, and that leads you to try out innovative formats. Sometimes you realize that users of your program need more background material than you've given them, and you need to find out just what terms your target audience does and does not understand. Whenever you go out on a limb with your designs—set challenging objectives for yourself and your work—you need to test.

2   Select the Test Points

A *test point* is an issue or feature of a document that might interfere with the efficient and effective application of a program to user's work activities. Test points fall into two areas: problems with content and problems with document design. To identify test points you need to have a sense of what kinds of difficulties your manual poses to the user. Software errors can occur at two levels: the level of accuracy of content and the level of document design. The problems that may block success in manipulating the interface or carrying out a procedure may stem from the fact that you recorded the steps for using the interface incorrectly. On the other hand, the problem may come from a miscalculation of what tasks actually apply to what work activities.

Select Procedures for Testing

Identify what tasks you want to test by looking at the points in your documentation set where you perceive that a mistake on either level could introduce a chance of user failure, or a cost of user failure. Remember that here you're testing the documentation, not the user.

You should seriously consider testing procedures that are overly complex (involves over 10 steps) or involve one-time activities (installation, configuration, etc.). Because of their complexity and uniqueness users run the risk of getting them wrong because they don't build up a learning curve for them. The same goes for highly abstract or technical tasks such as higher-level programming functions, advanced graphics processing, and importing files from other applications and so on. Here the user is often coming to the procedure fresh and can experience confusion in using the interface or applying the procedure.

In many situations the cost of user confusion or failure may be high. For example, certain tasks seem to provoke a large number of expensive support calls. Check with technicians and phone support personnel to identify these procedures.[1] Other costs of user failure stem from risks of file deletion, maintenance of data files, and storage. User mistakes in this area could cost time and money.

Look for specific places in your documents where a mistake on the user's part could cost time (by causing other errors) and reduce efficiency (by forcing the user to do it the hard way). Review your document goals and any important, information-oriented tasks. For example, a documentation system that supports information transfer and storage should be tested to ensure that procedures for those information-related tasks contain no errors. Such tasks include:

- Importing information from other programs
- Creating, naming, and formatting files
- Exporting information to other programs or other program formats
- Creating printouts and reports

Select Design Strategies for Testing

Identify what document design strategies you want to test by looking at the following design elements.

- **Terminology.** Identify what terminology you want to test by looking at the language in your documents. Test technical terms that might confuse novice users and also test subject-matter terms (relating to the user's workplace).
- **An index.** Test an index for consistent, recognizable terminology.
- **Cuing patterns.** Test icons and labeling graphics that might confuse; or test to make sure users can recognize images.
- **Headings/layout.** Test for headings too small to see, layout that hides key information.
- **Navigation.** Test for navigation that doesn't match the user's usage pattern when performing workplace actions.
- **Extraordinary document formats.** Test for special conditions. Waterproof, fireproof, or childproof documentation should be subjected to these conditions and tested.

3 Choose the Type of Test

The three types of tests indicated in Table 8.2 relate to the test points you identified in the previous stage. It may be that you will have to perform more than one type of test on a document.

TABLE 8.2   Match Types of Tests with Test Points

Test Point	Type of Test	Description
Tasks	Can-They-Do-It Test	Often called a *performance test,* this test requires users to perform a procedure.
Terminology	Can-They-Understand-It Test	Often called an *understandability test,* this test requires users to provide a summary of material they have learned, or provide definitions of key terms.
Document Design Strategies	Can-They-Find-It Test	Often called a *read-and-locate test,* this test requires users to use mocked-up portions of a manual—the index or table of contents—to find information on key topics.

4   ## Set Performance and Learning Objectives

Because you want your tests to measure actual behavior, you must come up with numbers that correlate with the kind of performance you want from your users. Performance objectives simply put numbers on that behavior. Often called *operational definitions,* performance objectives state, in clear terms, how long a procedure should take or what frequency of correctness we can expect users to perform software tasks. Often testers refer to performance objectives as *exit criteria* because performance objectives specify the criteria that a task must meet to exit from the testing situation.

You should also be aware that performance does not always mean getting a task done in the shortest period of time. Effectiveness of software use results from users mindfully following steps, bringing their often unspoken knowledge to bear on tasks, and considering consequences of their actions in the workplace. Thus, effective performance may take longer to accommodate the complexity of doing work with software. When you set performance objectives you should try to take these complexities into account.

A number of methods exist for coming up with performance objectives. With performance tests, you can simply measure your own speed or pace in actually performing procedures. Then, given what you know about your user, compare the time he or she would take. This little exercise can help you establish the optimal rate of performance, but only if you carefully consider the differences and similarities between your performance (knowing the technical aspects of the program very well) and the performance of your users (knowing the workplace application of the program very well). Other methods of setting performance objectives include surveying potential users to determine how long they would take to do a task, or piloting the test with actual user. But the trial and error method works for most cases. Above all, you want to arrive at numbers that you can measure and compare. Table 8.3 illustrates some types of performance objectives.

Learning objectives for tutorial documentation are similar to performance objectives for procedures, but differ because of the different purpose of tutorials. Because

TABLE 8.3   Ways to Set Performance Objectives

Kind of Objective	Description	Example
Time-related	Time taken to perform a procedure	The user can install the program in under five minutes.
	Time taken to find a topic	The user can locate the import function in under one minute.
Error-related	Number of errors made during a procedure	The user can perform the procedure with a 20 percent error rate.
	Number of times a passage gets re-read before comprehending	The user can paraphrase the meaning of *field variables* after one reading
	Number of tries in the index	The user can find the **Remove** function in under three tries.

the purpose of tutorials is to teach, the main objective of the lessons is whether or not the learner can perform the skill or skills from memory. Thus, when testing a tutorial your main question is: "Did the tutorial teach the desired skills." Your test for this skill will often consist of having the evaluator perform the test and then having the person perform the skill using sample data.

Objectivity and Testing

Objectivity in testing means that you try to set up the test in such a way that you don't prejudice the outcome too much, so that the procedures don't "pass" automatically or the document doesn't come out with flying colors to let you get on with the project. While no test can be 100 percent objective, pilot testing can help you avoid slanting the test in one direction or another, but first you should understand where a lack of objectivity can come from. A bias can creep into your test from a number of areas, mostly unintentional. Consider the following scenarios:

- **Work pressure.** You got the project late, you lost personnel during the work, other members of the development team sigh and turn away when you announce that you're going to need a week to test before you can sign off on the document. So you shorten, simplify, expedite in order to conform to everyone else's schedule and production values.

- **Pro-forma testing.** Your department always "tests" procedures, but you pretty much know how the results will turn out because you don't put a lot of originality into the documents in the first place. You use the same forms you've always used and don't really pay attention to the "results" because you know that nobody else will pay attention to them either. They only care about getting the test form signed and getting back to work.

- **Caring too much.** You have gone out on a limb with a new design of a fold-out, 3D, multicolor layout and you don't want to go back to the drawing board to come up with yet another killer design for your information. This one shows how original you really can get; it reflects the reason you should get a raise or a promotion. No way do you want the document to fail the usability test.

You may fall victim to these or other situations that cause bias to creep into your test. You cannot avoid these totally, but you can recognize them and pilot test first to make sure the test forms and situations don't provide you with only the evaluation you want. Also, rely as much as you can on numbers—calculations of frequencies, performance statistics, and such—to keep yourself "honest." Of course, you do have to interpret, but postponing your interpretation until you have some reliable data can help ensure objectivity. Also watch out for leading questions, such as "Don't you think this procedure makes you more efficient?" Such questions lean the user in the direction you want as opposed to the direction dictated by the usability of the information.

Above all, objectivity results from a right attitude on your part toward testing: it will help you improve your design. If you see testing in this positive light, you will want to make your tests as objective as possible so you will end up with better documentation.

5   ## Select Testers and Evaluators

The *tester* is the person who administers the test: arranges the meeting with users, sets up the test situation, records the test activities and so on. Often you will administer your usability tests yourself. You will act as tester. On the other hand, you may increase your objectivity if you devise test materials for someone else to administer. That way you help eliminate the bias you will probably have for the test to come out positive for your materials. Often you can trade favors with other writers or development team members to obtain unbiased testers.

The *evaluator* is the person taking the usability test. Selecting evaluators poses some interesting problems. If you began incorporating your users into the development process back when you did the user analysis (Chapter 5, "Analyzing Your Users"), then you may have already lined up some potential evaluators among those you interviewed then. Let's be practical. You may not have actual users, or even potential users at your disposal—ready to spend one to five hours helping you do your job—so you will have to compromise. In Table 8.4, you will find an overview of potential situations and some suggestions for coping with them.

6   ## Prepare the Test Materials

Depending on the complexity of your usability test, the written and other materials you supply for testers and evaluators can get very complicated. You will probably not require all these materials for your test, unless you plan to set up a testing lab, or temporary testing lab. For an informal field evaluation you would only use some of the

TABLE 8.4   Ways to Compromise on Evaluators

Evaluator	Characteristics	Suggestions
Actual users	A given user type (novice, experienced, or expert) A given degree of subject-matter knowledge	No compromise necessary.
Similar to your actual users	Same user type but lacking in subject-matter knowledge	Acquaint the evaluators with the major uses of the software. Have the evaluators read your user scenarios for background.
Not similar to your actual users	Different user type and lacking in subject-matter knowledge	Acquaint the evaluators with the basics of the software if they need to test advanced tasks. Remind them that they should assume the role of novice if they need to test basic tasks. Acquaint the evaluators with the major uses of the software. Have the evaluators read your user scenarios for background.

test materials. In Table 8.5, you will find a list of kinds of test materials and definitions indicating when you might want to use them.

Along with written test materials, you need also to pay attention to the location of the test and the kinds of hardware and software equipment you require. Table 8.6 lists and describes most of the materials needed for testing.

Pilot Testing

Pilot testing means that you test the test. It's a way of reviewing your test, trying it out in a kind of dry run, to see if your testing materials will gather the kind of information

TABLE 8.5   Kinds of Written Test Materials

Test Material	Description and Use
Evaluator selection survey	A brief list of questions that potential evaluators fill out to ensure that they meet your profile as typical users. The selection survey should include a user *Permission To Participate Form,* with provisions for permission from the user's employer when needed.
Instructions for evaluators	A one-page list of instructions to help evaluators understand their role and what they should do during and after the test.
Test schedule	A schedule of testing times and locations for testers and evaluators to follow.
Instructions for testers	A set of instructions telling the test monitors how to conduct the test and how to record information. May also include operational definitions of key terms such as "error," or "success."
Test lab procedures	A set of instructions for operating the testing facility that expresses its purpose and policy of fair treatment of evaluators.
Pilot tests	Mock-up versions of test forms to try out as a way of determining how well the test works and the suitability of the performance objectives.
Test forms	Actual tests written up as test scenarios (suggesting job-roles for evaluators) task descriptions (telling evaluators what to do) and/or procedures (portions of hard-copy and online documentation products).
Test subject materials	Actual copies of documentation—hard-copy and online tutorials, procedure, and reference manuals—that the test intends to evaluate.
Software overviews	Marketing or overview information that informs evaluators of the main uses and features of the documentation they will use.
Test results records form	Charts and tables with spaces for testers to record results of task performance times, error rates, location names.
Exit interview questions	A brief schedule of interview questions for evaluators after the test, designed to gather incidental but potentially important information.

TABLE 8.6   Kinds of Hardware and Software Test Materials

Material	Description and Use
Tape recorder	A portable mini-recorder used to record the evaluator's spoken comments during the test procedure.
Notebooks	Regular letter-sized notebooks for taking notes on the evaluator's behavior during the test and interview after the test.
Evaluator face camera	A VHS camera used to record and time the evaluator's eye movements from the manual or help screen, to the keyboard, to the screen, etc.
Evaluator keyboard camera	A VHS camera used to record and time the evaluator's use of the keyboard, mouse, light pen, digitizer, or other input device.
Screen camera	A VHS camera used to record and time the state of the evaluator's computer screen and use of online help.
Intercom	A portable walkie talkie or AC-Line intercom used to communicate with the evaluator during the test, if necessary, from an observation room.
Tables/chairs	Office furniture used to simulate the user's work environment.
Computer(s)	PCs, terminals, printers, modems, phone and network connections needed to simulate the user's computing environment.
Sample data files	Software-related files containing mocked-up data for use during the test.
Timing clock	A stop-watch or specialized clock for recording start, stop, and elapsed times for evaluator's test performance.

you want them to. It doesn't tell about the product, the document, but about how well your questions work and what kind of data you can expect. When you consider the ways that a misunderstood term or the omission of a detail can ruin the data you get, it just makes sense to administer a draft of your test to one or two representative users, or your office mates, and then revise it.

Pilot testing doesn't require a lot of extra effort, and it provides a wealth of information to help your final version work better. For one thing, you can try the test out on a very small group and get their response. Suppose you're asking users to record information on a form. Do a dry run to find out if the kinds of tasks you ask about actually get performed by your users. Or you can make sure that you have given clear instructions to your evaluators. In general, pilot testing can help you in the following areas:

- **Instructions.** Do the guidelines you give to evaluators allow them to perform the test correctly?

- **Terminology.** Do you use any terms the user can't understand?

- **Timing.** Can the user perform the test in the time you have allotted?

TABLE 8.7   Advantages and Disadvantages of Various Testing Environments

Environment	Advantages (Lead to Clear Results)	Disadvantages (Lead to Mixed or Ambiguous Results)
Field (user's workplace)	Irreplaceable similarity to actual work demands	Intrusive in the user's workplace Less control over interruptions
Lab (documentation, software, hardware usability lab)	Controlled and consistent testing Better recording equipment Trained testing personnel	Lacks similarity to actual user's environment More expensive
Combination	Pilot testing done in lab can lead to better field results	Requires setting up two test events (expensive)

7   Set Up the Test Environment

The environment for your test may range from the user's work environment (the field) or a controlled laboratory. Your best chance to learn about actual use in the context of the user's work and information environment comes from field testing. But, if you have one available, the laboratory offers you a greater degree of control and you may find it more convenient. Assuming you have a choice, Table 8.7 will help you decide where to perform your tests.

Researchers with experience in usability testing often recommend a combination of testing methods as a part of a complete usability program. Using different methods independent of one another can help ensure a clear understanding of usability. Besides providing information about tasks for a specific product, you can use field testing to help establish test points: features that you want to examine more closely in the lab. Conversely, field testing can help to validate results that you discover after having done strict lab research.[2]

8   Record Information Accurately

As you conduct the test, you need to use accurate methods of recording what you see and hear. Your observations serve to flesh out what the cameras and other pieces of equipment record. For this reason, you should observe some guidelines in recording information. Synchronize your timing with the recorders so that you know when an event happened. That way, if the voice recorder, say, indicates a pause in the tape, you can go back to the same place in your notes and see what you recorded the user as doing at that time. Take copious notes, even of things that don't seem, at the time, to pertain to the evaluator's performance. Later these details can help you relive the event and interpret the results. Unless it would crowd the scene, invite disinterested observers to watch your tests. Their per-

TABLE 8.8   Variables Affecting Test Results

Variable	Description	Example
Halo effect	One detail of the testing procedure affects the entire test	In testing at the CIA you discover that the presence of cameras makes evaluators nervous. Tester's disdainful attitude causes evaluators to screw up intentionally.
Results not clear cut	Widely varying results	All three evaluators use different methods of finding a key term.
Wrong performance objectives	Evaluators perform a flawed procedure flawlessly	Evaluators take one minute to perform a procedure you thought would take two minutes but which actually should only take half a minute.
Unexpected factors	Chance details about the evaluators cause difficulty in performing otherwise easy procedures	Users are used to having color in manuals and yours are black and white. The evaluators you picked have an emergency meeting so the department sends you three substitutes named Larry, Curly, and Moe.

spective can help you by reinforcing what you saw or filling in the gaps you missed.

9 Interpret the Data

Interpretation requires you to take into account all the elements that can go wrong with testing so that you get clear results. A number of variables can have the effect of clouding the data. Table 8.8 summarizes those some of those variables.

Interpretation requires, of course, more than just calculating the data and making the changes that you can justify by the numbers. It requires that you make common-sense decisions about your manual design, so that the changes make sense to you and to others involved in development. In fact, most of the results that require a change in the design—changing a cuing pattern, adjusting the format of a table of contents—should appear clear from the test. A good test shows you what you should have seen anyway, and what makes sense after the numbers direct your attention to it. It should reveal something about your users, a missing piece of the puzzle of their usage and application of the program that you didn't know, that helps you make an intelligent design choice.

On the other hand, numbers can have great persuasive force with some technical audiences. A software usability specialist once confessed that he liked to record a bug with a program interface at least ten times, so that he could show it to the programmers and they would believe him. In this case, the sheer numbers of persons having difficulty with a part of their design helped convince them of the need to change. You may find yourself in a meeting with development team members for whom a satisfying flourish of numbers can help support your case.

10   ## Incorporate the Feedback

Testing does you no good unless you incorporate the information into the design of the documentation product. Ideally, your testing produced such useful results that you could make the suggested change, and then re-test a few times with consistently positive results. However, you may only obtain partial results, in which case you analyze them reasonably, make the changes you think they imply, and get on with the project.

Discussion

In this chapter we study the usability testing of documentation products, which is a subset of usability testing of software and of products in general. A way of involving users in the documentation process, documentation testing consists of a series of structured inquiries that attempt to measure the effectiveness of various elements of manuals and online documentation. Testing of manuals differs from testing of online products in that when you test an online product you follow procedures for testing an actual software program.

What Is Testing?

Testing usually requires a tester, an evaluator, and the subject material (a manual or online system). Figure 8.3 shows the necessary components.

Testing resembles reviewing; it generates information about a draft of a manual or online system. But it differs in the kind of information it creates. Testing generates statistical information—data in the form of numbers about user performance tasks or use of document features. Consider this list of questions you might have about a manual or online system:

- How well do installation procedures work for a program?
- Does the index contain sufficient cross-reference information to enable novice users to find information they need?
- How clear are the definitions of the concepts in this manual?
- Can the user perform the task from memory after following the tutorial?

Testing, done correctly, can provide answers to these and other questions. First, in testing you set the performance criteria by using an operational definition, (e.g., "work well means can be performed with 100 percent accuracy at least 90 percent of the time"). In testing, you get real people to actually try the procedures out. It puts your document "to the test." Your proficiency in doing usability tests lies in your ability to construct a reasonable test situation and to interpret the results intelligently. Testing fits in with the task-oriented approach supported by this book, because it constitutes another way of building your design around software users.

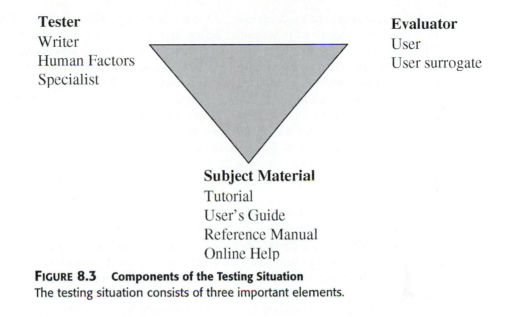

Tester
Writer
Human Factors
Specialist

Evaluator
User
User surrogate

Subject Material
Tutorial
User's Guide
Reference Manual
Online Help

FIGURE 8.3   Components of the Testing Situation
The testing situation consists of three important elements.

The Importance of User Testing As Part of User-Driven Design

User-driven design follows the idea that the technology should adapt to the person, rather than the person adapt to the technology. Software documents enable that adaptation. They show the user how the program can perform useful work as a tool in various environments, something the user might not see or appreciate without the documentation, and something the software product cannot usually do on its own. You can't have user-driven design for your documents unless you first study the user, then test to find ways to make the documentation adapt to the user's needs. User testing also supports task orientation because it helps the documenter build a clear picture of the user's work environment and encourages a broader view of software use than mere learning of functions. From a practical standpoint, testing can occur in a variety of forms as a part of your documentation project.

Example of a Field Test

Many writers use informal user-preference tests of document design features as a regular part of designing manuals for software programs. One writer, Stacie, wanted to do a *field test* at the design stage of her work on a desktop calendar program for the Macintosh. She identified the following test points for evaluation:

- Body text size (9, 10, or 12 point)
- Heading size (bold or plain)
- Cropped screens (versus whole screens)
- Cues for steps (italics, bold, plain)
- Page orientation (right/right or right/left)

To study these items she made up nine different mock pages with examples of different options she considered in her initial design. Her performance objective consisted of 100 percent, that is, a design feature had to get acceptance by 100 percent of her users before she would incorporate it into the document. In Stacie's test she assumed the role of tester, and she used three potential users as evaluators: a manager of an agricultural services company, a fellow writer, and a features writer for the local newspaper. These users represented the range of user type she had identified for the program (novice to experienced) and each possessed roughly the same amount of subject-matter knowledge of the program (using personal information management programs). Materials for this test took the form of a set of interview questions for the users, mocked-up pages showing the features, and a brief overview of the program's capabilities.

The tests took place in the user's workplace. After explaining the nature of her project to the evaluators, Stacie simply gave the three sheets to the users and asked them a series of questions about the page design features. Her results, when tabulated, showed that some of the test points won unanimous approval by her evaluators (thus passing the performance objective), while others did not. Interpreting her results, Stacie found that some of her original ideas (such as using italics to indicate steps in procedures) did not work for these users because they distracted or confused the users. The results allowed her to adjust the original design and taught her a lot about her users.

Example of a Laboratory Test

Recently I had the opportunity to visit an IBM usability lab in Richardson, Texas and to watch some usability testing of documents and interfaces for office automation products developed at the site. The lab consisted of two large rooms, one for the testers and one for the evaluators. The director of the lab had dual Ph.D.s in human factors and behavioral psychology. The assistant had training and experience in usability testing.

The test room consisted of a number of computers in office-like settings with video cameras strategically placed. One camera focused on the user's face, another on the screen, and another on the keyboard. The evaluator, a woman, sat at the computer and also spoke into a microphone connected to recording equipment in the observation room. She used manuals beside her on the table. A one-way mirror separated the test room from the observation room.

The observation room held a wealth of recording and processing equipment. The VCR cameras in the testing room fed into a bank of VCR recording equipment. The evaluator's face appeared on a number of display screens, above the digital timer. In fact, the tester showed how images of the evaluator's screen, keyboard, and face could all be imposed on one screen for a total view of the testing event. Microphones allowed the testers to communicate with the evaluator in case she needed help. The test director told me he often invited writers and programmers to visit the lab (and sit in the row of observation seats provided) or he would record the testing events for replay for product developers. Usually testing involved performing tests designed by documentation and software developers, and then preparing reports based on the results. An elaborate testing facility looks something like the one illustrated in Figure 8.4.

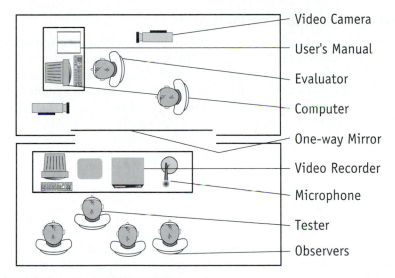

FIGURE 8.4   A Layout for a Usability Lab
Usability labs for software and documentation require equipment for viewing and recording user interaction with manuals and help products.

Clearly the two situations described above differ in a number of ways, ways that reflect the nature of commitment to testing within an organization or with a given project. Below we examine the two primary variables pertaining to the elaborateness of a testing effort.

Testing as a Corporate Priority

You may work for a company that places testing is a high priority. Such institutions, research and development organizations, or companies used to producing high-tech computer and other products, see the value of usability as a way of involving the user, client, or customer in the production loop and producing better products. In many situations, however, technical communicators find themselves having to argue for the value and benefit of testing. Often, documenters who take on the challenge of instituting usability testing programs in corporations find themselves faced with organizational cultures that resist a user-centered design of products and the shift in development processes it might take to institutionalize testing. Those working to develop a testing facility may find it difficult to obtain authorization to bring in the human factors professionals it often takes to start a usability testing program. In some instances, the cause of usability can become a battle ground for other forces within a company (factions supporting one or two strong personal leaders with separate visions for the company). In this circumstance, usability may receive short term support but fail to flourish because of the lack of a user-oriented infrastructure in place in a company.[3]

Testing As a High-Cost Endeavor

As you can imagine from the description above, setting up a testing lab requires a large initial investment. Not only does the recording equipment cost, but you have to include costs for room renovation, maintenance, and salaries for trained personnel to act as test monitors, testers, and evaluators. While testing on a limited scale can work for smaller projects, once you think about labs the cost goes up dramatically and you begin to need the buy-in of other important departments within your organization. Testing, at least the lab-type described above, requires resources in time and equipment that many companies just cannot commit.

The Advantages of Field Testing

Because of the high cost of testing and the resistance to user-centered design often encountered by technical communicators in traditional corporate cultures, many software documenters turn to field testing as a way to gain valuable information about the use of their documentation products. Field testing is simply taking your test materials to the actual users worksite and gathering data on the effectiveness of your documents. Field testing can make up a part of your ongoing, managerial attempt to discover and test ideas with actual users or can constitute a more modest step in your regular documentation process. Researchers have identified a number of things that a documenter can learn by conducting field research. Table 8.9 identifies some of them.

Remember that field research poses issues of time, politics, budgets, ethics, and legality that require you to proceed carefully. While most users will welcome you into their workplace, they will most likely do so only if you approach them with full authorization, treat them with respect, and conduct yourself professionally. Researchers who have done extensive field testing recommend that you follow the guidelines below.[4]

- **Do preliminary research into the company and users you want to test.** You can study the history of user feedback from technical support personnel, or

TABLE 8.9   Things You Can Learn from Field Testing

Topic	Explanation
Demographics	Information about age, gender, years in the profession, job titles, educational levels, specific individuals in fields
Environments	Office design, network designs, team structures, reporting structures
Information Access	Corporate information channels, sources of documentation/computer support, user communities
Information Use	Sharing, storing behavior, information systems
Software Use	Tasks, applications, training, upgrading, purchasing
User Satisfaction	Effectiveness of written and online documents, support

examine the trade literature for reviews of products like yours to get a sense of the kinds of companies you're investigating and their use of software. Contact any persons you know in the company to find out the ideal employees to contact and how to do it.

- **Construct a sensible field testing plan.** A well-designed field test should have a plan that includes the following: a schedule, list of resource personnel, questions that will guide the test, objectives of the test, and the authorization to conduct the test. A plan like this, in writing, gives you something to send to managers and supervisors who need to approve your testing activities.

- **Prepare to compromise.** Unless you have developed a track record with a certain company or group of users, you will find that your actual site testing will involve fewer people, or you will have less time than you thought you had, or you won't get the exact users you wanted. Flexibility in your plans and good research ahead of time can help accommodate some of these inevitable shortfalls.

Field testing should be a part of your ongoing involvement of users in your project. It helps build rapport and sends a message to users that you care about their software support needs. Additionally, because you focus on the users' environments, field testing enables you to support critical workplace tasks.

Methods of Field Testing That Emphasize Task Issues

While almost all of your work in field testing contributes to the user's ability to perform actions productively and proficiently, some methods specifically allow you to target information and communication behavior and thus give you specific insight into how to help users meet their work goals. Below you will find some of these tests described.

Modified Q-Sort Tests

Workplace efficiency using software relates in some ways to how people process information. In fact, people process more and more each day, especially those who work with information systems or manage and store their own information. The more we know about their thought processes and can support them, the more we can facilitate their information actions with a given piece of software. To perform their information tasks, users employ patterns; they group their tasks into action categories. To identify these categories, psychologists have devised methods called q-sorts: These operate like surveys by asking users to rank their preferences for items.[5] Q-sorts can help you identify users' patterns for the information tasks they perform in their work.

Q-sorts require users to respond based on a scale, but instead of ordering preferences on a single sheet of paper, users receive a set of index cards with one preference listed on each. They only have to sort the cards in the order of their individual preferences. For example, you could put real-world actions on the sort cards and ask users to rank them in order of performance or order of frequency of performance. The cards can contain statements such as "I transfer a document to a work-group storage area on the server" or "I download information from the main office." The actual test asks users to sort the statements into categories, such as the following:

- Perform every time the software is used
- Perform sometimes, but not every time the software is used
- Perform seldom, or only under special circumstances
- Never perform.

When you administer the test, you can present these guidelines to the user:

1. Take as long as you like to sort the cards.
2. Look at all the cards before sorting them.
3. Make a list of the card numbers in the pile after sorting.

Analysis of the results consists of calculating the frequency that certain cards fall into certain categories, as illustrated in Figure 8.5.

Results from q-sorts can help you establish users' general patterns in using a particular piece of software, or with regard to their perceptions of their actions. Those actions that receive a high percentage of frequency (i.e., are used often by many users) should be supported in your documentation. You can use the results to help you organize the operations of your software program to best reflect user needs.

Q-sorts not only work for establishing what software features and operations belong in the user's toolkit for using the software, but for establishing the importance of other elements of the user's psychology and work environment. You can modify this technique to measure the kinds of information indicated in the following list.

- **Work motivations.** Create statements of internal, external, and environmental motivation to perform tasks and have users arrange them in categories of importance.

- **Task Categories.** Use task names from the task list and have users arrange them according to perceived difficulty, centrality to their job success, relationship to group or team activities.

Categories of Frequency	Task types
Perform every time the software is used	Task A, Task Y, Task C
Perform sometimes, but not every time	Task D, Task F
Perform seldom, or under special circumstances	Task L
Perform never	Task B, Task M, Task O

FIGURE 8.5   **A Way to Record Results of Q-Sorts**
Organizing responses can help you see how to incorporate results into your documents.

- **Documentation types.** Create statements of use of various documentation media and types and have users arrange them according to categories of frequency of use or attitudes.

- **Document sections.** Present names of existing document sections (tables of contents, index, chapters, quick reference cards) and have users categorize them according to their perceptions of how frequently they think they would use them.

- **Tutorial lessons.** Present titles of existing tutorial lessons on cards and have users arrange them by task sequence, perceived usefulness, or perceived difficulty.

Vocabulary Tests Tap the User's Viewpoint

The language you use in the manual provides the key to the usability of the manual. Language evokes the schema or mental patterns in your users and allows them to read more easily and process instructions more efficiently. If you use the right kind of vocabulary in your manual you have a better chance to evoke divergent thinking, suggest information-related usage of a program, and achieve your goals of efficiency and effectiveness. Based on these observations, it makes sense for you to test the users' vocabulary extensively.

You begin identifying your user's vocabulary during the user analysis phase of the project (although you will learn new terms all along). If you have different user sets or groups, they will probably distinguish themselves by their different vocabularies. As you build lists of these vocabularies you should consider ways to test the user's knowledge of them. You have the following two kinds of vocabularies to consider with users: subject-matter terms and computer terms.

- **Subject-matter terms.** Subject-matter terms relate to the area of expertise reflected by the program. Professionals in engineering, accounting, physics, genetics, medicine, retail, and many other fields rely on the jargon of their profession to communicate with each other and to build their sense of professional identity. Because you want to evoke that sense of professionalism in the use of the software product, you will want to use the terminology to do it. Thus in retailing you'll refer to "cumulative monthly" figures in regards to business data or "SKUs" when you're referring to data items in a program.

- **Computer terms.** Computer terms refer to those terms associated with the software and hardware that the user has to manipulate in the use of the program: all the terms relating to processors, printers, interface cables, network protocols, hard drives, modems, keyboards, screens, files, operating systems, desktops, and windows. As with subject-matter terms: Use the wrong ones at your peril. Incorrectly using terms like "jumper-wires" or "register tables" can get you into trouble with users.

Create vocabulary tests to make sure that users understand any problematic terms, or that you can get away with using jargon in your information products. A couple of simple vocabulary tests you can employ are listed below.

- **Match definitions with terms.** Providing a scrambled list of terms and a list of definitions to match can help you see which terms users understand. This test has the advantage of putting the terms in a context of other terms so that the user doesn't have to rely on the active vocabulary, but can take cues from the context of usage as he or she would in a workplace context.

- **Ask for definitions.** Ask your users to provide short definitions for terms. This kind of test allows you to see how the user defines a term, so it can also help you find terms that the user just doesn't know and doesn't have the chance to guess at.

Scenarios Trigger Problem Solving

Scenario-based testing differs from the forms above because it allows the user to explore a product and documentation on his or her own, guided by goals set in a scenario. Scenarios—brief narratives of realistic work situations—suggest a situation to the user as a way of guiding the testing activity. As author Donald Norman points out, scenarios evoke a specifically human perspective to thinking.[6] According to Norman, human thinking differs from strict, logical, machine-type thinking in approaching problem solving from a story perspective, detailing how a person did one thing and then another, sometimes failing, sometimes succeeding, until achieving an acceptable solution. For example, when a person gets lost driving in a strange city, logical thinking might suggest that the driver should ask a pedestrian for directions. But people often will drive around trying first one street then another, until they arrive at the destination. Such trial and error characterizes human thinking.

In testing, you can easily end up with results that diverge from reality if you encourage too much logic, or machine-like methods of problem solving. Your carefully planned task sequence for setting up a spreadsheet, for example, might not appear logical to the user, who can easily see alternative sequences. And additionally, using stories in testing allows your user to bring in human emotions and biases, to consider personalities and human informational needs and preferences—something of the rich context of realistic workplace problem solving, the tacit knowledge that often marks the difference between success and failure.

Depending on your knowledge of the user's informational or organizational background, you can focus these scenarios on information tasks, or on more routine tasks. The example in Figure 8.6 represents a commonly used scenario.[7]

In the test associated with this scenario, the tester provided the user with a prototype of the software and a draft of the documentation and appointed an unbiased test administrator to observe the test. The administrator recorded how long the user took to perform the test and the errors the user made during the test. After the test, the

FIGURE 8.6   A Typical Workplace Scenario
This scenario allows the tester to observe the user applying the software to workplace tasks.

It is Monday morning, and you find a new computer on your desk. Your boss has left you a memo, asking you to set up the computer and enter product forecast data (provided on the attached sheet). Set up the spreadsheet and get a printout for your boss.

writer and program developer interviewed the user to gain information about reasons for the user's specific performance.

This kind of test can provide a large amount of information about all aspects of a document and product, including the following:

- **Task selection.** What tasks did the user choose to perform the task?
- **Access methods.** What access elements (table of contents, index, running heads, etc.) did the user rely on in performing the task?
- **Time to perform.** How long did the user take to perform specific tasks? the overall task? to recover from errors? to read background information?
- **Success rates.** How often did the user achieve the desired results? how often did the user fail or get sidetracked into necessary but non-relevant tasks?
- **Assistance needs.** What elements (user's guide, quick reference cards, online help) of the documentation set did the user depend on for help?

As you can see, a scenario-based test can reflect on many parts of a manual, encourage the kind of divergent thinking you want to encourage among users, and give a wealth of information about your document.

How to Interpret Test Data

The question of interpretation of test data relates to the difficulties of understanding numerical representations of survey results. Interpreting data of any type presents a challenge. For one thing, you have constructed operational definitions of un-definable things, so, by definition, these definitions contain flaws. For example: you can say that a good cup of coffee would contain 1teaspoon of sugar, 1 oz. of cream, and be served at 109° Fahrenheit. Many coffee-drinkers would agree that a cup of coffee with those characteristics (defined in this way) would classify as "good." But what kinds of bias would cause us to suspect such a generalization? Regional, for one thing: Many drinkers in the southern United States drink their coffee black. Age, for another: many younger drinkers of coffee require much more sugar. Test bias, for yet another: the test may not have been given to enough subjects. *Tester* bias, for another: the test may have been administered by representatives of the dairy industry, known to favor the inclusion of milk products in coffee.

Whenever we make generalizations about data, we assume that what some specific examples show as true necessarily represents the whole. Thus, if nine out of ten of our procedures meet the acceptable performance objectives, then we assume that the tenth one will also meet acceptable performance objectives, or that our documentation will lead to efficient and effective use. But the problem stems from the fact that we have given a number to something (operational definitions) that is innumerable, and such generalizations necessarily contain flaws.

Interpreting test results means converting the data you obtained into document design changes. Often if you just naively make changes based on the data, you may overlook a bias you had in your test that would invalidate the results. For this reason you need to stay aware of biases, and make sure that any changes to the documentation reflect what you, your other team members, and often your users see as reasonable and based on common sense.

Remember the Testing Paradox

Besides bias in data, the other problem with interpretation has to do with the testing paradox, stated as follows:

> The **earlier** you test the **weaker** the results but the **easier** it is to make changes; the **later** you test the **better** the results but the **harder** it is to make changes.

Consider the design stage of your documentation, when you can make vast, sweeping changes in all aspects of the document design: page layout, type size, style, and font. You can include or exclude tasks, specify all kinds of special job performance aids, design or delete innovative formats. Even as your project progresses, you can make changes to task formats and other elements with little cost. Changes this early in the project might cost you $10 to $12 each. But as your project progresses, those changes begin to get much harder to make, because one change in design principle means changes in many instances of that design in the documentation set. Sure, your word processing and help authoring systems can mitigate the difficulty of making such changes because you can adjust formats and style sheets. But you still have to consider the labor and time it will take to make even slight changes. The cost for each change may escalate to $45 to $50 per change. By the time your project has reached the later stages, where you have whole sections completed, edited, and ready for review, the cost for even slight changes may reach $100 or more.

Now consider the quality of the information that your testing reveals. When you did your design testing you obtained pretty good results, but you had to use mocked-up pages to do it. To what extent did those mocked-up pages affect the validity of the results? Often you can't even get good results until you have the entire book or online system completed. You certainly can't really check the comprehensiveness until that stage. Consider the example of a 350-page manual. You really can't check whether the gutter margin is too narrow until you print the whole book. And by the time you print the whole book, the cost for changing the binding seam has gone through the roof. The situation with testing resembles that of performers: there are some things that you just can't rehearse until you get on the stage. But by the time you get on the stage the cost of messing up has increased because of the presence of the audience.

The testing paradox puts extra stress on you as a tester. Design your early tests as carefully as possible to ensure that you can make changes while the cost remains low. Detect major flaws early in development before you build them in and can't remove them easily.

Distinguish between Problems with the Documentation and "Problems with the Product"

Documentation has a special relationship to the software product (program) itself: it helps make the program usable. But what about programs with inherently bad or poorly designed interfaces? Can documentation make up for bad system design? And if it can, to what extent should documenters feel obliged to make up for bad system design?

Your first obligation lies in distinguishing between product deficiencies and documentation deficiencies. Baker identifies three characteristics of testing that seem to indicate a problem with the usability of a product rather than the documentation. These include:

1. An expert can do the task in much less time than the neophyte
2. Documentation has been debugged (corrected), yet users still struggle with tasks
3. Subjects understand instructions but object to the procedure itself[8]

Often the problem will lie with the documentation, but sometimes users left alone with just the product experience difficulties. Usually, Baker points out, users can figure out how to use a feature after a few tries without documentation, but if they continue having difficulty, then the problem may lie with the product.

In many development situations you will simply have to make the best of the system because you can't change it. This situation presents an advantage: With a set interface you don't have to play catch-up with last minute product changes. But most documenters consider themselves experts in adapting technology to human users and so welcome the opportunity to contribute their expertise both in document design and interface design. Where possible, you should offer your expertise in interface design as part of the overall development team's attempt to make the technology usable.

Glossary

evaluator: the person taking the test.

exit criteria: the performance standards that you set, so that when the procedure meets these standards you can leave the test situation for that procedure.

field test: a usability test conducted at a user's location or workplace.

operational definition: a quantified definition of something that you can't otherwise quantify because of its inherent vagueness. Defining "efficiency" in terms of time (under ten seconds) and effectiveness (50 units per hour) means you have a definition with which you can operate.

pilot test: a preliminary test of the materials (questionnaires, surveys, test forms, etc.) associated with a usability test.

tester: the person conducting the test, sometimes called the *test administrator.*

test points: specific elements in the documentation that you want to learn about such as problems with specific procedures, specific terminology, or special elements of document design.

testing paradox: a conundrum that identifies the trade-offs between testing early in the manual development cycle and testing later in the development cycle.

usability lab: a facility equipped with computers, software, and recording equipment for use in testing the usability of software and documentation products.

☑ Checklist

Use the following checklist to remind yourself of the important elements in planning and executing usability tests.

Test Plan Checklist

❑ What document or section are you testing? Describe briefly.

❑ What is the overall objective or purpose of the document or section?

❑ Describe the phase of development for this test (design, development, field evaluation).

❑ What elements of the test (test points) will you be testing (tasks, terminology, design features)?

❑ Describe the test type (performance, understanding, read and locate).

❑ Describe the performance objectives for the user tasks and actions.

❑ Name and describe the evaluators. In particular, tell how closely your evaluator resembles your actual users and what compromises you will make in using him or her instead. What extra background will you need to supply for these evaluators?

❑ Name and describe the testers. Tell what materials you need to provide for conducting the test.

❑ What gathering methods will work best for your test (talk-aloud protocol, unobtrusive observation, interviews, watch and take notes).

❑ Describe any special test materials you will have to write for this test (instructions authorization forms, performance record sheets, product information, scenarios).

❑ Describe the test environment (user's site, neutral site, testing lab, other).

❑ Narrative: tell what will happen, step-by-step, during the test. Try to envision it.

❑ Explain what you intend to do with the results, especially naming who will take responsibility for reworking the documentation.

Practice/Problem Solving

1. Test Report

Find a manual or help system that you think could use some improvements in its procedures or step-by-step documentation. Select a relatively simple task, such as changing the date and time display in Windows or Creating a New Folder.

Performing the role of both tester and evaluator, use the Procedure Test Form in Figure 8.1 to test a procedure from the manual, and then write a brief, one-page report telling the results. Discuss these topics in your test report:

- Objectives of the test
- Expected performance of the procedure
- Test structure and activities
- Results of the test
- Recommendations for improving the procedure

2. Deciding on Kinds of Tests

Considering the advice in this chapter, what methods would you use to test the following kinds of documents or sections. Can you test them in more than one way?

- Reference page containing images of all the menus of a program (to accompany an engineering drawing program)
- Three-part quick setup and installation fold-out card (to accompany a program to teach students principles of microbiology)
- Brochure giving an overview of how to use color in documents (to accompany a desktop publishing program)
- Online index (as part of a help system for a financial analysis package)

CHAPTER 9

Editing and Fine Tuning

Editing, like reviewing and testing, challenges you to create an attitude conducive to productivity. The examples show how to apply proofreading marks to a page and how to use style sheets to maintain editing consistency. Like other elements affecting document design, editing requires you address the needs of knowledge workers in a high-tech workplace. The guidelines help you develop good editing attitudes and adapt the types of edit to the needs of your project. The discussion links editing tasks with users' information-task concerns and covers editing as a management concern.

How to Read This Chapter

- If you're working on a project, here you need to change gears and approach your document from a different angle—that of the editor. Read the Discussion section first to put some distance between your mind and the project. Then edit, using the symbols in Figure 9.1.

- If you're reading to understand, it doesn't make that much difference which order you follow here. The Guidelines in this chapter offer practical tips rather than a specific sequence, and they offer an interesting overview of the function of editing in the information workplace. The Guidelines and the Examples will give you the information you need to complete a documentation plan.

Examples

A Page with Editing Marks

Figure 9.1 shows two versions of a *marked-up* page that illustrate the use of copy-editing marks in a manuscript. As you can see, the editor has made the corrections clearly and has made them easy to find. To facilitate ease of reading, editors often use red, blue, and green pens. You might have to draw some lines or use both margins to avoid clutter on the page, or you might have to rewrite certain passages and paste them into the manuscript or print them on separate pages and attach them. If you're

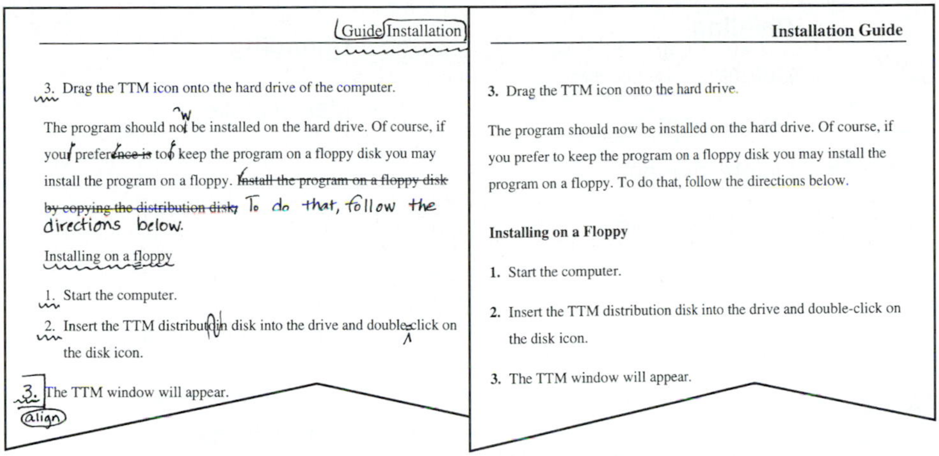

Guide Installation

3. Drag the TTM icon onto the hard drive of the computer.

The program should not be installed on the hard drive. Of course, if your preference is to keep the program on a floppy disk you may install the program on a floppy. ~~Install the program on a floppy disk by copying the distribution disk.~~ To do that, follow the directions below.

Installing on a floppy

1. Start the computer.

2. Insert the TTM distribution disk into the drive and double-click on the disk icon.

3. The TTM window will appear.

Align

Installation Guide

3. Drag the TTM icon onto the hard drive.

The program should now be installed on the hard drive. Of course, if you prefer to keep the program on a floppy disk you may install the program on a floppy. To do that, follow the directions below.

Installing on a Floppy

1. Start the computer.

2. Insert the TTM distribution disk into the drive and double-click on the disk icon.

3. The TTM window will appear.

FIGURE 9.1   Sample Marked-Up Page and Corrected Page
This example shows how editors use editing marks to include instructions to the writer in a document.

editing a help system, you might have to print it and edit the hard copy or use an electronic annotating feature to put notes into the manuscript.

The symbols in Figure 9.2 represent standard ones used throughout the industry. The example shows how you use them on a draft of a print manual or a printed version of an online help system. As you will see in other parts of the chapter, you base decisions about changes on a variety of authorities: standard style guides, in-house style guides, specifications for a specific document, and, more often than not, common sense.

Style Guide for a Software Company

The example in Figure 9.3 shows a sample page from an in-house style guide for a software company, or for any computer hardware and software manufacturer. Such a company would employ a staff of about twelve writers and editors, perhaps even a full-time indexer. For these writers, the style guide provides general information about documentation at the company, and tells how to write and format procedural tables. The document itself demonstrates the overall format it describes, and it also refers to an online version of the document so writers and editors can see what commands to use to achieve the exact same format.

Guidelines

The steps involved in editing and fine tuning are shown in Figure 9.4.

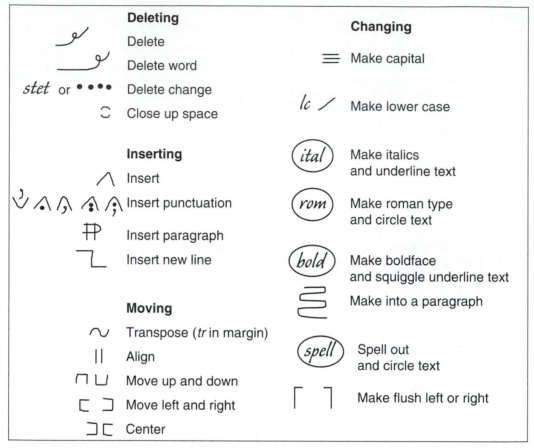

FIGURE 9.2   A Copyeditor's Marks
These copyeditor's marks help you standardize the instructions you give to writers about text.

1 Establish Project Guidelines

Each manual or help project is going to differ from all others you've done because of the uniqueness of the user's situation, and the content (interface and terminology) associated with a program. When you start a project you should make sure you and your writers and other editors understand your roles and the goals of the editing process. In some organizations people edit their own work; in other organizations they have editors assigned to projects. You may find yourself in one of three situations:

WRITER AND EDITOR'S ROLES COMBINED. In this situation you would perform all these edits yourself, preferably systematically so as to counterbalance your tendencies to overlook your own document's weaknesses. Knowing which levels to perform at what time helps you organize your editing work.

WRITER SUBMITTING TO AN EDITOR. In this situation you would specify the kind of editing you want done on your manuscript. Depending on your efforts and skill as a

460-006-000   General Guidelines
Issue 1, April 1992   Page 2-1

ACRONYMS

General   The following information applies to acronyms and their usage:
- Acronyms can be used in section names, sideheads, table titles, bulleted lists, etc., if they are already coined or are coined in the text after the first appearance.
- Coin acronyms upon first usage each time they appear in a table or figure, or use a legend.
- An acronym list is required and should always be the last subsection in the first section or the introduction of a document.
- Always index acronyms or abbreviations that are defined in the document's acronym list under their fully spelled-out form.

Placement   The standard header for an acronym list is as follows:
- All Caps
- Bold
- In the right column
- Preceded and followed by short nonbold lines

Rulers   The ruler for the acronym list is as follows:

```
          19          35                    70
          |           |                     |
   10......1.......W2.........3....W....4.........5.........6........R
          |           |
   Acronym begins    Meaning of Acronym begins
```

Lines   The acronym list header is preceded and followed by a short nonbold line. There are no other lines associated with acronyms.

FIGURE 9.3   A Page from an In-House Style Guide
These instructions to writers tell how to format and handle acronyms for company manuals. Style guides help maintain consistency among writing staff members.

writer, you may not require substantive edits. You may be revising existing documents where the organization is set and all you need is a copy edit and proofreading. Establishing what kind of editing to perform helps you give the right instructions to the editor.

EDITOR OF THE WORK OF ANOTHER WRITER. In this situation you would ask the author to specify what type of edit the work requires, and then perform just that edit. You would only perform the required edits, or you might examine a manuscript and let

FIGURE 9.4   Guidelines for Editing Software Documentation

> 1. Establish project guidelines.
>
> 2. Understand types of editing.
>
> 3. Plan your editing tasks.
>
> 4. Develop the appropriate editing forms.

the writer know how what kind of editing a document requires. Knowing what kind of editing to perform helps you estimate how long you will take and how much it will cost.

Once you have some idea of what role you will play in the editing of documents, you can identify the objectives for your project. In a recent presentation at a WinHelp conference, writers for the Windows 2000 operating software documentation faced the challenge of editing over 2 million words in over 20,000 source documents.[1] With such a large amount of information to work with and so many programs all needing to work together, it was important for these editors to recognize ahead of time the need for consistency in how users experienced all these products. What hurdles in terms of use or understanding do you have to overcome with your project? What special production demands do you have to meet? Do you have to schedule a product with the release of other products? Are you trying something different with your project that you haven't done before?

The following are examples of the kinds of editing goals you might want to set for your project.

- Consistency in how the user perceives or experiences the document: styles of graphics, use of frames and panels to present information consistently, graphics of standardized resolution and size, consistency of voice and tone.

- Consistency in the purposes of the information in the documents: making sure that documents teach, guide, or support according to the user's needs.

- Applicability to multi cultural or cross cultural readers: using standardized English or specialized vocabularies to ensure easy translation and readability by English speakers in foreign countries whose native language is not English.

- Correspondence of tasks and activities in a manual or help system: making sure that the activities and actions described in the help system match the activities and actions in the user's workplace.

- Smooth interaction among editors and writers, content specialists, managers and all involved. Editors play a key role here.

I find it useful to use the four categories of editing, each applying the kinds of edits described above, but each providing a different kind of service to a document. Below I describe each kind or level of editing. These levels can help you both in planning your editing and in clarifying the kinds of editing services you may offer to a client.

In some ways you will perform these kinds of tasks on your own documents if you have control of the production process. Most likely you will work in a development team with specific responsibilities for the documentation assigned to an editor. In that case, a clear idea of the kind of editing you will provide helps all team members understand and appreciate your contribution.

2 Understand the Types of Editing

Editing requires a lot of very hard work, partly because of the overall abstractness of the task. You can make the task more manageable by understanding ahead of time the basic kinds of editing work that you will do or that you will request of others for your project. As you will see from the description below, the *types of edits* (sometimes called *levels of edit*) correspond roughly to stages in the writing process. The managerial edit sort of covers the entire process because it concerns itself with planning and monitoring standards for the process, but the other three levels reflect how editors and writers work together to build documents. They start by looking at substantive issues of content and task orientation and progress through stages of refinement in which the document content, language, graphics, and other details become more stable, correct, appropriate, and ready for distribution.

Clearly, all documents require the edits listed in Table 9.1, regardless of whether they get done by the writer or a designated editor. The first edits on a document have to do with managing the production itself and as the document grows toward its final version, the edits focus more on the ideas and their expression.

TABLE 9.1   The Four Types of Editing

Type	Goal	Explanation
Managerial	consistency in document based on consistency in the process of document creation	Editing documentation plans, project reports, product specifications, outlines
Substantive	consistency in the overall information elements of a specific document	Editing outlines, rough drafts, organization and structure of content, appropriateness for specific readers
Copy	consistency in expression and presentation of information from the word sentence level up	Editing sentence-level style and organization, clarity of phrasing, mechanical (grammar and punctuation) consistency, format consistency and readability
Proofreading	consistency in production of copy-edited text in final versions of documents	Editing for consistency and correctness in all terminology, spelling and mechanical correctness, format consistency, consistency of the document against copy editor's marks

Managerial Editing

Managerial editing concerns itself with the documents and their planning and production rather than their actual format and content. This kind of editing doesn't relate specifically to the language of the document, but more to how the document gets produced. To do this edit you track and coordinate all the production processes, and the relationships with other documents. A managerial edit requires involvement all during the documentation process, from the documentation plan onward. This kind of edit requires you to make sure that all the document specifications and plans are consistent and informative so that the writers and developers can follow them easily. To do this kind of editing you don't necessarily concern yourself with what's in the documents, but more whether they conform with the requirements and whether they're on time. Your communication is not so much writing on pages (as in the other forms of editing) but in writing memos and messages to remind writers of deadlines, list specifications for documents, announce changes to style guidelines, and so forth.

A managerial edit, sometimes called a production edit, is done by managers and editors from an editing group working with writers on a team. The managerial editor looks at the documentation and design plans and checks them for consistency and accuracy. He or she may also check them for conformance with company policy and conformance with other documentation plans created by the company. The managerial editor also examines review and test plans to make sure they conform with the documentation plan and design plan for the project. In addition to this the managerial editor sets the styles for documents and then reviews them (along with the copyeditor) for conformance with the style guidelines. In most instances, the managerial editor is in charge of maintaining the *style guide* and periodically reviewing it based on usability testing results and other information about the manuals and help. Managerial editing usually adds between 15 to 20 percent to the cost of a project.

Substantive Editing

Substantive editing, sometimes called "developmental editing" involves editing language and information. To do this kind of edit you work very closely with an author to address the overall organization and structure of a document, as well as the clarity of ideas. The substantive edit is done earlier in the development process and used as a way to get all the information elements together and in an effective order. This kind of editing looks at sections and task orientation to make sure that the topics in the document conform to the actions the reader will undertake using the software. During the substantive edit, you also examine the sentences as expressions of the information in the document, making sure that each expresses ideas clearly and forcefully.

All documents should receive a substantive edit in some form or another either by the writer or an external editor. Less skilled writers or writers just beginning in software documentation need this kind of edit because the task of planning of a manual or help system is so large that you can lose sight of the purpose when confronted with the task of editing. Often documentation gets written or sketched out by subject matter experts such as programmers or engineers. In case these persons have little training in software documentation, you will want to require or suggest a substantive edit. With substantive editing the editor gets highly involved in all aspects of the document's structure and ideas. The editor suggests re-organizing, developing new sections, deleting others, and indicates where particular types of information would be helpful. This kind of editing is informed by a full knowledge of the user and the user's need for information.

An experienced editor can usually substantively edit two to three pages an hour but the pace of your work can vary greatly depending on the quality of the writing in the document. Often a substantive editor will look over many versions of a document as a way to help a writer focus the content and get all the information elements structured well. For this reason it's hard to give a benchmark on the time it takes to do this kind of editing and your best bet is to try out a few pages and keep track of your times to use as benchmarks in the future.

The following are some of the things you look for in a substantive edit.

- Overall organization of the document meets user task needs
- Ensuring fluency of one sentence to another within paragraphs
- Crafting parallelism in steps and lists
- Deleting for conciseness in sentences
- Ensuring proper use of description, elaboration, examples, screen-focusing statements
- Clarifying definitions of abbreviations, acronyms, and symbols
- Making sure all elements are in the right order
- Ensuring that titles, introduction, and appendices contain the right information
- Checking divisions of information are logical and consistent
- Maintaining the correct emphasis on certain elements, such as information-related tasks
- Minimizing redundancy and repetition
- Omitting irrelevant or inappropriate material
- Finding instances of missing information

Copy Editing

The copy edit concerns itself with editing for grammar, mechanical style, and format: finishing work to get the document ready for the final draft. By the time you do this kind of editing the overall content of a document has been set so the editor doesn't need to worry too much about this. In this type of editing you pay attention to all the surface-level elements of the words, sentences, paragraphs, pages, and books. The copy edit can range from very light to very heavy: essentially, you address details of prescribed styles and conventions and those conventions accepted in the profession. For example, you make sure that bulleted lists are all done with the right kind of symbols. Or you read the introduction to a procedure to make sure it's not too long and that it addresses the reader's needs accurately for that procedure. Basically, when you copy edit you ensure clear ideas and conformity with style guides for usage, list and table format, footnote formatting and placement, page layout, and other visible elements of the document.

Copy editing is done on documents that writers have already tested and subjected to user and other reviews. Copy edits occur also in organizations with set practices and forms for documents, where you can pretty much assume coherence among the document parts. This kind of edit doesn't necessarily assume that you know a great deal about the reader of a specific document. For instance, if you're copy editing a manual for engineering readers and you come across the following sentence: "Calibration of pneumatic cylinder bore size is done using already prescribed air pressure and force required." You might recommend dropping the "already" as a way to simplify the sentence and reduce redundancy. But you don't necessarily need to know about mechanical engineering or your engineering readers. You do, however, need to know a lot about users in general and principles of good technical writing to do this edit.

An experienced editor can usually copy edit eight to ten pages per hour. Copy editing takes longer than substantive editing because of the necessity of getting all the sentences and paragraphs correct, all the format done just right.

The following are some of the things you look for when copy editing:

- Correcting spelling
- Correcting subject/verb agreement
- Fixing sentence fragments
- Fixing incomprehensible statements due to missing information
- Checking the suitability of screen shots for publication
- Checking typography: type styles, leading, column widths, headings, indentations
- Checking for correct style: continuity instructions (so users don't get lost between pages), positioning of figures and tables
- Testing progress indicators and navigational aids
- Running headers and footers
- Fixing margins, spacing, rules, fonts, page numbers, binding, tabs
- Checking mechanics and punctuation related to content ("CPU" versus "cpu")
- Regularizing word compounding ("online" versus "on-line")
- Checking the form and construction of numerals and terms ("drive A:" versus "A: drive")
- Checking the form and use of acronyms and abbreviations
- Testing cuing patterns (bold, script, color, etc.) for specific words or content

Proofreading

Proofreading is the last stage you go through before shooting or printing the production copies of a document. Proofreading looks at all elements of a document: format, mechanics, graphics, punctuation, and style to make sure that everything is as perfect as it can get. It entails making sure that all the changes suggested during the copy edit were done, so often you're comparing one document to a previous version. As the next stage in the production process is printing and distributing, proofreading is the last opportunity to catch typos, spelling errors, and everything you didn't catch in copy editing. It also gives you the chance to catch errors that were introduced after copy editing. For example, using the example above, let's say you struck the word "already" in the sentence, but left an extra space between the words "using" and "prescribed" like this "using prescribed." Proofreading would catch that error and delete the extra space.

Because of the tedious and detailed nature of proofreading you often do it with a partner, with one person reading the document aloud and the other following along (sometimes switching roles.) It is very hard to prescribe or predict how long proofreading takes in terms of pages per hour because of the nature of the document (how many errors it actually contains) and the length of the pages. As with copy editing it's best to do a few pages as a test and then try to estimate based on that.

The following are some of the editing tasks you perform when proofreading.

- Checking for consistency in the table of contents, matching the text pages and page numbers
- Checking lists of tables and figures, matching the tables, figures, and screens in the text
- Checking that navigation and routing sequences specify the correct location of the necessary information within the document or help system
- Cross-referencing that tutorial lessons, user's guide, reference, and online support is consistent
- Checking that screen shots and figure numbers are unique and consecutive
- Checking that numbered or lettered sequences, as in steps, are correctly labeled
- Checking that the spine copy, bleed tabs, and index pages are consistent

Editing Help Documents

When you're editing help documents the same levels of editing apply but because of the significant differences in the process of help development and the technical features of help documents, you have additional editing problems to address.

MANAGERIAL. With help systems you need to account for the extensive planning and building of the system, testing on the target hardware, and working with programmers. Often you're dealing with help systems written by programmers, so your content won't arrive until after the programming is underway or complete. Help system development requires you to coordinate with phases of software development or else you might not have a functioning system to work with. With some help systems you have to turn a document around quickly to reflect product changes. Additionally, you need to control versions of the help system, usually using the same system of version control used in product development. Finally, you may have a number of contributors to the help system as the product unfolds, which means that you need to set styles for interface, content, graphics, and other elements clearly.

Substantive. Substantive editing for help systems means that you have to set and maintain standards for organizing into electronic books, using linking and image mapping, using graphics and icons, and organizing navigation strategies. You need to work with authors to experiment with ways to organize and present text in ways not afforded by print documents. For example, you need to edit "alt text" that appears when users move the mouse over graphics. Or you might need to help authors decide what kinds of information goes into pop-up windows. Often you print versions of help systems and work with authors on the hard copy.

Copy. With copyediting you also edit paper printouts of help files. Much of what you look for in this stage of editing parallels what you look for in print documents: punctuation, grammar, style, sentence organization, parallelism in lists, consistency of graphics, and so on. If you examine the document online, however, you can get a sense of how the information elements work together. For example, any given topic of information may have associated with it a glossary term, a related topic, a further explanation or any number of other information. So when the user clicks on these additional information elements what results may or may not be consistent with the original topic. So working online during copy editing can help you detect some of these kinds of problems at the sentence level.

Proofreading. In her book on *Editing Online Help,* author Jean Hollis Weber emphasizes the importance of working onscreen for what she calls "production" editing.[2] The reason for this is that elements of the help system such as links and mouse-overs need to be checked on the type of monitor that the user will employ in reading the documents. The editor needs also to make sure that all the corrections made in the copy editing stage have been carried out and work technically.

3   Plan Your Editing Tasks

Planning for editing should begin at the kick-off stages of a project, but often editors get brought into a project later as a last-minute quality control measure. The best situation for the editor is for editing to be planned out from the start, so that the tasks and role definitions of the editor are spelled out in the documentation plan (see Chapter 6, "Planning and Writing Your Documents"). When the editor comes on the job late you may find yourself having to do some retro planning in order to establish a productive editor/writer relationship and a workable schedule.

Scheduling depends on the kinds of editing work you will be doing on a project. Depending on how many writers you are working with and your editing role, you will have to schedule two things: 1) time for going over documents as an editor, and 2) meetings with authors and developers about your work. Editing is a very hands-on activity.

Scheduling Editing Work

Scheduling editing work allows you to budget in the time you need to complete your editing tasks and match your activities with others on the documentation team. The time it takes to complete editing tasks depends on a number of things: the quality of the work you receive for editing (which is, in turn a factor of the quality of your writers) and the nature of the material you're editing. For instance, as Collins points out, editing indexes can take up to a third of all the time devoted to indexing,[3] where as editing procedures may take up a smaller percentage of time devoted to writing procedures. On the whole, as Hackos and others point out, editing can take roughly 15% of your entire project time,[4] but that time may be distributed in different ways depending on the type of edit and the material. Most editing time is spent during the implementation phase (doing substantive and copy editing) and during the production phase (doing proofreading). Table 9.2 shows an overview of editing tasks and the times it might take to complete them.

The advice of experts in editing is that you track the time it takes you to perform editing tasks in the organization in which you work.[5] However, depending on the kinds of information you're working with, Rude, following Lasecke, offers the general guidelines in Table 9.3 for performing substantive editing, copy editing, and proofreading (essentially all editing tasks) on various kinds of print documents.

When you're editing online materials or editing online help the time it takes you to complete tasks may change somewhat because of the extra time it takes to work within an electronic medium and the extra tasks that face the editor of online help. David Dayton notes that generally editing electronic files using mark-up tools such as those found in Microsoft Word is often reported to be take less time because of two reasons: 2) you can more easily incorporate changes into the text and 2) tedious tasks

TABLE 9.2   Estimates of Editing Times for Print Documents

Task	Time	Activities
managerial editing	10–15% of all activities	overseeing production, scheduling edits, meeting with writers
substantive editing	6–8 pages per hour	reading drafts, suggesting organization and content
copy editing	1–3 hours per page	correcting sentence clarity and structure, correcting grammar and regularizing mechanics
proofreading	5–10 pages per hour	verifying changes against copy edited pages, checking for consistency in layout and graphics

TABLE 9.3   Guidelines for Types of Print Documents

Type of Topic	Guideline Hours
Step by step procedures	4–5 hours per procedure
Glossary terms and definitions	0.75 hours per term
Reference topics	3–4 hours per topic
Error messages	1.5–2.5 hours per message
Graphics and screen shots	0.5 hours per screen[6]

(like replacing all instances of one word with another word) in a document go faster.[7] How much faster, however, is unreported in Dayton's survey.

As for editing online materials, the editor has to look at a slightly different set of tasks. Online editing requires working with linked text, and checking and correcting broken links. Also, the tools for editing online documents change, depending on the format. Most editors of WinHelp files work on paper printouts, so changes have to be manually put back into the document. Those working with Adobe portable document files can use the editing tools available in that program.

Technical editor Jean Hollis Weber has attempted to set out general guidelines for how long it takes to edit online help. Table 9.4 breaks down the time estimates she uses (while counseling editors to record their own times based on their clients and projects.) The estimates in this table assume a standard help topic containing 100 words, while some types of help (pop-ups, definitions, etc.) may only contain half this amount.

As you plan your editing activities, identify the tasks you expect to have to fulfill and clarify them in the documentation plan (Chapter 6, "Planning and Writing Your Documents"). The following discussion summarizes the nature and types of tasks that you may want to specify in the plan.

Managerial Editing Tasks

If you are doing a managerial edit, then you should plan to attend meetings and edit documents such as the documentation plan, test and review forms, the style guide for the project, and all the documents associated with the project as well as the product. In fact, your role as managerial editor should be spelled out in the documentation plan already. At the beginning of a project you will be concerned about making sure that all the documents pertaining to management are consistent with one another, and then during the project you will be concerned with keeping project documents up to date and communicating with authors, reviewers, testers, clients and other persons associated with the project. Above all, communication is the key. Judith Tarutz points out that, in the role of managing editor, "Failure to communicate effectively and often will harm the project faster and more devastatingly than any other error."[9]

TABLE 9.4   Time Required for Editing Online Help

Task	Time Required
Develop design specifications, style guide, templates	40–80 hours (note 1)
Edit substantively, including some rewriting	2–4 topics per hour
Edit table of contents	2 hours (note 2)
Copy edit	6–12 topics per hour
Light copy edit (skim help, correcting obvious errors in spelling, grammar, punctuation, consistency and completeness)	12–30 topics per hour
Check links against specifications (Do they go to the right place? Are they useful links?)	50–70 links per hour (note 3)
Edit index (500 entries)	1 hour quick check 4+ hours detailed check, no fixes 5–20+ hours to fix problems
Quality or production edit	60–100 topics or more per hour (note 4)

Notes:

1. The time required depends on whether you're adapting or updating an existing specification and style guide, or starting from scratch; whether a lot of negotiation with other stakeholders (and revision of the specifications) is needed; and to some extent whether the project uses one author or several. You'll need to allow time to revise these materials after the first edit of sample material or the first review of a prototype design.

2. You may need more time to edit a very large project or a badly organized table of contents, especially if more than one writer is involved, or if the editor needs to make detailed suggestions on how to reorganize the contents.

3. Depending on the complexity of the linking system used in your project, the time required could be considerably longer than suggested here.

4. Depending on the complexity of the project and the number of tables and graphics, the time required could be considerably longer than suggested here.[8]

Managerial editors need to schedule the following events:

- establish styles for print and online documents
- a meeting to review the documentation plan
- occasional memos to communicate updates to the documentation plan
- reminders of deadlines, drafts, reviews, and test activities
- periodical updates to the project style guide

Substantive Editing Tasks

Substantive editing requires an additional editing pass at the outline or very rough draft stage of a document. The substantive editor checks documents as they are being developed and advises the writer on how to organize and design the content of a document according to the reader's needs. Typically substantive editing requires that you review the outline, and, depending on the project, sections or whole drafts before they go to reviewers. Often you will have to schedule a meeting, conference with the author as part of substantive editing. These conferences, sometimes by phone, sometimes in meetings with more than one author, usually take about an hour and offer you the opportunity to review your comments on the rough draft, explain why you made the comments, and help the writer understand the tasks that he or she needs to undertake in working on the document. These meetings also give you the opportunity to resolve any conflicts between your view of how the document should be organized versus the author's view.

Substantive editors need to schedule the following events:

- review the documentation plan and style guide for the project
- deadlines for outlines and rough drafts
- deadlines for returning outlines and rough drafts
- meetings to discuss editing comments and suggestions

Copyediting Tasks

Copyediting usually only requires one session per draft and is done after the document is completely written in draft form. Copyediting, because it doesn't delve into structural or organizational areas as with substantive editing, doesn't require you to schedule meetings with authors. Once you have done your copyediting you send the document back to the author who then incorporates your corrections and produces the final draft. Copyediting usually takes longer than substantive editing because you may have to make many passes through a document looking for one feature after another.

Copyeditors need to schedule the following events:

- start and end dates for editing sessions on drafts (usually one session per document or one per document section)
- meetings with writers after drafts

Proofreading Tasks

Proofreading, as mentioned above, is a matter of double-checking things and making sure that the work of previous edits is in place. As the last pass over a document, the work of proofreading can often get tedious and so scheduling two persons to work together can often help. The issue is not just the boredom of proofreading but the necessity of needing another pair of eyes to catch mistakes. If you had to go from one document to the other you might miss something, or if you have to scan the document for errors you might loose concentration after a while because of the monotony. Usually what you do when you proofread is have one person read the copy-edited version and one person read the proof version. Working in tandem can help ensure that you don't let errors slip. Then if you come across a matter of discussion, you have someone to talk to.

Proofreaders need to schedule the following events:

- arrange proofreading sessions with another editor
- scan documents for grammar, spelling, headings, graphics, figures, tables, layout, notes, table of contents, index[10]

Don't Confuse Editing Tasks with Other Tasks

While you edit you should try to see your editing tasks as separate from your writing tasks. This allows for greater efficiency in the process. Also, if you edit the work of others, you should identify tasks that you don't have to do. The tasks listed below don't really fall into the editing category.

Don't supply missing material: procedures, definitions, explanations. Editors do the work of forming and shaping a document, not the research that provides the information in the document. As an editor, you will find many areas during the substantive edit that require more information, especially if you have the readers' informational needs in mind. Mark these for the writer to address.

Don't supply missing screen captures. If you think a procedure needs a screen, indicate so, but don't capture it yourself. Screens require careful planning, in most cases, and the right software.

Don't write more than short passages. Often you will write short passages because the original sounds so garbled or misdirected that you just can't stand it. Often it will not fulfill the purpose of the heading. But editors don't write longer passages.

Don't edit a manuscript more than once. Theoretically, you should only edit a document for one level of editing once. That should catch most mistakes. Of course, on the flip side, you need to catch all the errors and problems the first time. With a particularly needy manuscript you will want to spot-check it afterwards.

4 Develop the Appropriate Editing Forms

Once you have laid out an editing plan, you should turn to developing forms to help you carry it out. Because editing requires you to establish relationships with other persons on the documentation team or in your organization, you will find that creating editing forms, or using existing forms, can help you regularize your procedures and communicate with others more clearly. Table 9.5 summarizes the main kinds of forms you might use as part of your work in the role of the editor.

When you planned your documents, you should have planned what styles you would follow during the writing. Thus, when you make up any editing forms for a project, it's a good idea to consult your documentation plan (Chapter 6, "Planning and Writing Your Documents") first, to update the project plan, and more important, identify the styles it specifies and reuse as much of the original information as possible.

One of the most important forms you can create is the style sheet. A style sheet is a way of recording information about your editing conventions as you go along. The reason a style sheet works is because it allows you to make a decision and then refer

TABLE 9.5   Forms Used in Editing

Editing Form	Description	Purpose
Editing cover sheet	A cover sheet to attach to documents you edit, containing names, titles, schedules, and specifying the level of edit preferred.	Allows you and your client to easily see the terms of the edit. Acts as an informal editing contract.
Editing checklists	Lists of items with check boxes at the left.	Act as reminders of elements to examine in editing.
Editing review form	Contains information about the document and editing activities. Asks questions about the writer's agreement with editing suggestions and elicits constructive criticism about editing.	Allows writers to critique editing practice. Adds to quality control of editing groups. Gathers constructive criticism.
Policy document	Overall policy statement regarding scheduling, style guidelines followed.[1]	Helps your client understand the editing relationship and accept suggestions. Contributes to your professional image.
Style sheet	A single page divided into 9 alphabetized squares where you record conventions for a specific document.	Helps you remember the changes you made at other places in a document. Enables you to maintain consistency within a document.

back to that decision later in the editing process. Figure 9.5 shows an example of a style sheet. All you have to do is create a sheet with a four by two grid and divide up the alphabet into the cells you create. Then, as you're working on a document you can record your styles.

Style sheets are not a substitute for a style guide, which specifies styles for a document ahead of time. What a style sheet does is allow you to keep track of the smaller decisions about format and mechanics that you make along the way as part of copy-editing.

Develop a Style Guide

A style guide is a document that contains all the rules and conventions for a specific project, for a publications department, or sometimes for a whole company. An invaluable tool for editors and writers alike, the style guide fulfills a number of purposes. First, it can regularize the production of documents, so that writers and editors follow uniform styles from one document to another and within documents. This uniformity creates more consistent reading for users of the documents. Aside from this primary purpose, the style guide can also set standards that the members of the documentation team can follow, thus increasing their productivity and making them more comfortable as writers. Even the most hard-nosed writer still likes to know that his or her work conforms to styles before the editor gets hold of it. A style guide can help writers avoid mistakes by regularizing procedures for things like lists, tables, rules, etc. For instance, if

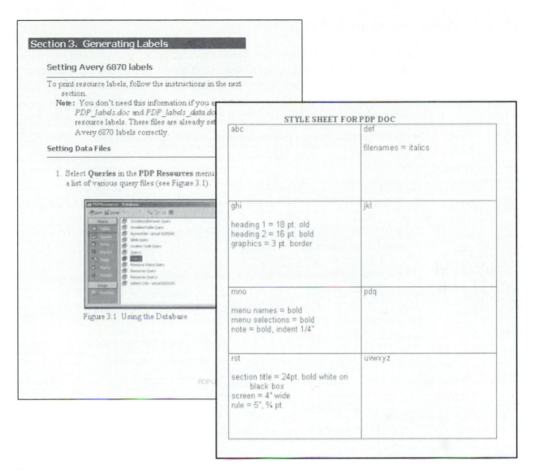

FIGURE 9.5   An Example of a Style Sheet

you specify in a style guide that rules are to be created by using the "Borders" command, then writers will use that method instead of using the line drawing command, which can cause irregularities in printing. Finally, a style guide is useful in training of new writers. Figure 9.6 shows an example of a style guide.

To put together a style guide is no small undertaking. To begin with you can look to the existing style guides in your organization, perhaps one for programmers. You may be able to use their format. You can also examine manuals and help from the past. Bill Sullivan suggests writing a style guide in bits and pieces, one email at a time, until you have amassed a body of topics that you can then organize into a departmental or company style guide.[11] So it's a good idea to see the style guide as an evolving document and many style guides encourage writers and editors to contribute.

You should also consider constructing a style guide for a specific project. Such a guide contains exceptions from the usual style guide and reflects the demands of a specific product or specific user. Usually you can construct a project-specific style guide from information in the documentation plan and make it available on a web site or internal document.

How to Write for Translation

This section provides you with specific rules about how to write for translation, with practical examples. The topics in this section are the most common causes of translation errors. You can use the following basic grammar guidelines to avoid most situations that cause difficulty for translators and readers.

Topic 1	Use of gerunds—the -ing form of verbs.
Rule	Do not use gerunds where a translator can mistake them for adjectives.
Example	Clicking Programs opens the Programs menu.
Rewrite	Click on Programs to open the Programs menu.
Exceptions	1. You can use the gerund in section headers, for example: *Using the Menu Editor*. 2. You can use the gerund in the present continuous tense, for example: The application is running. 3. You can use the gerund for the term *settings*, for example when you describe controls in a dialog. 4. You can use the gerund for the adjective *following*, when you introduce a list or a procedure. 5. You can use gerunds preceded by an article, for example *the*.

Topic 2	Use of passives.
Rule	Use the active voice wherever you can. If you do use the passive voice, analyze the text to see if you can say the same thing in the active voice.
Example	The Menu Editor is started from the Main Menu.
Rewrite	You start the Menu Editor from the Main Menu.
Exceptions	Sometimes you can not avoid the passive voice. If the sentence does not make sense in the active, then use the passive but limit the occurrences.

FIGURE 9.6   An Example from a Style Guide

5   Conduct Editing Sessions

Some kinds of editing occur at different phases of the document developing process, but at some time you'll have to sit down and perform your edits. How and where you do this is highly dependent on your specific situation, but a few guidelines might get you started.

The number one requirement for good editing is having no distractions. Editing requires concentration that you can't get if you're constantly having to answer the phone, respond to email beeps, or having visitors at your cubicle. Some editors I know

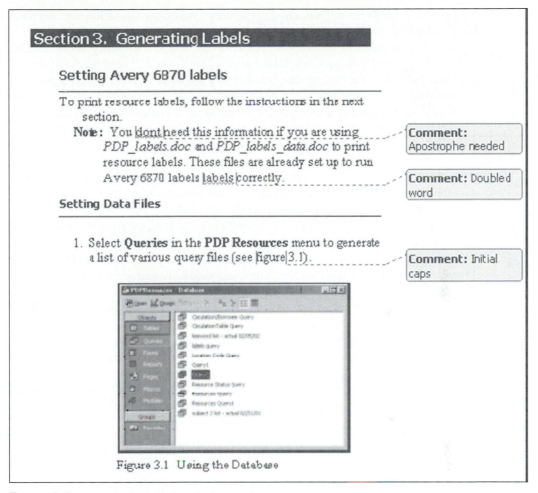

Figure 3.1 Using the Database

FIGURE 9.7   **A Page Showing Marked-up Soft Copy**

wear earphones (the larger kind) during editing to help drown out office noise. If you're editing on a computer, so-called "soft-copy" editing, you'll need to understand the change tracking and markup capabilities of a program. With Microsoft Word, for example, you can easily mark a text using the reviewing features of the program, as shown in Figure 9.7.

It is a good idea to have a checklist handy while you're editing, especially for copy editing and proofreading. Hansen recommends this for when you're editing your own writing.[12] Part of the reason for this is that we all have blind spots. Among others, my blind spots include the misspelling of certain words ("ballance," "recieve," "proceedure"). Also, because we created a document, we tend to feel protective of it, or we want it completed without extra effort, or we want it to reflect our view of the user. For these reasons, seasoned writers know that you need special tools, like checklists, to edit your own work. It resembles the quip of lawyers: "A lawyer who defends herself has a fool for a client." An editor who edits her own work has a client who doesn't believe editing will do any good.

Besides the issue of blind spots, you also have fatigue to worry about. Editing requires hard, often tedious work. Often when editing you will read the first part of a page and then, after not finding many challenging mistakes or points of consideration, continue reading without really comprehending the text—just moving your eyes over the page. I have done this many times. It's hard to remain attentive while editing.

To mitigate the effects of your own biases and human weaknesses in editing, consider two techniques that help ensure productive editing.

EDIT WITH A PARTNER. Ask another member of your documentation team to help you with proofreading. Typically you would use two versions of a manuscript, with one person reading aloud and the other person reading along. Professional copyeditors use this technique to ensure against tediousness and to balance their blind spots.

SHORTEN EDITING SESSIONS. Don't try to do a whole document at once. Do a task at a time, or section at a time, so that you pace yourself, and shift tasks frequently. First read for sentence coherence, then for mechanics, then double-check the format, then move on to the next task or section. This shifting of tasks helps you keep attentive.

The kind of reading you do as an editor differs from reading you do as a user or a writer. Often it means not reading at all, but looking, stifling the urge to understand in order to pay attention to editorial details. Untrained editors often lack these skills. Judith Tarutz recommends developing the kinds of reading skills that help you find typos and other mistakes quickly.[13] You might identify some methods such as those listed below:

- Flip test (ten seconds per manual): get the overall organization and layout of pages, check for glaring inconsistencies

- Skimming (six to ten pages per minute): spelling, mechanics, punctuation

- Reading selectively (two to three minutes per page): spot-checking for tone, grammar, complete sentences, transitions, parallel sections

- Reading analytically (five pages per hour): missing information, technical inaccuracies, paragraph organization and unity

- The long look (one to two minutes per page; errors will not appear until after the first minute): omissions in title pages, tables of contents, indexes, headers and rules out of place, misaligned graphics, tiny extra spaces in front of some words, double spaces between words (like this), stupid mistakes

You can use these techniques selectively, or apply them all to the same document at different times. I find, for example, that when I read selectively I often find patterns of mistakes I can then catch by skimming through the rest of the document.

You can use these estimates of pages per minute in planning your documentation project. Chapter 6, "Planning and Writing Your Documents," discusses where you can use this kind of estimating information.

Discussion

Writers and editors share a lot of the same characteristics: they work on development teams and work in the same environments. They may work in the same publications department or work group and often have the same reporting structures. They concern themselves with the same documents and share the same goals: to produce usable documents to support work. But to do good editing you also need to know the differences between editors and writers.

Editing Graphics

The editing of graphics in software documentation requires that you make and monitor decisions about the nature of illustrations in manuals and help. Illustrations, which are often lumped as a group called "figures and tables" (Rude, 114) make up one of the most important parts of any software document. Illustrations in software manuals and help are primarily screen shots, but can also include icons, buttons and other interface elements, clip art used to adorn and exemplify ideas, and, increasingly, animations of procedures showing pointers selecting from menus and the resulting changes in program screens.

As part of the typography of a manual, the graphics are often produced by the graphic artist assigned to a project, which means that the editor has to learn how graphic artists do their work. What programs do they use to capture and label screens? What text do they work with in deciding the content of captions, callouts, and labels? Answers to these questions can help you decide how to mark up their work, whether you can use software to do it or whether you will work from printed copies (which is usually the case.)

Decisions about standards for graphics in manuals and help are set in the documentation plan, so you can consult that document as the beginning of your planning to edit graphics. However, along the way writers and graphics artists will introduce variations in to the illustrations and so the editor's job becomes one of editing for appropriateness and consistency.

As a substantive or developmental editor, your job is to evaluate the appropriateness of the illustrations for the given manual or help system. Rude notes that, "illustrations must match content and purpose in quality, taste, and seriousness."[14] You should examine graphics for their type (whole screen, partial screen, diagram, icon, etc.) to make sure they meet the standards set for the document. You can also edit for the location of the graphics. For example, most screen shots should follow closely the command element in the step so that illustrations come in when the user moves from having "done" to witnessing the result.

Sometimes you can have too much correlation with the graphics so that the illustration looks more like a dart board than a screen shot, with arrows drawn from the steps to the image. As a rule of thumb writers should avoid drawing lines from the step to the screen shot as this introduces clutter. In such instances it's often best to set a standard whereby graphic artists set a single, highly visible arrow in the screen shot that corresponds simply to the command in the step. Similarly, writers often like to use numbers to correlate with the elements of an illustration. This technique works only if it is used sparingly so as not to cause too much hopping back and forth from image to text, trying to keep track of numbers.

As a copyeditor your job is to edit for consistency of a number of elements of illustrations. As mentioned above, usually you will work on printed copies of pages or screens to do this work. The marks you use for illustrations as Rude notes, are the same as those used in marking regular text.[15]

- **Captions:** captions for figures usually are placed underneath the figure on the left and are capitalized and numbered (as in "Figure 1: The CalendarMaker Main Screen"). Only if you have chapters would you use "Figure 1.1," with the "1" indicating the chapter. Often if you have a short document you don't need to indicate "Figure" at all but just name the screen with a descriptive title like "Calendar-Maker Main Screen." Make sure capitalization is consistent.

- **Linking with image maps:** image maps (images in web pages that allow users to click on hyperlinks associated with regions) should operate consistently and correctly in an html help document.

- **Callouts:** Callouts (boxes or labels with text) should be aligned along a vertical margin or horizontal rule and should point with the correct sized line and arrow head to the right portion of the screen.

- **Resolution:** illustrations should show a consistent resolution and, when they are to be read, should be clear and legible. You can also check for the size of image files (in kbytes) to make sure you don't have one or two that are so large as to slow down the entire help system.

- **Size:** screen shots should be of a consistent size given the type of graphic. Icons should also be sized consistently and often will have a border around them to set them off from the page or screen clearly.

- **Location:** Screens should be located after the step to which they refer or in a coordinate position in the left margin. This way the user learns where to look after executing a command to see what happened, or learns where to look for an illustration of menus or toolbars.

- **Cross references:** cross references from the text to the illustration should be clear and easy to follow and consistent throughout.

Writing Versus Editing

Usually you write first, then you edit or get edited. Some writers edit while writing, stopping to fix a typographical error or dress up a table before writing more sentences. I do it as a way of lingering over a passage, of slowing down writing so the words that appear on the screen have time to sink in. But this kind of fiddling with text while you write doesn't pass for systematic editing, which can't really happen until you complete the whole document. Writing also differs from editing because editing requires you to form a different relationship with the text than you had as a writer. You need a distance from text when editing, so you should approach it "coldly," after a period of time. Similarly, if you edit someone else's work, you approach it with a different attitude from the one the writer had. Even though you may have editors helping the development team from the start and you consider matters of style all through the early stages of writing, you can't escape the feeling that when you write you make the document and when you edit you shape something already made. When you turn from writing to editing, you turn from issues of content to issues of consistency, integrity, shape, and form.

Table 9.6 gives you a broad overview of the differences between writing and editing. For writers who have to turn from one activity to another, such a comparison can help you avoid a confusion of roles that can lead to unproductive work and, ultimately, to missing your documentation goals.[16]

While you edit you should try to see your editing tasks as separate from your writing tasks. This allows for greater efficiency in the process. Also, if you edit the work of others, you should identify tasks that you don't have to do. The tasks listed below don't really fall into the editing category.

- **Don't supply missing material: procedures, definitions, explanations.** Editors do the work of forming and shaping a document, not the research that provides the information in the document. As an editor, you will find many areas during the substantive edit that require more information, especially if you have the readers' informational needs in mind. Mark these for the writer to address.

- **Don't supply missing screen captures.** If you think a procedure needs a screen, indicate so, but don't capture it yourself. Screens require careful planning, in most cases, and the right software.

- **Don't write more than short passages.** Often you will write short passages because the original sounds so garbled or misdirected that you just can't stand it.

TABLE 9.6   A Comparison of Writing Versus Editing

As a Writer You . . .	As an Editor You . . .
Concentrate on generating, collecting, testing useful information	Concentrate on document standards, production processes, printing, schedules
Become expert in a specific program, interview experts, learn all the features	Become expert in writing standards, company policies, stylistic guidelines
Work on one project at a time	Work on multiple projects at a time
Compose	Correct, check, compare
Maintain direct contact with readers	Maintain indirect contact with readers through writers and document specifications
Work on one kind of document: user documentation	Work on many kinds of documents: internal publications, style guides, brochures
Are very familiar with a specific product or technology	Are very familiar with the company and a variety of products
Start with a product and produce a document	Start with a document and produce an information product

Often it will not fulfill the purpose of the heading. But editors don't write longer passages.

- **Don't edit a manuscript more than once.** Theoretically, you should only edit a document for one level of editing once. That should catch most mistakes. Of course, on the flip side, you need to catch all the errors and problems the first time. With a particularly needy manuscript you will want to spot-check it afterwards.

Before desktop publishing, editors formed important links between the writer and the document and the production personnel (illustrators, typists, typesetters, printers). The editor read the text and gave instructions to these persons on how to manipulate—reformat, set, draw—the information according to the author's intentions. With desktop publishing, the editor's role often changes to giving those instructions back to the writer who often, in the role of typesetter or illustrator, will make the changes.

Editing for Cross-Cultural Readers

Editors often have to accommodate cross-cultural readers by setting standards for and checking for language, graphics, and make-up characteristics that could render a manual easier to use by non-native English speakers. Cross-cultural readers are persons in or from foreign countries who speak English as a second language. Short of translations, readers in international locations, whether they be English speakers (such as the British, whose English is significantly different from the American Standard) or persons with some knowledge of English living in the Middle East or Far East have to adjust for translation.

In addition to editing for potential translation, editors often have to also accommodate writing that will be "globalized" or "localized." *Globalized language* is writing that has been stripped of features that make it special to one culture or another. For instance, the use of special vocabularies to create "generic" text is a characteristic of globalized language. It is so general that practically anyone with a rudimentary knowledge of English can understand it. Such language has little regional flavor. For a globalized version of a manual you might include time formats using both a 24-hour clock time (17:10), as is used in Europe, the Middle and Far east, and a 12-hour clock (4:40 p. m.) as is used in the western hemisphere and North America.

Localized language has a much more narrow focus on a target audience in another country or generally from a national or corporate culture other than the one you're used to. Language that has been localized is specifically designed for use in a certain region. For example, to localize a manual for British readers you would use the £ symbol for currency, or if you were localizing for the European Union you would use the Euro symbol €. Not only symbols, but elements of language also differ so much from one group to another that you need to know the kinds of grammar and syntax to look for.

Simplified English or "plain English," represent attempts to write in ways that are straightforward and uncomplicated so that non-native speakers of English can

```
Special English is a subset of the English language developed by the
United States Information Agency for world-wide news broadcasts on
the Voice of America.  Using only this limited vocabulary of
approximately 1400 words, it is possible to talk about current events
and everyday activities.  Special English is an example of a "controlled
language," i.e. a carefully engineered version of a natural language.

The list below gives the vocabulary of Special English.  Thanks to
Stephen Tice who sent us this list in 1991.


a               an              able            about
above           accept          accident        accuse
across          act             activist        actor
add             administration  admit           advise
affect          afraid          after           again
against         age             agency          ago
aggression      agree           agriculture     aid
aim             air             aircraft carrier air force
airplane        airport         alive           all
ally            almost          alone           along
already         also            although        always
ambassador      amend           ammunition      among
amount          anarchy         ancient         and
anger           animal          anniversary     announce
another         answer          any             apologize
appeal          appear          appoint         approve
area            argue           arms            army
around          arrest          arrive          art
artillery       as              ash             ask
assist          astronaut       asylum          at
atom            atmosphere      attack          attempt
attend          automobile      autumn          awake
award           away            back            bad
balance         ball            balloon         ballot
ban             bank (shore)    bank (money)    bar
base            battle          be              beach
beat            beauty          because         become
bed             beg             begin           behind
believe         bell            belong          below
best            betray          better          between
big             bill (legisl.)  bird            bite
bitter          black           blame           blanket
```

FIGURE 9.8   An Example of a Controlled Language

follow easily. According to the Boeing, Inc. definition of simplified English, the aim is to create sentences that:

- Use the active voice
- Use articles wherever possible

- Use simple verb tenses
- Use language consistently
- Avoid lengthy compound words
- Use relatively short sentences[17]

As an editor you need to train yourself to detect sentences that exhibit these characteristics and offer simplified alternatives.

Often simplified English will use what is called a "controlled language:" a language with a limited vocabulary that has been tested for readability among non-native speakers of English. Figure 9.8, on page 291, shows an example of a controlled language.

Editing for Translation

Editing for translation means that the text of your manual or help system needs to be checked for characteristics that allow for language to be easily rendered into another language. Many translators or translation services use automatic translating software as a first pass on a document, and then correct the errors that occur in that step. If you have created the document using the principles of simplified English in the first place, then there is less of a chance for error using the machine translation. I want to emphasize that writing for translation is very complicated work, too complex to address properly here. However, a few guidelines for the editor can help you see the general point of writing in this way. Andrews offers the following considerations for writing documents that you intend to translate (or be used by non-native English speakers.)

- Manage terminology. Use a glossary to keep track of words with specialized meaning in the target culture.
- Use short sentences, active verbs, and simplified syntax.[18] Simplified syntax refers to avoidance of subordinate clauses not introduced by relative pronouns. For instance, the sentence: "The following menu shows the selected database," is hard to translate because of terms like "selected" and "database," which both are in the position of object of the verb "shows." An easier sentence to translate would be: "The following menu shows the database that you selected," because the subordinate clause is introduced by the relative pronoun "that." It gives the signal.
- In addition to these specific rules for software manuals, you should check documents for elements of simplified English.
- Check graphics for colors and icons that have different or opposite meanings in other cultures.
- Check for appropriateness with workplace cultural conventions in the target audience.
- Keep commands short to facilitate clear translation.
- Avoid complicated pronoun references. Pronouns can often be substituted for the antecedent for clarity and to avoid the gender elements in foreign languages.
- Learn the terms for the international interfaces and keep menu names and titles as consistent as possible.

Problems with Editing Online Systems

When you edit online help systems you face essentially the same concerns as you face when editing print: consistency, clarity, and integrity. But you have to deal with the different medium. I've made a list of some of those differences for you here.

- **Heavy emphasis on editing the index.** Users don't know online systems as well as they do books, so you have to edit search keywords and navigational aids carefully.[19]
- **Different production process for planning and scheduling.** Your online help system will usually be built long before your manual will, and you have to coordinate with programmers for much of it.
- **Format of material: multimedia, hypertext.** Because you're dealing with electronically presented text, you can't scribble on it in the same ways as you do on paper. You can't distribute it as easily, in some cases, because of the equipment that readers need to view the draft material. Sometimes you end up printing and editing the draft.
- **Editing for completeness and accuracy of topics given the initial topic list.** In online help systems you should have an updated topic list to work from, so you can make sure all topics have been written and that they all meet their original specifications.
- **Print screens, then edit them, but also edit on the screen.** Customarily you print your help topics and then work off the paper version. While you can see the same words this way, you don't see the words the same (through a screen). You can't get their impact as electronic text, can't scroll them, can't see the same colors or highlighting as on a screen. These subtle differences can affect your editing.

How Do You Know What's Correct?

People expect the editor to know the correct spelling, usage, grammar, punctuation, and capitalization of all matters of language. For example, should you tell users of a network to "Login to the system," or "Log in to the system," or something entirely different? Try as they might, editors have yet to disabuse the general public, especially writers, of the myth that somehow they know the correct usage in this and many other cases. In truth, language, especially the language of high-tech computers, changes daily, and often you have difficulty finding standards of usage, spelling, and other matters. As an editor, you need to know where to look to find answers to questions like this.

Answers to questions of correctness depend, in large part, on how widely a particular spelling or usage has gained acceptance. To find correct answers to these matters you should start at the top of the hierarchy of authorities and work your way down. Clearly, the higher an authority you can cite, the better your chances of gaining acceptance among those you wish to convince. Basically, you can consult the five levels of authority illustrated in Table 9.7.

Like all activities associated with tailoring technology to users, editing begins with a clear idea of the needs of the people who will put the manual or help system to productive workplace activities. Commonly, novice editors see their work as making documents conform to style guidelines or hunting for mistakes in grammar and mechanics. True, editors rely on the *Chicago Manual of Style* or their company style

TABLE 9.7   Levels of Authority in Editorial Decisions

Level from Highest to Lowest	Description	Example
General Style Guides: *Chicago Manual of Style* *The American Heritage Dictionary*	All elements of punctuation, mechanics, grammar, spelling, and style.	"Use a hyphen to join multiple-word unit modifiers: 'easy-to-learn software'."
In-house Style Guides: *CONVEX Style Manual*	Standards unique to a company, the subject matter, kind of product, and contro-versial matters where style is adopted for consistency.	"Always place CONVEX in all caps. The exceptions are those specific product names that use the word otherwise. Examples: ConvexOS, ConvexAVX, ConvexRTX."
Document Specifications for a Specific Document	Standards unique to a specific document or product.	"Refer to the product as 'DOANE On Disk.'"
Common Sense	Standards used because they appear logical.	"Use *online* rather than *on-line* as a spelling because the word gets used so frequently the unhyphenated spelling will make it easier to type."
Consistency	Standards used in order to achieve consistency.	"Use *on* instead of *upon* consistently for no particular reason other than to avoid arguments."

guide, but they inform their best decisions as to level of detail, sentence structure, and language by reviewing information about the software user.

When you come to editing tasks, the thorough job you did in Chapter 4, "Analyzing Your Users," will serve you well. In particular, you should pay attention to the tone of language the user employs and the level of formality he or she expects. All the elements of language that you use should reflect some aspect of the user's characteristics, as Table 9.8 illustrates.

Editing encompasses more than finding mistakes. The good editor should see his or her job as molding all the details of document design, from pages and binding down to punctuation and mechanics, as the fine tuning of communication guided by user characteristics. Table 9.8 indicates how some of the areas of editing relate to user characteristics.

So you can look to guides and to the users, but you should also refer to the documentation plan. It will contain the specifications for the document that the document designer laid out at the beginning of the project. As an editor your job is two-fold: 1) to make sure that the design specifications are carried out in the draft of the document as it was written, and 2) to ensure consistency in the document, so that the same format, language, and other design features are the same throughout.

TABLE 9.8 Editorial Areas Reflect User Characteristics

Editorial Area	User Characteristic
Tone	User type: novice, experienced, expert
Level Of Detail	User's knowledge of computers
Vocabulary	Level of subject-matter knowledge, user type: novice, experienced, expert
Sentence Structure	User's education/demographics, learning preference
Examples	Learning preference, subject-matter knowledge
Graphics	Learning preference
Organization	Usage pattern
Content	Information and task needs

TABLE 9.9 Ways to Take a Constructive Attitude toward Editing

Editorial Relationship	Problematic View	Enlightened View
Editor and Writer	Editor as grammar police: "Make me bleed."	Editor as partner: "Help me communicate better with users."
	Editor as outsider: "When the document's done I'll send it to the editor."	Editor as team member: "The editor and I will work together on this document."
Editor and Subject Matter Expert	Editor as window dresser: "Make it look pretty," or "Dot the i's and cross the t's.	Editor as adder of value: "Help me avoid embarrassing technical flaws and inaccuracies and improve the document."
Editor and Self	Editor as writer: "I can see all the mistakes because I wrote it."	Editor as alter ego: "I've got to shift roles to see my blind spots as a writer."

Take a Constructive Attitude

Editors have to deal with other people, often in a relationship fraught with problems from the start. Consider the quips people make about editing: "A person's greatest urge is to edit someone else's writing," or "Commit random acts of editing." Tell someone you work as an editor and see how long it takes them to make a joke about watching their language. Like it or not, people see editors as grammar police.

So take an enlightened view. See yourself not as an adversary, but as a partner sharing the goals of good communication with the writer. Table 9.9 summarizes the kinds of relationships editors can have with others, and indicates the way the tenor of certain relationships depends on how you interpret them.

Seeing yourself as a partner with the writer can take you a long way toward having a satisfying experience as an editor. Roger Masse, among others, has examined the relationship between editors and writers and suggests that we see it as a dialog rather than a confrontation. He suggests that working to achieve the components of a

TABLE 9.10   Elements of a Constructive Editorial Dialog

Quality	Description	Technique
Genuineness	Being yourself and expressing what you think and feel, not what you think you ought to express	"We have a problem with the style guidelines here." (not *"You* have a problem. . . .")
Accurate, empathetic understanding	Comprehending and understanding the other person in a relationship	"You've done a lot of work here."
Unconditional positive regard	Affirming the other person as a partner in dialog	"Thank you for working with me on this document."
Presentness	Being consciously and actively present in a dialog and concentrating on the other person	"Explain what you want to do with this passage." "Let's convert your verbal explanation into prose."
Spirit of mutual equality	Seeing the other person as an equal	"I like what you've done here." [Find areas you really admire in the document and say so at the start.] "What do you think?"
Supportive psychological climate	Communicating without preconceptions	"So what I hear you saying is . . . [accurate reflection mirroring of the other person's ideas]"

dialog, shown in Table 9.10, can ameliorate some of the problematic elements of the editorial relationship.[20]

Consult Standard Style Guides

You will find yourself needing to check general style guides and reference materials from time to time, especially when controversies occur over points of mechanics or style. The following list presents examples of style guides you may have at your disposal.

GENERAL STYLE GUIDES AND DICTIONARIES. Guides that give broad ranges of information on mechanics, punctuation, and usage, especially technical terminology. Examples include the following:

The Chicago Manual of Style, 14th ed. Chicago: University of Chicago Press, 1997.

Technical Writing Style, Dan Jones. Boston: Allyn & Bacon, 1998.

Publication Manual of the American Psychological Association, 4th ed. Washington, DC: APA, 1994.

Scientific Style and Format: The CBE Manual for Authors, Editors, and Publishers, 7th ed. Chicago: Council of Biology Editors.

Technical Editing, 3rd ed. Carolyn Rude. Boston: Allyn & Bacon, 2002.

Science and Technical Writing: A Manual of Style. Philip Rubens. New York: Henry Holt, 1994.

The GPO Style Manual, 28th ed., Washington, DC: U. S. Government Printing Office, 1984.

IN-HOUSE STYLE GUIDES. Guides that embody the specific conventions for text and format, and all the style elements for a specific company. Examples include:

The Digital Style Guide of Digital Equipment Corporation, Burlington, MA: Digital Press, 1993.

GNOME Documentation Style Guide
http://developer.gnome.org/documents/style-guide/

Rochester Institute of Technology Style Guide
http://www.rit.edu/~932www/style_guide/index.html

Ray Johnson's "TCL Style Guide" http://dev.scriptics.com/doc/styleGuide.pdf

SPECIALIZED REFERENCE WORKS. Reference books that contain specific conventions for publishing and documentation in specialized technical areas cover broad conventions within fairly narrow definitions of fields. You will find specialized guides like those listed below of great value in establishing project conventions.

IEEE Standard Dictionary of Electronic Terms.

Electronic Computer Glossary, Alan Freedman, The Computer Language Company, 1995.

The Computer Glossary, ed. Alan Freedman, 6th edition, 1993.

IBM Dictionary of Computing, 10th edition (August 1993), McGraw-Hill.

IEEE Encyclopedia of Computer Science, 3d edition (1993), IEEE Press.

McGraw-Hill Dictionary of Scientific and Technical Terms. 5th edition, 1994.

Galaxy, a guide to Internet-based reference materials and dictionaries,
http: //galaxy.einet.net

Usability Glossary http://www.usabilityfirst.com/glossary/main.cgi

The Free Online Dictionary of Computing, http://www.instantweb.com/~foldoc/

Council of Science Editors http://www.councilscienceeditors.org/

Glossary

controlled language: vocabularies of words used to maintain consistency of terminology in manuals and help used for non-native English readers.

globalized language: a form of writing using generic language that lacks, as much as possible, conventions or characteristics that associate it with any specific culture.

localized language: language that uses conventions and characteristics that tailor it to a specific culture.

mark-up: the system whereby editors insert corrections, comments, and other information to guide the writer of a document.

simplified English: language that uses short sentences, active voice verbs, consistent language, limited technical vocabularies, and simplified syntax to communicate clearly to non-native English speakers.

style guide: a book or booklet-length collection of conventions of grammar, punctuation, spelling, format and other matters associated with written and online text.

style sheet: a page or document divided into alphabetical sections and used to keep track of conventions used for a specific document.

types of edit (levels of edit): a system for identifying the methods used and results of edits on documents broken down into four categories: managerial, substantive (or developmental), copy edit, and proofreading.

☑ Checklist

The conventions covered in this checklist cover general conventions rather than specific guidelines for a company. Also, I could not include every convention in the book, nor would I want to here, because you can find them in standard reference guides.

Editing Checklist
Mechanics and Punctuation

❑ Abbreviations: Follow style guidelines, use consistently, explain clearly. Example (if this is what you had decided): "3.5in." rather than "3.5-inch" or "3.5"."

❑ Acronyms: Check for consistent use, explain unusual terms used for the first time. Example: use "DOS" but spell out newer terms to avoid confusion, as in "Virtual instruments (VIs) have three main parts. . . ."

❑ Capitalization: Check all trademarks and product names for consistency and conformity to style guidelines ("eHelp," "Microsoft Corporation.").

❑ Cues: Check all cues such as glossary terms, key-press sequences, and menu choices for consistency and conformity to style guidelines.

❑ Hyphens: Check all compound adjectives preceding the word modified for proper hyphenation. Example: "end user documentation" should be "end-user documentation," but "documentation for the end user" is correct.

❑ In-text definitions: Follow style guidelines for use of bold and/or italics, and (possibly) explanations in parentheses.

❑ Quotation marks: follow style guidelines for usage around words or Chapter titles, to name two possibilities. Example: . . . see Chapter 3, "Program Manager."

❑ Spelling: use a spell checker, spell special terms according to style guidelines.

❑ Steps: Number/bullets consistently and consecutively.

❑ Trademarks: Check with the trademark owner for legal restrictions on the use of trademarks and names. Always say "Use the Excel^{tm} spreadsheet program to . . ." rather than "Use Excel^{tm} to . . ."

Language and Usage

☐ Consistency of voice: Use active and passive voice consistently. Example: "You need to check the results of this step to be sure the stream network has been thinned to one pixel width. Once this has been accomplished, continue to the next step." The second sentence should begin "Once you have accomplished this" to keep it parallel with the active voice in the first sentence.

☐ Fragments. Check for sentence fragments where they would be inappropriate.

☐ Paragraph length: Check for unusually long paragraphs. You can probably subdivide any paragraph longer than 1.5 inches into subtopics.

☐ Parallelism: Check for parallel grammatical structures in all lists. Example:
The options described below allow you to do the following:
  • Choose a logarithmic scale.
  • Scale limits defined. [Should be "Define the scale limits."]
  • Set the minimum step of the scale to be an integer.

☐ Performance orientation: Make sure all instructions and explanations indicate performance of a task in the workplace. Example: "Grass 4.0^{tm} has a number of tools provided to help find and solve problems that can occur in digitizing." [should be "Grass 4.0^{tm} provides you with a number of tools to help you find and solve problems . . ." or "Grass 4.0^{tm} provides a number of tools to help you find and solve problems . . ."].

☐ Unbiased gender references: Check for appropriate gender representation in language and examples. If names are used, for example, include women's names and men's names in approximately equal proportion, even if the activity has traditionally been performed more commonly by one gender than the other.

☐ Precision of diction: Check for precise technical terms. Don't say "use" when you mean "install." Don't say "boot" when "start" would be clearer to the uninitiated user. Whenever you use jargon you should do so with proper explanation. Other jargon terms include "download," "site," "hub," "black box," "tower," "fiber," "backbone," and "log." This standard goes double when you're writing for translation or cross-cultural communication.

☐ Sentence length. Check for unusually long sentences and re-write as two or more sentences.

☐ Unclutter cluttered sentences. Example: "Do not install other desk accessories either in the system file in the Server Folder on the startup volume or in the System file in the System folder on the AppleShare Server Installation disk." This should be rewritten as more than one sentence to separate similar-sounding terms.

Organization

☐ Completeness: Check for the following parts (does not all apply to all manuals):

☐ Title page	☐ Acknowledgements	☐ Notetaking page(s)
☐ Copyright page	☐ Trademark list	☐ Revision history page
☐ License	☐ List of Tables	☐ Index
☐ Table of Contents	☐ List of Figures	☐ Appendices
☐ Foreword	☐ List of Screens	☐ Advertising insert

☐ Organization: Check for the following overall organizational schemes for manuals and sections:

❑ Degree of difficulty (beginning, intermediate, advanced)
❑ Sequence of use (starting, processing, analyzing, printing)
❑ Tasks (task a, task b, task c)
❑ Job-related topics (topic a, topic b, topic c)
❑ User types (novice, experienced, expert)
❑ Program areas (area a, area b, area c)
❑ Alphabetical (a, b, c)
❑ Order of menus (menu a, menu b, menu c)

Format

❑ Lists: Follow style guidelines established for the project. Example: "Colon preceding list, no periods for fragments; square 10-pt. filled bullets align with left text margin; 1/4" indent to text; text returns wrap to same 1/4" point."
❑ Tables: Follow table guidelines established for the project. Example: numbered titles above in bold, rules above and below first column heads, column heads left-justified, rule below last row, initial caps in row heads.
❑ Page consistency: Check that all pages have the same top, right, bottom, and left margins.
❑ Headers and footers: check that headers and footers reflect the information in the sections and contain the correct text.
❑ Pagination and page breaks: Check for correct right/left pagination and page breaks above individual tasks and sections.
❑ Page numbering: Check front material ("i, ii, iii," etc.) and main section pages for consistency, correct sequences, and format.
❑ Vertical spacing: Check for consistent spacing between the lines and paragraphs, and between rows in tables.
❑ Horizontal spacing: Check for consistent spacing before lines, between words; check centering of figures and lateral alignment of headings and marginal icons.
❑ Type styles and fonts: Check for the correct type styles and fonts for body text, cued text, user-typed information, figure and table titles, callouts, header and footer text, and headings.
❑ Orphans and widows: Check for orphans (abandoned at the bottom of a page) and widows (isolated at the tops of pages).
❑ Rules: Check line thickness and length of rules in tables, headers, footers, columns, and section breaks.

Illustrations

❑ General: Check for presence of all titles.
❑ Unneeded elements: Check for clutter in screens, too many arrows and lines.
❑ Cross-references: Check for clear, consistent references to visuals in the text.
❑ Appropriateness: Check that screens and cropped screens reflect the appropriate elements discussed in the text.
❑ Elements correctly identified.
❑ Aligned correctly with text.
❑ Accurate figure numbers.

❑ Accurate and complete captions.
❑ Legibility: make sure all screens are legible.

Tables

❑ Units of measurement in column heads.
❑ Placement appropriately in the text.
❑ Accurate table numbers.
❑ Punctuation of elements consistent.

Practice/Problem Solving

1. A Style Guide for Editing Online Help

Editing online help requires attention to individual topic modules, examining them for elements that we know make online help work. These might include:

- Information access
- Navigation
- Overall help design
- Keywords (equivalent to index)
- Completeness/accuracy of topics
- Image fit, orientation

Imagine that you have just started as a publications manager of a software firm looking to expand its staff of writers and to shift documents online. You realize the value of a style guide for an expanded staff, many of whom haven't written help before, so you start taking notes. Considering the challenge of editing online material, make a list of the kinds of topics you want to address in your style guide and how you will create consensus in the guide.

2. Study the Work Habits of an Editor

Find the editor in your company or an editor of software documentation at your university. Set up an interview with him or her and see what you can find out about the editing habits discussed in this chapter. Using email and the Internet you may find a cooperative editor willing to answer brief questions in an email message. Consider exploring these questions:

- How does the editor's work differ from that of writers of the documents?
- What forms, such as style sheets and checklists, does the editor use?
- What different demands do different media, such as print, online, and multimedia put on the editor?
- How much say does the editor have in establishing guidelines for his or her own organization?
- What kind of training does the editor see as most helpful for preparing editors?

3. Resolving Controversial Usage

Often technical editors get drawn into discussions of editing specific yet controversial usage. Such editors may have to refer to authorities to resolve the discussion. Consider the lists below. How would you resolve the controversy, and what authorities would you refer to for help?

The user *who* initiates the first search procedure on the network	The user *that* initiates. . . .
Upon opening the file, you will see table headers.	*On* opening the file . . .
You will find tables of commands in the *Appendixes*.	. . . in the *Appendices*.
Procedures are found in the "User's Guide."	. . . in the *User's Guide*.

4. Create a Mock Style Sheet

Study a manual or help system (with which you're familiar) for the conventions it follows. Mock up a style sheet for the document following the example in Figure 9.5, "An Example of a Style Sheet". What elements of the style of the document would you want to record as you wrote it? Can you find any inconsistencies in the document?

PART III

The Tools of Software Documentation

CHAPTER 10

Designing for Task Orientation

This chapter presents tools and techniques for responding to the characteristics of software users. The chapter suggests a problem-solving approach to document design, allowing the writer or designer to apply those techniques that adapt technology to human use through a manual or help system. The chapter covers guidelines for designing to meet user needs, and then explores techniques for responding to each of the elements in the user analysis checklist. Next it presents a document designer's toolbox in a discussion of each element.

How to Read This Chapter

All readers should read this chapter from start to finish. Along with Chapter 11, "Laying Out Pages and Screens," it functions as a key statement of how I articulate the relevance of design concepts to task orientation. These represent just the beginning. With the ever-broadening definition of the page and the book—including things like htmlHelp and other *hypertext* technologies—the technical writer in the software industry needs to know a growing number of page and book design features. This chapter gives you the basics of book design, and Chapter 11, "Laying Out Pages and Screens," gives you the basics of page design.

- To those working on a project and anxious to get started, practical tips for design are discussed in Guideline 5: "Review the User Analysis."

- Those readers interested in broadening their understanding will appreciate the psychological discussion of users under the heading "The Design Problem." And if you are in the midst of planning a project, the "Solutions to the Design Problem" for print and online help systems acts as a handy resource, a gallery of features of manuals and online help systems.

Examples

Figure 10.1 and Figure 10.2 present some elements of document and screen design.

Guidelines

Guidelines for designing documents are listed in Figure 10.3 and described in the paragraphs that follow.

1   Create a Table of Contents

Think of document design as a sequence of stages beginning with a goal (meeting users' needs) and ending with a solution (a manual and/or help). The user analysis will suggest overall goals for your documents. What workplace tasks do they use the

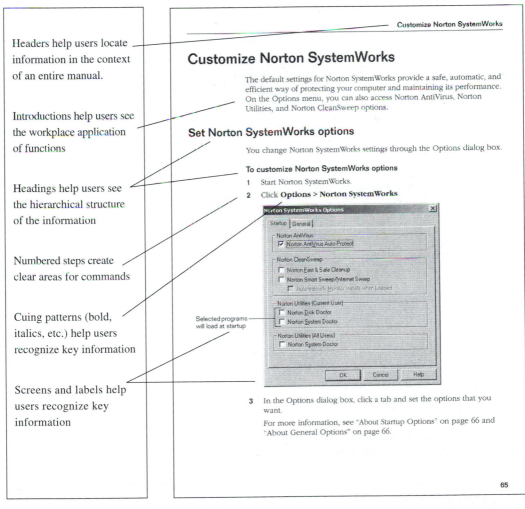

Headers help users locate information in the context of an entire manual.

Introductions help users see the workplace application of functions

Headings help users see the hierarchical structure of the information

Numbered steps create clear areas for commands

Cuing patterns (bold, italics, etc.) help users recognize key information

Screens and labels help users recognize key information

Customize Norton SystemWorks

The default settings for Norton SystemWorks provide a safe, automatic, and efficient way of protecting your computer and maintaining its performance. On the Options menu, you can also access Norton AntiVirus, Norton Utilities, and Norton CleanSweep options.

Set Norton SystemWorks options

You change Norton SystemWorks settings through the Options dialog box.

To customize Norton SystemWorks options
1   Start Norton SystemWorks.
2   Click **Options > Norton SystemWorks**.

3   In the Options dialog box, click a tab and set the options that you want.
    For more information, see "About Startup Options" on page 66 and "About General Options" on page 66.

65

FIGURE 10.1   Elements of Document Design
This page from the Norton SystemWorks *User's Guide* shows how elements of document design help users gain access to and manage information.

Norton SystemWorks basics

About Startup Options

Startup options let you select which SystemWorks programs start when you launch Windows. When you check the checkboxes next to these programs, they load whenever you start Windows:

- Norton AntiVirus Auto-Protect
- Norton Fast & Safe Cleanup
- Norton Smart Sweep/Internet Sweep

    If you select Norton Smart Sweep/Internet Sweep, check Automatically monitor installs when loaded to have Norton CleanSweep automatically monitor program installations.

- Norton Disk Doctor
- Norton System Doctor

About General Options

General options let you choose to display information about programs when they launch and also let you schedule One Button Checkup.

Option	Description
Display program splash screens	Displays the graphic window for each program when that program is started. Uncheck this checkbox to bypass the graphic and open the program's main window.
Display program introduction dialogs	Displays a brief description of the program every time you start the program.
Norton Tray Manager Windows 98, Me, NT, and 2000	Collects the taskbar icons for memory resident Norton SystemWorks programs into one icon.
One Button Checkup Scheduling	When this option is checked, Norton SystemWorks' One Button Checkup runs at the scheduled time. For more information, see "Schedule a One Button Checkup" on page 51.

66

Headings help locate and identify information

Lists help users decide how to apply program functions to workplace tasks.

Tables give users an overview of information to support decision making

Page numbers help users navigate among abstract concepts

FIGURE 10.1   Elements of Document Design *(continued)*

program for? What levels of usability and support for information-related work do you want to see in your users? What do they need to understand and how do they need to perform? Your overall goals will suggest organizational strategies because these goals are usually expressed in terms of sequences of user actions. The sequences (based on typical use scenarios or workplace actions) require you to structure the operations of the program (opening files, using program tools, saving work, etc.) in innovative ways. Thus, the main design element is the outline of the information in your documents, because the outline embodies your most innovative and user-oriented ideas.

A number of organizational strategies already exist for structuring information about a program interface. These strategies or methods are presented in Table 10.1.

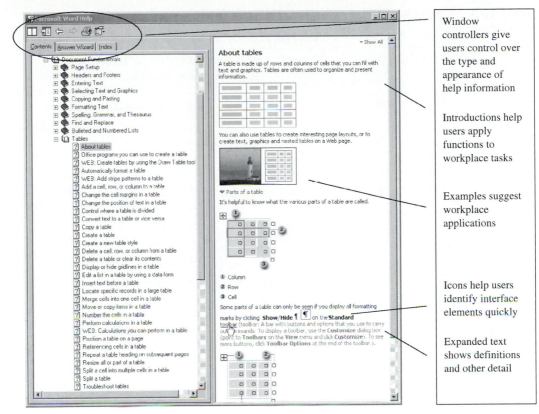

Window controllers give users control over the type and appearance of help information

Introductions help users apply functions to workplace tasks

Examples suggest workplace applications

Icons help users identify interface elements quickly

Expanded text shows definitions and other detail

FIGURE 10.2   Elements of Online Help Design
This screen shows some design features used in online documentation that enable users to access and manage their information.

FIGURE 10.3   Guidelines for Designing Documentation

1. Create a table of contents.
2. Match the user analysis with information design strategies.
3. Acknowledge productions constraints in document design.
4. Test and review the design.
5. Follow a design process for online help.

TABLE 10.1   Typical Ways to Organize to Meet Users' Task Needs

Method	Structures	Used For ...	Examples
Degree of Difficulty	Beginning Intermediate Advanced	Tutorials	From *HBOOK Reference Manual:* (Makes histograms based on scientific data from particle physics experiments) *"1 and 2-D Histograms—Basics"* *"Advanced features for Booking and Editing Operations"*
Sequence of Use	Starting Processing Analyzing Printing	Tutorials Standardized usage software: word processors and spreadsheets	From *Getting Started with PV-Wave:* (Allows scientists and engineers to visualize and analyze technical data) *"Importing ASCII Data"* *"Creating a Date/Time Variable"* *"Using Mathematical Functions"* *"Reformatting Datasets"* *"Displaying Data as an Image"* *"Creating 3-D Surface Plots"* *"Printing Your Results"*
Jobs or Tasks	Job a Job b Job c	Programs with distinct users: administrators, clerks, executives	From *Municipal Court System Ref. Manual:* (Describes parts of the citation entry system for a city court system) *"Entering Citations"* *"Processing Receipts"* *"Setting Trials and Hearings"* *"Performing Batch Runs"* *"Handling Warrants and Capes"*
Job-Related Topics	Topic a Topic b Topic c	Large systems Fewer user types Identifiable topics	From *DELPHI: The Official Guide:* (Software to connect to an online information service) *"Business and Finance"* *"Computing"* *"Entertainment and Games"* *"Groups and Clubs"*
User Groups	Group a Group b Group c	Educational programs Networks Groupware	From *Ear Trainer:* (Software to teach note and interval recognition to voice students) *"Teacher's Guide"* *"Student's Guide"*

TABLE 10.1   Typical Ways to Organize to Meet Users' Task Needs *(Continued)*

Method	Structures	Used For ...	Examples
Major Program Areas	Area a Area b Area c	Distinct program elements	From *MSGRAPH:* (Software to draw graphs from table data as part of a word processing program) *"Working with a Datasheet"* *"Working with a Graph"*
Alphabet	Task a Task b Task c	Experienced users Expert users	From *WordPerfect:* (Word processor) *"Align"* *"Append"*
Order of Menus	Menu a Menu b Menu c	Special menus	From *Pegasus Mail: Assistance Features* (Electronic mail system) *"Context-sensitive help (<F1>)"* *"Listing users on a server (<F2>)"*

This table can help you find ways to organize your tasks, and in turn, provide a basis for organizing your manual or help system. It covers eight recognized methods of organizing tasks that you can use for individual documents or for sections of documents.

Once you have decided on the overall organization of the manual or help system you can turn to the next problem which is how to design the information so that the user can find it and use it most efficiently. This process is sometimes referred to as interaction design, where you are designing the way the user interacts with the system and your information.

2   Match the User Analysis with Information Design Strategies

The following suggestions are from the user analysis checklist presented in Chapter 5. These questions give you a way of summarizing the important characteristics of your users. Specific design suggestions follow each element of the checklist. You should remember, however, that designing to meet user characteristics requires creativity and a willingness to test new ideas.

Use this stage of the process to think globally about how you will design your users' access to information. These strategies are described more completely in the design guide that follows in the discussion section. If you are unclear about any of the techniques suggested here you can read ahead in the design guide and then come back this section. It would also help at this stage of your document designing to use the "Document Design Planning Guide" in the Checklist section of this chapter. As a planning tool it can help you sketch out your ideas and experiment in your thinking about how your user could best use the interface information you're providing.

Design for Different Groups

Possible users of the program may vary depending on the operations that each one employs and their typical tasks. As you design for different groups, consider including all of the following elements:

- **Navigational aids.** Navigational aids make sure user groups get to the information pertinent to their needs. These can include special statements directing users to sections of documents, lists of figures and tables, headers and footers, and a whole host of "information access" devices. These are discussed in detail in the discussion section below.

- **Scenarios.** Scenarios give each group a role model. These examples of usage help users identify themselves and their main workplace activities.

- **Icons.** Icons identify information for each group. Icons are a handy way to catch the user's eye and direct them to short cut information for advanced users or extra details for beginners.

- **Metaphors.** Metaphors (stadium, showroom, floor plan, school room) for different users. Metaphors make implicit relationships to the workplace explicit so users can see and feel like the document is familiar to them.

Design for Specific Program Issues

Program issues can range from difficult interfaces to a need to supply special workplace information to perform special tasks. For instance, some programs have to support transfer learning from previous releases or from competing systems or software. Some require hardware knowledge or have installation difficulties. To help your users meet the difficulties you identified, design your documentation to include:

- **Job performance aids.** Job performance aids cover technically difficult or repetitive tasks. These stand-alone documents (often one or two pages each) can focus users on key workplace applications of the program. Example: "Entering Information about a New Patient"

- **Background information.** Extra background information to meet special needs. Sections of background information can help users feel like you're making sure they don't fly blind. Example: "How to Use a Light Pen" or "The Concepts Behind this Program."

- **Special forms.** Tear-out forms or printable documents can help users collect information in the field for later inclusion into the document. Example: "Automobile Record Form," "House Inspection Form," "Patient Information Form."

Meet the User's Task Needs

A lot of complaints are passed along by developers from their customers, but the most common complaint is that the documentation doesn't tell the users how to do things. It just describes the system and defines the terms and fields on the screen. Follow the suggestions here and in the rest of this book to make sure you don't get that complaint about your manuals.

- **Illustrations.** Show photographs, drawings, or clip art of users performing familiar tasks.
- **Layout design.** Make the document fit the user's desktop.
- **Examples of usage.** Include introductions explaining examples of use of the program.
- **Special document sections.** Provide a "getting started" section with three or four useful examples.
- **Tips.** Include performance-oriented elaborations and introductions.

Meet the User's Information Needs

Meeting the user's information needs requires you to understand how users manage information within a job setting. Then you "simply" write a manual that tells him or her how to use your program to support that work. The following strategies work well for meeting these needs.

- **Explanations.** Certain functions can lead to storage and re-use of information, but only if the user understands their use and their importance. Explain why file naming and directory structuring of program files can help retrieve reports and files.
- **Examples.** Examples that illustrate workplace uses of information the program generates can help users see how to manage their own work. Show program data imported into a word processing or database program.
- **Meet efficiency goals/command summaries for efficiency.** Both keyboard and mouse users appreciate avoiding irregular and time-consuming behaviors, so you should provide shortcuts, quick-key combinations, and pre-existing macros liberally. Unfortunately, many users may not see the relevance to their own goals, so include an introduction that explains the use of these shortcuts.
- **Problem solving.** Workers see themselves as problem solvers. Figure 10.4 shows how a writer's knowledge of examples of problems faced by a variety of user types allowed for an emphasis on problem solving with the program. Encourage problem solving by suggesting options, encouraging creative solutions and thought-provoking suggestions, and generally letting the users know that the writer has anticipated their problems and can show them how to use the program to solve them.
- **Emphasis on information management and communication work.** Identify functions that relate to information management and communication: file naming, saving files under other names, configuration of the program output, translating

PV-WAVE P&C is well suited for a variety of engineering and scientific tasks. The following table summarizes PV-WAVE P&C's capabilities.

Application	Capabilities
Test Engineering	Visualizing and analyzing data from analytical instruments.
	Evaluating results from quantitative experiments in fields such as cryogenics and materials testing.
	Visualizing vibration, heat transfer, stress, and emissions test data.
Space Exploration and Astrophysics	Studying geodynamics and seismology of planets.
	Analyzing data from satellites.
	Determining the composition of comets, meteorites, and other celestial bodies.
Computational Fluid Dynamics	Identifying flow patterns such as shock waves, vortices, shear layers, and wakes.
	Applying CFD research to aeronautics, automotive design, weather forecasting, and oceanography.
	Analyzing data from thermal dynamics, fluid dynamics, and nuclear reactions.
Medical Imaging	Displaying and analyzing bio-science imagery, including NMR/MRI, X-ray, CAT, and electron microscopy.
	Planning medical treatments, such as targeting areas for radiation therapy.
Earth Sciences	Simulating meteorological conditions and making predictions.
	Analyzing well logs to locate mineral deposits.

6   Getting Started with PV-WAVE Point & Click

FIGURE 10.4   A Way to Support Task-Oriented Work
This example from *Getting Started with PV-Wave* enables the user to see how the functions of the programs support critical decision making and professional goals for a variety of user fields.

files to other formats, importing information from other programs, printing, report generation.

Match the User's Computer Experience

To suit the predominant type of user, consider the following characterizations and how you can incorporate them:

- **Novice.** Tutorial covering basic functions, definitions, full screens, sample files and templates, keyboard templates, automatic installation, humorous and reassuring tone, second (elaborate) table of contents

- **Experienced.** Problem solving support, transfer learning by references to commands in other programs, short-cuts, special section on "Ways to Use the Program Efficiently," error recovery techniques, full index

- **Expert.** Highly structured reference, troubleshooting support, command reference, quick reference card, full three-level index

Enhance the User's Subject-Matter Background

Take advantage of the user's knowledge of the subject in your documentation, to enhance the usability of the software.

- **Special glossary of background terms.** Learn this by studying the user's job and workplace. What terms do they use to indicate what? Put these in a special glossary. It doesn't have to take up a whole section, but just a part of a page. In online help you can allow users to jump to definitions.

- **Index entries linking background terminology to program functions.** Identify the users' terms for special workplace concepts and include these in the index, along with mention of the relevant program functions.

- **Special booklets/sections describing background concepts.** PageMaker software now comes with a booklet explaining the elements of page design for desktop publishers. This may have reduced the frequency of "over-designed" documents that were being produced by those new to desktop publishing.

- **Elaborate examples with explanations of key concepts.** In the program Glider Design, the documentation includes a section on the theory of flight.

Leverage the User's Workplace

Your documentation can help the new user build the necessary sense of community. Try to incorporate the following:

- **Getting help from co-workers.** Introductory passages/scenarios suggesting help from other users

- **Suggestions for support groups.** Special sections on in-house users groups or in-house technical support

- **Descriptions of network use.** Suggest ways that users can share troubleshooting information and on local area networks or on the Internet.

Meet the User's Learning Preferences

Meeting learning preferences has to do with your choice of *media* and the design of documents for those media. The characteristics of the three predominant media vary; the following list includes some of the most usual ones.

- **Instructor learning.** For the instructor: lesson plan, overheads, sample files, quizzes, background information, equipment list, schedule/sign up sheet, video tape; For the learner: workbook, diskette, notes pages
- **Manual learning.** Tutorial manual, or elaborate "Getting Started" section, list of learning objectives, sample files
- **Computer-based learning.** Programmed computer-based training modules, sample files

Meet the User's Usage Pattern

To meet the user's usage patterns, first determine which category is most appropriate:

- **Regular usage.** "About This Manual" section, interrelated examples to support incremental learning, organization around beginning, intermediate, advanced functions
- **Intermittent usage.** Troubleshooting to help with error recovery, fully detailed procedures to support on-demand learning, help systems, reminders of the interrelationships among program features, support for problem solving (tips, scenarios)
- **Casual usage.** Quick cards, keyboard templates, scenarios suggesting mental models of program usage and application

Your user analysis will also contain suggested document features that you should try out with potential users as part of the design process.

One way to organize your thinking in terms of content design is to use a design planning guide that shows your user characteristics and design solutions in the same place, and then pencil in your ideas. Table 10.2 shows a partially completed design planning guide. A complete version of this table is shown in Table 10.6. The planning guide allows you to fill in your ideas as you think through how you want the document to look. If you're working on a team, discuss your designs until you get a sense for what you can do. After you have set up a "wish list" using the matrix you can adjust it based on your production constraints (discussed in the next guideline.)

3 Acknowledge Production Constraints in Document Design

After you have decided on what design features you would *like* to have you need to determine what features you can *afford* to have. In any situation, whether a corporation with a large publications department or a smaller, one horse company, the envi-

TABLE 10.2   An Example of How to Use a Design Matrix

User Characteristics	Design Solutions		
	Visual Elements	**Textual Elements**	**Structural Elements**
	cues, rules, headings, labels, icons, illustrations, progress indicators, color,	layering, advance organizers, scenarios, parallel structures, hierarchies, document overviews, examples, explanations	job performance aids, organizational strategy, lists of figures, tables, screens, table of contents, index, links, browse sequences
User Groups:	*icons for advanced users*	*none needed*	*glossary for beginners*
etc.	*etc.*	*etc.*	*etc.*

ronment in which we work poses constraints on the design choices we make. These constraints allow you to adjust your "wish list" and settle on something that might be a compromise but which will actually work in your situation. Constraints set useful limits.

In Table 10.3 you will find some of these constraints posed as questions one should ask as part of the design effort.

Knowing what your limitations are can help you decide on what you want to put into your design plan—the one that goes into your documentation plan. At the stage that you have a design plan, you are ready to undergo some kind of reviewing or testing.

4   Test and Review the Design

In the design testing and review phase you evaluate through reviews by clients and sponsors and you test problematic areas in a lab or field test. As noted in Chapter 6, involving users in the process of designing documentation allows you to eliminate a number of alternatives that would not work for your users. For example, one manual puts the index at the beginning of the document. This idea seems like a good one because users claim that they use the index more than any other part of a document to find information. Had the designers of the document tested its organization with users ahead of time, they would have realized that the "index in the front" needed some kind of explanation and justification. As it turned out, users rejected the document because it didn't fit their expectations of having an index in the back. Other design and layout elements require testing, such as legibility of help typography, suitability for layered access, special navigation schemes using color, and other design innovations.

To do testing of your design you can follow these steps.

1. Mock up pages with access elements on them and field test them. Give users sample pages and find out which one they prefer. Preferences mean a lot in manual design. Find out your users' definition of "cluttered" pages, and avoid them.

TABLE 10.3   Production Constraints

Type of Constraint	Description
Writing Tools	What design capabilities does your word processing or publishing programs offer?
	How can we optimize the design of documents by using a computer network?
	What graphic programs or other graphic technologies (such as scanners, plotters, and printers) can we use?
Production Tools	What type styles and fonts does the laser printer support? Can we use color reproduction?
	How can different binding methods make manuals more accessible?
Human Resources	How much do members of the writing team know about user- or task-oriented design?
	How can we use management know-how to achieve the most usable design?
	Do we have clerical or other basic help available to allow for more complicated design solution?
	What testing facilities can we use to try out new design ideas?
Budgets	How can we fit all the design ideas into the existing budget?
	What neat design ideas will we have to sacrifice or work around to meet cost and printing constraints?
	What design ideas will work when we have limited testing and review budgets?
External Considerations	How do other departments in the company design their documents and to what extent should we design like each other?
	What legal constraints—in terms of copyright/trademark protection and adequate warnings—should we build into the design?
	How do competing software products design their documents?

2. Consult the chapter on testing for ways to do quick usability tests. Many field testing techniques, like summary tests and vocabulary tests, can yield data to help you decide between one design and another.

3. Decide on a design. Decide on a design with confidence, having followed a process based on logic and experimentation.

At the heart of any task-oriented documentation you will find step-by-step procedures called tasks organized so that the user can follow them to perform useful work. In terms of the design of your book or help system, you'll have the job of organizing those tasks so that they match the pattern of activity your user will understand as logical. And what's logical for some users won't appear too logical for others, precisely because they do different kinds of work.

5   ## Follow a Design Process for Online Help

The design of online help should parallel the process of designing for print, however the process must be adjusted to accommodate 1) the technical differences between print documentation and online documentation (for example the different *authoring environment* for online help) and 2) the different features of online documents versus pages (such as links, graphics, scrolling and so forth). The next section looks at the design process for online help and identifies some of the challenges it poses.

Identify and List the Online Help Topics

At the heart of any online help system you will find topics, which can be defined as follows:

> **Topic:** *an identifiable body of usable information associated with a user activity*

Examples of topics include the following:

- *Steps* for performing a specific program operation, such as "open a file" or "select a database"
- Definition of a *command*
- Definition of a *term* relating to a program
- A labeled *screen* with explanations of its interface elements

Figure 10.5 shows the types of topics available with the Sevensteps help authoring system. Topics in a help system don't necessarily differ from those in print manuals or tutorials. The difference between print and online, however, lies in how you present this information and how the user gets to the information.

From the user's point of view, the topic represents the final destination. The topic tells the user what key to press, what steps to follow, or what a term means. It provides the information the user needs to get back to the task of putting the program to productive work. To start designing a help system, you need to identify user activities and then write topics that support them. If you have done the research needed for a print manual, you have half the battle won because you already know the topics and know your users.

You can research information for topics by reviewing the information you collect during the user analysis. The Sevensteps program identifies the following seven kinds of topics.

- **About.** Offers introductory information about a program or introduces groups of topics.
- **Module.** Describes the modules of an interface to give the reader a clear understanding of the structure of the program.
- **Action.** Describes situations in which a user would use a part of the interface to achieve a workplace end. This kind of topic includes step-by-step instructions for performing operations with the program.

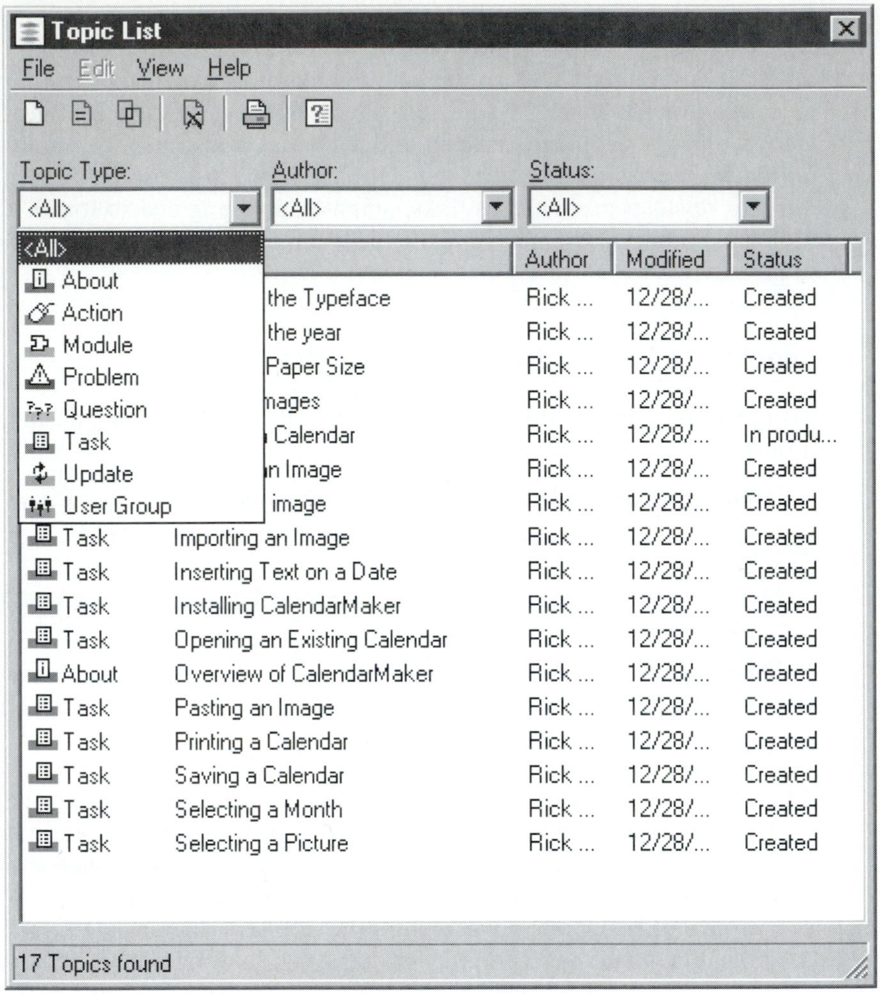

FIGURE 10.5   An Example of a Topic List
This authoring environment helps you set up topics of different types.

- **Problem.** Describes solutions to problems—bugs or errors—that the user might encounter with the program. This topic could also be used to illustrate varieties of user cases: problems in the workplace that the program can help solve. This kind of topic requires a great deal of knowledge of the user.

- **Question.** Describes questions that the user might ask about the program and provides answers or ways that the user might answer the question.

- **Task.** Identifies the workplace activities that the program supports. This very important type of topic is based on user goals and identifies procedures or operations that the user can perform to achieve the goal.

- **Update.** Describes new features of a program or application to the users. This kind of topic is used when a new version of the program is released.

- **User group.** Describes the types of users with similar ways of using the program. User groups are described in Chapter 5.

Other kinds of help topics relate to the actions the user has taken with a system and can be programmed to appear in response to those actions. Help author Cheryl Zubak points out two types of these automatic topics.

- **"Best practices."** The best practices topic opens when the user has performed an operation that is potentially complicated and which the user analysis indicates might require further guidance. Sometimes these topics carry the heading, "Did you know . . . ?". These topics will often suggest actions and contain links to further information or program commands to keep the user working.

- **Corrective.** Corrective topics open when a user has reached a dead end, such as a search that didn't turn up any hits. In this case the topic provides "corrective" action, such as suggesting better ways to phrase a search string or use of keywords. Sometimes these topics appear just as part of the interface, as in when you type "dogweed.jpg" as a search string. If the system finds no hits it might provide the following corrective advice: "Did you mean *dogwood.jpg?*"[1]

Once you have identified your topics and classified them into activity groups, you can move to the next step in designing online help: determining how they relate to one another.

Determine the Interconnected Elements

Those interconnections among topics based on user activities make up the heart of a help system because they allow users to follow familiar patterns of activity. In most help systems the relationships between one topic and another are called "see also" links. "See also" links go two ways: any given topic has links to it and links from it to other topics. In Figure 10.6 you can see three "see also" links at the bottom of the "Create a Calendar" topic. These links cause the help system to display three related topics: "Pasting an Image," "Inserting Text on a Date," and "Importing an Image." For the writer the trick is to know the user well enough to know what other operations might be associated with creating a calendar, and then create the links to those operations.

Using the help authoring system, the user can easily create links *to* these other topics as well as links to "Create a Calendar" from other topics. Figure 10.6 shows how this is done by listing the topics *to* and *from* any given topic.

In fact, you can see an online help system as one that allows you to identify interconnections and provide ways for the user to make the most of them.

For the document designer, the challenge lies in discovering the related topics within a help system and then making them available to the user through hypertext links. But keeping track of related or interconnected topics can get complicated. So to help systematize your planning for the kinds of relationships among topics, I constructed the Topic Design Form (Figure 10.7). This form provides a handy way to keep track of how a topic relates to other topics. Figure 10.7 shows how the topic design sheet would look filled out for the procedure for creating a custom report.

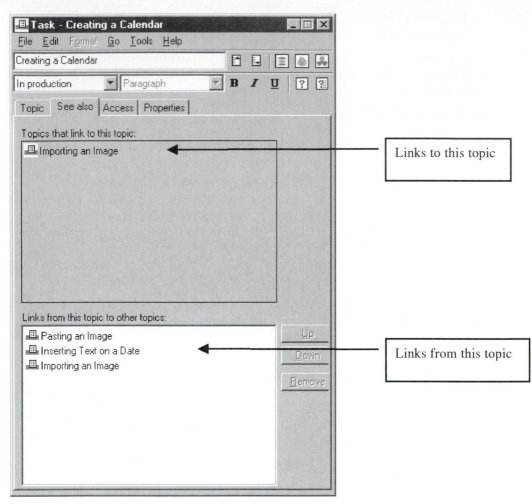

FIGURE 10.6   Links Showing Connections from One Topic to Another
Links in a help system go in two directions: links *to* and *from* a topic.

Later when writing the help system, I could refer to this sheet to see which other topics related to the task of creating a report. The related topics recorded on this sheet will appear later as the familiar hypertext links or pop-up glossary definitions that make up the connections in a help system. Depending on the system you design, you can make up your own design sheet.

Decide What Design Features to Use

Design features the software capabilities that you can build into a help system that allow the user to find and use the information it contains. Design features are the

Topic Design Form

Program:	CalendarMaker

Topic Name:	Creating a Calendar
Type:	How to…
Category:	Creating a Basic Calendar
Related Glossary Terms:	"Calendar"
Search Keywords:	Create, File Menu
Index Terms:	calendar, creating
Related Topics:	Pasting an Image, Inserting Text on a Date
Browse Sequences:	⇒ Importing an Image

Topic Name:	Importing an Image
Type:	How to…
Category:	Creating a Basic Calendar
Related Glossary Terms:	"Importing"
Search Keywords:	importing, images
Index Terms:	images, importing
Related Topics:	Using the Clipboard, Editing an Image, Pasting Images
Browse Sequences:	⇒ Changing the Typeface

FIGURE 10.7   Topic Design Form
This topic design form can help you organize the relationships among topics.

electronic interface elements of a help system: the buttons and links that you make available to the user for finding information. Indexes, tables of contents, headers, and so on, make up the design features of print manuals. One of the main differences between print manuals and help lies in the number of features available to you with the online media. Whereas the print manual requires linear reading of pages and the table of contents affords searching for topics, the online system allows you to get at the information in many more ways. Table 10.4, "Features of a Help System," lists and explains the features of a help system. Use it to help you determine what features you want to use in your system.

You will find a more complete overview of help elements at the end of the Discussion section.

TABLE 10.4   Features of a Help System

Feature	Explanation
Hypertext Links	Embedded codes that take the user to another help topic. Usually hypertext links appear in green, and the cursor changes to a pointing hand when over it. Usually users need only click once on a hypertext link to jump to new information.
Button	A feature of the help interface that causes the program to display another topic, a search dialog box, a history of topics examined, or other element of the help system. Buttons usually look like buttons—have raised edges and depress when you click on them. Like hypertext links buttons.
Hot Areas	Portions of an image (a screen or a diagram) that the user can click on to see more information. Usually the cursor changes to a pointing hand when the user moves it over the hot areas. Hot areas resemble labels on diagrams, only they operate electronically and on demand.
Browse Controllers	Forward and backward buttons that allow the user to follow a pre-determined sequence of linked topics. The sequence may cover procedures in a sequence, or sequences of background information.
Pop-ups	Small screens usually containing definitions that the user can view and then click anywhere to close.
Context Sensitivity	The ability to get help on the user's current screen or menu simply by pressing a designated help key, usually F1. The help program knows which screen to present by tags put into the program, indicating which screen or menu the user currently has displayed on the screen.
Balloons	A kind of help that displays, in a cartoon-like balloon, information about screen elements. The user selects balloon help from a menu or a button, and then moves the cursor to whatever part of the screen he or she wishes to learn about.
System Affordances	A range of buttons and capabilities provided by the help program. With Winhelp, for example, the help program used with Windows products, you can use a back button (to take you back to a previous screen), a history button (to show you the topics you've viewed in the current session), a search screen (to search the help system for keywords), an on-top button (to keep the window on the screen while the user performs the procedure), and an index (an electronic index system). Other help programs, such as HyperCard or Apple Guide offer a different set of system affordances.

Discussion

We live in an age of accommodation of writing to readers. All functional documents—not only those supporting software—require and allow a great deal more reader involvement than before. Readers no longer have to follow a strict linear form in reading. Consider web documents, in which readers determine the order in which they read: The electronic form allows them to restructure the document according to their needs. In software documents, users also need this kind of freedom. They need to select parts of a manual that fit their needs, that help them perform their tasks. The manual should reflect a design that matches their needs and tasks. Because computer support comes in pages and screens—a linear format—many readers have a problem finding information and mentally restructuring it according to their task needs. Documentation designers call this the design problem.

The Design Problem

The design problem of software documentation resembles that for instructional materials in general. It has to do with the user's needs for support versus the document's organization and content. Essentially the problem results from each reader needing to apply the computer system to a multitude of tasks. And because of the linear bias in most information presentation—pages or screens in a row—users expect information organized linearly. One book, many users. To meet the goals of efficiency and effectiveness we need to accommodate the one to the many.

The Difficulty Adapting Printed Books

Consider this example: An attorney writing a brief needs to shift a quoted passage to the right five spaces. The user might ask, "What do I do to make the text shift to the right with this word processor?" The user has a whole manual to search through to find an answer. She may not know what terminology to use to start: Is it an "indentation" problem, a "tabs" problem, a "page layout" problem, or a "margin" problem? As a rule, she will start looking in the wrong place, or at a random place in the book. Many users will simply open a book and thumb through it, looking for an answer. Others will use the table of contents or the index. She might try exploring the help facility, looking under "indent" or "margin" or "page layout." She might ignore the documentation altogether and use the space bar to force the text to the right, thus avoiding using the books at all. In fact, the information exists under the heading of "temporary indentation."

In terms of the design problem, the manual needs to accommodate her dilemma: The answer occurs in only one place in the manual, but there are many ways she might try to find that answer.

Documentation that overcomes the design problem depends in part, as we have seen in this example, on the vocabulary that users bring to their computing tasks. But the problem really involves more than vocabulary. In fact, it represents one of the most complicated and least understood aspects of software document design. So before discussing some of the design, let's examine the problem further.

Accommodating Groups of Users

You must constantly consider the degrees of experience among groups of users. Each group will react to your manual or help system differently. This variety means that you cannot just write one manual; you have to write two or three manuals in one book, sometimes satisfying two or three sets of expectations and needs on the same page. You may categorize users in many ways, as we will study further in subsequent chapters, but the two ways we will take up here have to do with categories based on users' degrees of experience with computers (novice, experienced, expert) and users' professional roles (installers, managers, operators).

A common way to define users distinguishes between those just beginning to use computers and those who have either intermediate or advanced experience. Below we will examine how these groups' psychological differences and their informational needs pose challenges to manual designers.

ROLES BASED ON COMPUTER EXPERIENCE. The experienced user often has more patience with a program, more confidence than a novice user. The novice user will often blame a difficulty on the program ("This word processor doesn't allow you to move text to the right . . . imagine that!"). A more experienced user, on the other hand, will show more persistence ("This word processor must have a way of moving text to the right; they all do, I just haven't found it yet."). In this example, the novice user would perceive the software design as weak, whereas the experienced user would give the design more slack.

Novices have less experience with documentation—both manuals and online help. They don't know conventions of documentation (such as tutorial, guidance, reference) and can get lost looking for things in the wrong sections. They don't know terms manuals use, such as routing systems, don't know how to use help, can't identify their level of knowledge, and don't know where to go to look. Experienced and expert users know many manual and help conventions and can use them with confidence.

In our example of shifting text to the right, the novice user would more than likely see the problem as one of margins, because of the similarity of the word processor to the typewriter. In this way, previous experience would cause her to look under "margins"— a mistake, because for the task of shifting text to the right, most word processors have an "indent" feature. In fact, only when one has a degree of experience with word processors does one know to look under "indent." The experienced user would recall the different terms for the task, and look in both, or other places, until finding the right term. The document designer needs to use a variety of features such as glossaries, cuing, graphics, and type styles to catch users of limited experience and get them on track.

ROLES BASED ON PROFESSIONAL ACTIVITIES. Another variable regarding user types has to do with what we call role-based user groups. To determine role-based user categories, we need to know what kinds of jobs the user performs with the software. Examples of role-based types include installers, operators, evaluators, decision makers, troubleshooters, and so on. Unlike the types based on degree of experience that we examined earlier, here the types describe the users' jobs and their relationship to the software. The manual set for an operating system, for example, segregates users into role-based user categories: users, system managers, and programmers. Each of these categories of users needs to use different functions of the program and requires different levels of support. The writer, in this case, should segregate the information into the appropriate sections and then use adequate cuing to make sure each user type doesn't wander into the wrong section, like a hotel guest lost in the kitchen.

What defines performance, achievement, productivity, or efficiency for computer users in the professional workplace? You need to know what motivates professional employees, then tap into that motivation as a stimulus for the user to learn to use the program.

Matching the User's Problem-Solving Methods

Another variable affecting the design problem concerns how people solve problems using computers. The more we know about how people get the program to do what they want, the better we can make their use efficient. Consider the way beginning artists typically solve the problem of drawing the human face. A common error among art students is to draw the eyes in the upper portion of the face, because people tend to see the eyes at this position, above the nose, above the cheek bones, above the mouth. In fact, the eyes fall about in the middle of the face. Knowing that people perceive the face the way they do, mentally putting the eyes in the upper half, the good art teacher will caution students about this, will watch for this common difficulty, and will design the instruction to counteract it (by having students actually measure the placement of the eyes, and so on). So it is with the designers of text instruction systems we call software documentation. The more we know about the common errors people make using computers, the more we can anticipate those errors, and design documentation that will counteract them.

In this regard, the current research into software use can help. Areas like cognitive science and **human factors** research have, during the last ten years, looked at how people use computers to find ways of making computers more usable, of making people more productive in using them. What do they have to say that might relate to the design problem?

Cognitive scientists show us patterns that people follow in solving problems with using manuals. Sullivan and Flower, two researchers at Carnegie Mellon University, used a **protocol method** to study manual users: they recorded all the keystrokes, pauses, and reading times of manual users to determine not how people read, but how they use manuals.[2] Their research led them to the following five conclusions. Each of their conclusions is presented below, followed by some design advice you can use to help anticipate and overcome the design problems involved.

1. No one carefully reads more than two sentences at a time. Busy professionals just don't have time for more than this. They value work time more than computer time.

   Design Advice:
   - Keep paragraphs short.
   - Arrange information into tables and lists wherever possible.
   - Put important information at the beginning of each paragraph.
2. Most of the users begin to use the table of contents before they read the manual. Professionals learn early in their careers that planning ahead, even for short tasks, pays off.

Design Advice:

- Make the table of contents complete.
- Use both an abbreviated and an elaborate table of contents for complex material.
- Use chapter-by-chapter tables of contents.
- Make table of contents headings task-oriented: "Setting up a schedule" rather than "Schedule."

3. Most users go to the manual or help only after they have failed to perform a task. At this point you will find them hasty, stressed, and impatient.

Design Advice:

- Be sensitive to a user's state of mind after failure.
- Make descriptions of error recovery clear and complete.
- Emphasize getting back to real-world tasks.

4. Most readers do not read the introduction first, nor do they generally read all of it, even if the introduction is only three short paragraphs. Most professionals consider introductions "useless information" and want to skip directly to the steps for doing.

Design Advice:

- Replace the "introduction" with useful information about user needs, special document features, or helpful routing information.
- Replace the "introduction" with material designed to get them applying the system right away.

5. Most readers do not read any section in its entirety. Reading by professionals almost always connects with some real-world task that helps them filter information.

Design Advice:

- Avoid the injunction to "Read this manual before proceeding." Instead tell them which sections to go to for particular tasks or problems.
- Make sure that all descriptions of tasks and functions contain complete information for performing that task.
- Don't expect users to remember on page 50 what they read on page 33. Repeat important information if necessary.

A Design Guide for Printed Documentation

So far we have examined the elements of the design problem, investigating how variables such as user experience, user problem-solving strategies, and production constraints can alter the nature of the design of documents. Now turn to some of the solutions to that problem: ways you can design user task orientation into your products, while still recognizing the constraints of users and production tools.

Navigation

Navigational aids are elements of a document that tell the reader where to go next for what kind of information. Typically, you would place a navigational statement or section at the beginning of a manual, listing the types of users and indicating what parts of the manual you have designed for each. This section, usually under the heading of

"About this manual . . ." distinguishes between levels of expertise (novice, expert, or power users) or users' professional roles (system administrators, terminal operators, sales persons, etc.). It tells them what portions of the manual contain information specifically for them.

Figure 10.8 shows how you can help users navigate to specific topics of interest to them. You will need navigational statements like this one to direct traffic to sections

---

How This Guide Is Organized

This guide is divided into five chapters, followed by appendices, a glossary, and an index. Use the following table as a general guide to find the information you need in the Wisconsin Package documentation.

If you're interested in	Turn to
An overview of the VMS operating system	Chapter 1, Getting Started
Starting the Wisconsin Package	Chapter 1, Getting Started
Specifics about using sequences	Chapter 2, Using Sequences
Databases	Chapter 2, Using Sequences
Learning about the basic concepts you will need to use the programs of the Wisconsin Package	Chapter 3, Basic Concepts: Using Programs
Learning how to run specific programs	**Program Manual**
Finding out about different data files, for example scoring matrices and restriction enzymes	Chapter 4, Using Data Files **Data Files** manual
Learning more about Wisconsin Package programs that produce graphic output, how you can use and manipulate that output, and what graphics configuration languages are available	Chapter 5, Using Graphics Appendix C, Graphics
Printing ASCII or graphic files	Chapter 3, Basic Concepts: Using Programs Chapter 5, Using Graphics
An alphabetical list of GCG commands and programs	Appendix A, Short Descriptions
Basic guidelines for using the text editor EDT	Appendix B, Text Editor Basics
Differences between VMS and UNIX	Appendix D, Command and Keystroke Differences Between VMS and UNIX
Definitions of some terms found in this guide	Glossary

FIGURE 10.8   Navigation Statements
These instructions from the Wisconsin Sequence Analysis Package *User's Guide* help the user understand where to go to find the right information for specific uses or experience levels.

of your print documents. You will also need to describe the user tasks that these sections support.

In large computer systems, role-based user types often need specific manuals of their own, describing the program tasks needed to fulfill their roles. In this case, a separate section in the main manual performs the navigational function. For example, the Novell NetWare's manual set contains separate manuals for supervisors, users, system administrators, and so on. The "Guide to the Manuals" contains elaborate definitions of user types, under the heading of "What Type of User Are You?" These definitions help the user determine whether he or she is a "Novice, Experienced User, Supervisor, or Installer." The user who is a "User (Novice or Experienced)" is directed to manuals specifically designed for a user's needs (and conveniently marked with a "U" on the spine). The supervisor's manuals have an "S" on the spine, and the installer's manuals have an "I" on the spine. This navigational system helps the user make the most of a very complicated six-manual document set (a documentation set not made for the casual reader).

What can you do to be sure, first, that you really need a navigational system, and, second, that it will work? For one thing, you should make your decision to include navigating as an organizing feature only after you have examined your user's tasks carefully. These tasks will show you what jobs the user will need to perform, and in what environments. In later chapters we will study how to construct a "pattern of use" for your manual. A pattern of use tells how you imagine your users applying your manual in the optimum work situation. This pattern of use can help you decide whether or not to use navigational directions as a method of overcoming the design problem and increasing the access to information. The clearer you can see a definite pattern of use among a majority of users, the more likely that a navigational system will work.

Cross-References

Cross-references point to other sections or chapters containing related information. For example, in a procedure where you give guidance support on how to print a document, you would include a cross-reference to the section of the manual containing procedures for setting up the printer. That way, if the user needs to adjust the printer settings, or hasn't set the printer up yet, he or she can easily flip to the appropriate page.

The difficulty lies in the hassle of including page numbers of all your cross-references. Most word processors, such as Microsoft Word, can handle these page references automatically. When you create a reference, you insert a field at the point of cross-reference that points to a bookmark in the target section. Then every time you update the fields of the document, the program automatically updates the cross-reference, inserting the new page number of the target section or information. So much for the mechanics.

In some cases, however, writers use an easier solution to this problem, by inserting generic references to items in the task list that the user knows will occur in the table of contents or index. Thus a reference like "See the section on Setting Up a Printer in this manual for more information" does not require a specific page on which the target information resides. The trade-off, of course, requires that the user take an extra step by looking the page number up in the table of contents.

Given the specific constraints on your project, consider that to a user the best approach means you include the information they need at the point they need it. When you can't do this, consider the ease-of-use scale in Table 10.5 in deciding how much trouble you want you user to go through.

Running Headers and Footers

Running headers and footers such as those shown in Figure 10.9 consist of the page numbers and text information that occupy the top and bottom lines of a page. These may include any number of the following elements:

- Chapter and section names and numbers
- Book titles
- Graphic cues and icons
- Task names
- Color to indicate sections

The following paragraphs discuss these elements.

CHAPTER AND SECTION NAMES AND NUMBERS. Chapter and section names and numbers help the user locate a specific page within the overall scheme of a manual. Because they occur at the head or foot of each two-page spread, they are always at the user's disposal. Names of chapters and sections should correspond to levels of support,

TABLE 10.5   Scale of Ease of Use of Types of Cross-References

Scale of Ease of Use	Reference Type	Example
Easiest	References to specific pages	"(For more information on setting margins see Page Setup, page 7–32.)"
	References to near-by information	"(Make sure to read the Tip for Efficient Use that follows this procedure.)"
	References to sections/chapters in the same book	"(For definitions of commands see **Chapter 4: Command Reference.**)"
	References to other books in the documentation set	"(For help understanding error messages consult the *System Error Messages Reference Manual,* vol. 6 of your Novell documentation set.)"
	References to manuals for other programs the user uses	"(See "Configuring Your System" in the manual that came with your operating system.)"
Most difficult	References to other sources of information (recognized professional handbooks, standards books, phone books, etc.)	"(See the *IRS Tax Schedule for 19XX* for information about tax rates.)"

1. Open the Diagram window.

2. Place the Sequence structure (**Structs & Constants** menu) in the Diagram window.

The Sequence structure, which looks like frames of film, executes diagrams sequentially. In conventional text-based languages, the program statements execute in the order in which they appear. In data flow programming, a node executes when data is available at all of the node inputs, although sometimes it is necessary to execute one node before another. The Sequence structure is the LabVIEW way of controlling the order in which nodes execute. The diagram that the VI executes first is placed inside the border of Frame 0, the diagram to be executed second is placed inside the border of Frame 1, and so on. As with the Case structure, only one frame is visible at a time.

3. Enlarge the structure by dragging one corner with the Positioning tool. To create a new frame, pop up on the frame border and choose **Add Frame After** from the pop-up menu.

Frame 0 in the previous illustration contains a small box with an arrow in it. That box is a sequence local variable which passes data between frames of a Sequence structure. You can create sequence locals on the border of a frame. The data wired to a frame sequence local is then available in subsequent frames. The data, however, is not available in frames preceding the frame in which you created the sequence local.

4. Create the sequence local by popping up on the bottom border of Frame 0 and choosing **Add Sequence Local** from the pop-up menu.

The sequence local appears as an empty square. The arrow inside the square appears automatically when you wire to the sequence local.

5. Finish the diagram as shown in the opening illustration of the *Block Diagram* section.

Tick Count (ms) function (**Dialog & Date/Time** menu). Returns the number of milliseconds that have elapsed since power on.

Random Number (0-1) function (**Arithmetic** menu). Returns a random number between 0 and 1.

Multiply function (**Arithmetic** menu). In this exercise, multiplies the random number by 100. In other words, the function returns a random number between 0.0 and 100.0.

FIGURE 10.9 Headers and Footers Help Users Navigate
These headers and footers from the *LabVIEW for Windows Tutorial* contain elaborate information to help users keep track of their location within a manual.

such as "Beginning Tutorial" for teaching support, or "User's Guide" for guidance support. Generally, names work better than numbers because they describe tasks more accurately. Numbers, however, can correspond with the index and table of contents as points of reference. With chapters or sections you will usually use sequential numbering and restart numbering with each section. The number "5-1" refers to the first page of Chapter 5; the number "6-1" refers to the first page of Chapter 6, and so on.

BOOK TITLES. Book titles, like chapter and section titles, allow the user to see quickly where the current page falls within the entire scheme of the documentation set. Again, like chapter titles, book titles can refer to levels of support (e.g. *Tutorial or Learning System X*) or to role-based user types (e.g. *Supervisor's Manual, User's Manual*).

GRAPHIC CUES AND ICONS. Graphic cues consist of icons, product logos, company logos, and other images that help orient the user to the overall design of the documentation. Figure 10.10 shows how you can include them in headers and footers as a way of increasing the user's awareness of design on the particular page, so that he or she does not have to search around, or, worse, keep a finger in the table of contents while examining specific pages. Graphic cues have the advantage of showing with pictures. This means that they can use the rich evocativeness of pictures to refer to things like program tasks, user types, sections, and other manual design elements. They make great labels.

Only your imagination limits the styles of graphic cues that you can include in headers and footers. Often programs and companies have logos that you can include in headers and footers. On a more specific level, you can put generic clip art icons for teaching (lectern), guiding (pointing hand), and reference (open book) to reflect levels of support. Another form of graphic cue, the progress indicator, tells the user how far along in the section he or she has come (see Figure 10.11). Progress indicators consist of timelines representing the total length of the section and markers separating the portion already covered and the portion remaining. In tutorial documentation, for example, a progress indicator would show graphically the lessons completed and those left to go.

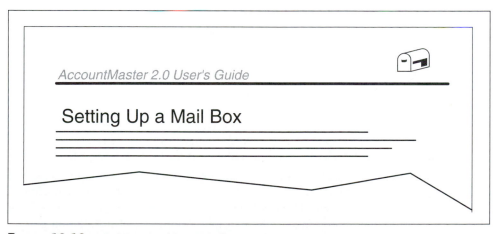

FIGURE 10.10   An Icon Used in a Header
Icons used in the header allow users to identify important information easily.

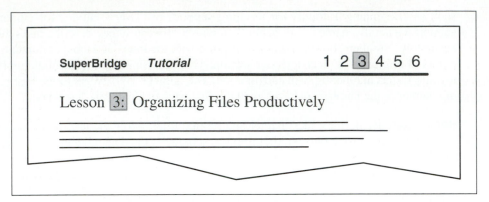

FIGURE 10.11   A Header Using a Progress Indicator
Progress indicators help users keep track of their location within a larger structure of a document.

TASK NAMES. Often in user's guides you will find the tasks that the pages describe indicated in the header or footer, as shown in Figure 10.12. For instance, "Using the File Management Function" or "Saving a Document" would appear in the header. The advantage here lies in the way this technique accommodates the user's technique to read quickly, scanning among pages to find a function. As we saw earlier in this chapter, many computer users become impatient with tables of contents and indexes, preferring to scan the manual searching for information to perform some real-world tasks. Putting names of program tasks, user tasks, tutorial lessons, or data tables (in reference sections) can help such users make sense of the overall manual structure at the same time as they relate to information on separate pages.

WAYS TO USE COLOR. You may also indicate sections by color bars in the headers and footers. Like graphic cues discussed above, these show sections using primarily visual information. They help reinforce the pattern of use of your manual because they tell the user where a certain page or section falls within the big picture.

Color can unify your documentation system and allow users to key on specific information. Below you will find some ways you can use color in your document design.

- **To identify specific sections of a document.** Color used in *rules, bleeds,* and as a highlight to section names can help users remember where they saw what kind of information. It's easy to do this in an online system because it's easy to use color in the non-scrolling region as a background for sections, such as yellow for step-by-step, green for command reference, blue for glossary, and so forth. The advantage here lies in not having to use a lot of color but still getting the benefit of identification.

- **To match pocket guides with user's guides.** If you have pocket guides associated with various documents, you can use the same color highlights for the pocket guide as for the document. The Thomas-Conrad Corporation's writers featured this kind of color linking in their award-winning four-volume set of installation guides for computer network adapters and cards. Each pocket guide has the same color scheme as its corresponding installation guide, even down to the colored spiral binding of the installation guides.

FIGURE 10.12   The Use of Task Names in Headers and Footers
Putting task names in the headers and footers (here shown in the footer) helps the user remember task names as well as navigate the document.

To cue specific kinds of information. Color can work very well for identifying specific types of information within your document layout. You can cue tips in blue or cautions in yellow. You can set the background color of all your demonstrations the same or box your elaborations in an eye-catching color. As with the other aspects of color use, cuing relies on color not as a decoration or to give a realistic feel to images, but to link to the structure of information in your document.

Layering

Layering refers to having two versions of information on the page at once, to satisfy more than one type of reader. For example, if you have elaborate steps for the novice user, you might also include on the same page an abbreviated version of the steps for the experienced user. Follow these suggestions for layering:

- Put keyboard equivalents next to mouse instructions (or vice versa).
- Put commands in the table of contents along with the terms.
- Put advanced instructions or definitions in tables alongside instructions for intermediate users.
- Use one column of instructions for beginners and one for advanced users.

At Texas Tech University we found ourselves faced with the need to provide layered information to support both the new and experienced users of our Technical Communication Production Lab. We wanted the new users to learn a number of basic word processing functions, and we wanted the experienced users to refresh their memories when they returned at the beginning of the semester. The solution that one of our design project teams came up with involved a specially designed manual with a layered, two-column format.

We put the steps in the left-hand column and the explanations in the right-hand column. At the top of the left-hand column we put a hare and at the top of the right-hand column we put a tortoise. The users actually figured it out. Experienced users just had to follow the steps at the left to complete their editing exercise. Beginning users could follow the explanations in the right-hand column to figure out how to follow the steps and to get help if they got stuck. Both users ended up with a completed real-world task (the edited document) that met the objective of preparing them for the start of a semester's work in the facility. The page made a good example of how you can layer information for users on the same page.

Headings

All manuals use headings, and all users expect them. Not all headings work equally well. In the example of the tortoise and hare tracks described above, the heading work fell to the icon. Icons make good heading elements, because of their visual nature. But most of us use text headings: short phrases indicating, at a glance, the contents of a section. Often we make headings too short: "Margins" or "Save." The problem here lies not in the intrinsic arrangement of the information: You can have a task-oriented organization, but still diffuse the usability potential by not focusing the information clearly. The problem here lies in the fact that the writer failed to make the headings part of the manual's system. Better headings would read "Setting Margins" or "Saving a file."

Don't forget that headings, besides focusing on workplace objectives, also make up part of the visual nature of pages. Below we discuss the way all text elements contribute to the look of a page. Headings do this. Usually we print them in larger, darker type so they catch the reader's eye. Good design of headings, like those in Figure 10.13, should support this function and make them fitting and easy to see. Easy-to-see often means less complex in font and style.

Follow these guidelines for including headings in computer manuals:

- Support workplace applications with elaborate headings: "How to Use Advanced Functions."
- Use a consistent font and style for headings.
- Use a sans serif (plain) font, in bold.
- Make headings task-oriented, as in "Promoting Arrays," or "Using Loops with Function Calls."
- Use appropriate graphical cues in headings, such as check marks, pointing hands, warning triangles, or note pencils.
- Make headings parallel in grammar.

Advance Organizers

Advance organizers (Figure 10.14) tell the user the structure and organization of the information that will follow. In tutorial documentation we see them in the form of statements that describe a lesson's structure. When expressed in terms of skills,

Chapter Heading	18 pt. bold non-serif	Use to indicate chapters.
Section Heading	14 pt. bold non-serif	Use to indicate sections that relate to groups of tasks.
Topic Heading	12 pt. bold non-serif	Use to indicate subgroups of tasks.
Task Name	10 pt. bold non-serif	Use to indicate task names.
Subtask name	10 pt. non-serif	Use to indicate groups of steps within tasks.

FIGURE 10.13   Headings Show Hierarchies
Headings indicate the organization and structure of information within a manual.

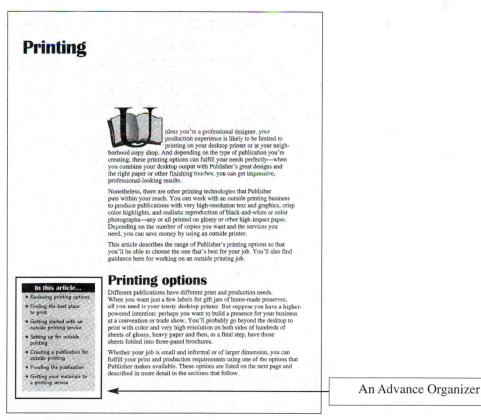

FIGURE 10.14   An Advance Organizer
This *advance organizer* shows users the organization of text and helps them understand the information.

these statements let readers know what kinds of activities they will encounter and their relevance to real-world tasks. In procedures and reference documentation advance organizers consist of introductory passages showing the organization of sections and telling how the user should read them for best advantage.

This advance organizer helps users anticipate what will occur in the chapter. Advance organizers can do any number of the following things:

- Introduce the user to the layout of the pages.
- Suggest real-world uses of the tutorial lesson.
- Introduce the focus example of the lesson.
- Tell approximate time commitments.
- Tell ways the user can structure the tutorial.
- Explain previous knowledge needed for the tutorial.
- Refer users to other documents.

A common and very effective type of advance organizer is the *graphic overview.* Often cartoon-like in design, the graphic overview uses images, arrows, menus, and icons to create an image in the reader's mind of how the elements of the program fit together.

Follow these guidelines for creating advance organizers:

- Put them in front of the information they reflect.
- Use them consistently: Users will look for them if they know they exist.
- Make them relate to user's work/task.
- Make them fit with other advanced organizers in style or design.
- Keep them short: People may not read them anyway.
- Make them functionally redundant: Don't include information readers won't get elsewhere.

Document Overview

When users open a document for the first time, they need to be introduced to its setup and how to use it to find information. Usually you perform this task using a section explaining "How to Use This Manual." Below you will find a list of the kinds of information you can include in this section.

- **Audience.** What kinds of users the manual is designed for and how it can help them use the software system.
- **Content.** What kinds of information (technical, instructional, reference, etc.) is contained and how it can be used.
- **Organization.** What comes first, second, third, and so on in a document, and how the sections can be used.
- **Scope.** What hardware and software the program works on, and what kinds of computer systems or peripheral devices (printers, monitors, etc.) are required.
- **Navigational information.** Where various user types need to go to get the information they need.

FIGURE 10.15   Parallel Headings
Parallel headings reinforce users' recognition of
information structures. These are parallel in terms
of the document structure, and they "match" in
typographic style and grammar (both verbs end
in "ing").

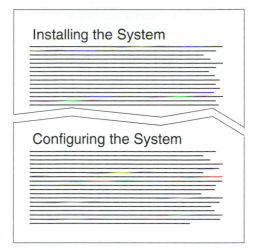

Parallel Structures

A good software manual contains useful patterns to help the user identify information easily. It repeats itself in a way that does two important things for the user. First,
parallelism reassures the user that the writer has sorted out the important information. Second, it creates patterns of expectation so the reader learns how to use the
document. Parallelism puts similar things into the same grammatical or spatial relationship. You can use many techniques for structuring information in parallel ways.
Figure 10.15 shows examples of these from the smaller elements to larger elements.

Cuing

Cuing refers to the technique of including visual patterns to make a certain kind of
information memorable. We respond to visual cues often when driving: An octago-
nal sign means stop, triangles mean caution, and so forth. Your reader needs the
same cues when navigating your document. The psychology of doing this is pretty
straightforward: If you always include warning signals in a double box, then the
reader will know to pay attention to them without having to dig the warning out of
the rubble of text.

Software documenters can use many elements as cuing devices in manuals:
icons, rules, fonts, styles, and others described below. The many cuing devices in
software manuals include what we call "notational conventions": styles of type, use
of brackets, boxes, highlighting, and so on for certain kinds of information (commands, keys, input text, etc.). On the one hand, notational conventions make up the
pattern of text the document uses; on the other hand, from a cognitive point of view,
they add to the schema of a manual and greatly increase its usability.

CUING WITH ICONS. Figure 10.16 shows simple icons that can lead readers to the information they need.

CUING WITH RULES. Cuing with rules means using solid or gray-valued lines to indicate the hierarchical structure of information in a manual. As shown in Figure 10.17,

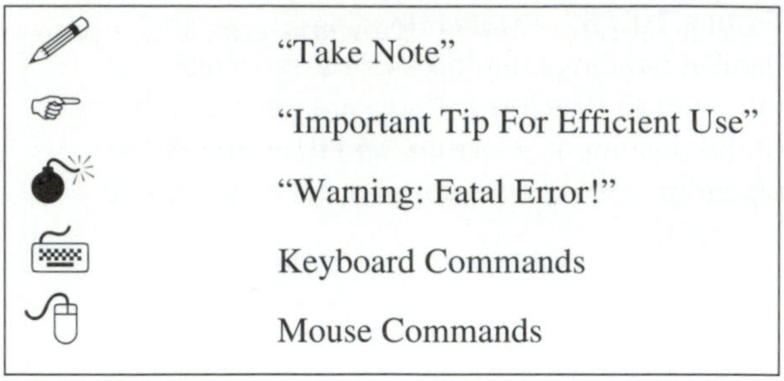

FIGURE 10.16   Cuing with Icons
Simple icons like this abound in
software documentation to help
readers identify information quickly.

✏	"Take Note"
☞	"Important Tip For Efficient Use"
💣	"Warning: Fatal Error!"
⌨	Keyboard Commands
🖱	Mouse Commands

1. To open a file, chose **Open. . .** from the **File** menu
2. To give the name of the file type `test.doc` and press <Enter>

FIGURE 10.17   Cuing with Fonts
Changing to a different font for types of information creates easy patterns
of recognition for users.

usually the thicker lines indicate higher—more general—information. You can use
rules with headings to indicate hierarchies.

Headers and footers can also contain rules that indicate, by their relative width of
line, the hierarchy of the entire document. The thinner the line, the more detailed the
information. Rules also help with layout, as we will see in Chapter 11, by helping the
reader's eye define the page.

CUING WITH FONTS. In the example in Figure 10.18, the menu commands appear in
bold and the information the user should type appears in Courier font.

CUING WITH CAPS. The use of capitalization styles is an important way to build parallel cuing structures into a document. In fact, you can choose from a number of capitalization styles which, when combined with bold letters, larger type sizes, and color,
can give you a wide range of possible patterns. But don't overdo it. The best cuing
scheme is one that does the job elegantly and simply.

Writers and companies use capital letters in many innovative ways in the software industry. Often you see capital letters in the middle of names of programs, like
"WordPerfect" or "SystemWorks." While you may not have the opportunity to take
these kinds of liberties with capitalization, you can follow these conventions:

- All caps   CHAPTER OVERVIEW
- Initial caps   Using Bookmarks, Cross-Referencing
- Down-style caps   Bookmark text, Bookmark names

In many manuals you will find initial caps used in first-level headings and down caps
in second- and third-level headings. This arrangement is logical because the number
of capital letters diminishes as the user moves into the lower levels of the text.

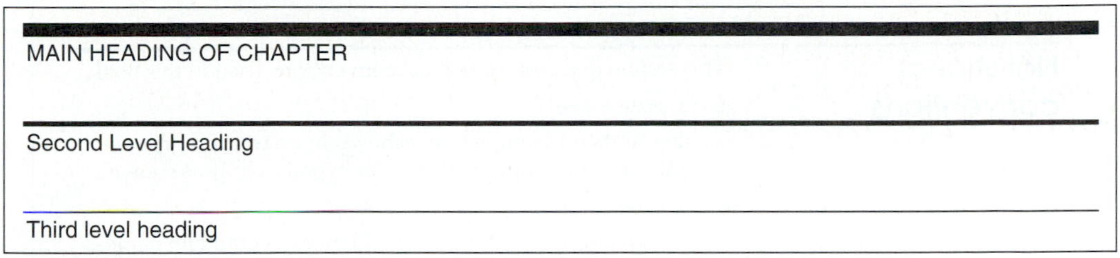

FIGURE 10.18   Cuing with Rules
Varying the thickness and style of rules adds to the ease of understanding document structure.

Figure 10.19 shows an example from a manual illustrating how you might explain the notational conventions that make up the cuing pattern you decide on for a page or screen.

Indexes and Tables of Contents

Indexes and tables of contents make up the two most important user tracking and navigational devices in any manual. Users consult them more than any other part of the manual. The table of contents describes the contents of the document from a task perspective: It is often arranged in order of typical use, and often task by task. The index, besides being the collection place for all the important terms used in a document, also contains abbreviations, synonyms, slang terms, substitute words, and user questions: all the terms the user might use, along with the terms the user should use to get at the right functions of a program.

Often a manual will present the table of contents in more than one form: an extended form including a great deal of detail for the uninitiated, and an abbreviated form for the advanced user. You can also include a table of contents at each section or chapter to remind the reader of the contents without sending him or her back to the front of the book. Its layout, as we shall see in the next chapter, uses dots or "leaders" to make it easier for the user to follow from the terms on the left to the page numbers on the right. Some newer formats for tables of contents put the page numbers next to the entries, allowing readers to connect page numbers and topics easily.

The index provides a meeting place of all the users of a program. It includes all the synonyms for program terminology that users would employ in their workplace and, through "see" and "see also" references, can put users in touch with the right information in the manuals. Users who migrate from one system to another, as in switching from one word processor to another, find the index especially useful because in a well-designed index they can find terms they know with references to terms they need to know. The index also presents a number of opportunities for the writer to use some of the cuing and parallelism tools discussed earlier in this chapter. The well-done index will contain lists of keystrokes and lists of commands and will present them in the font and type style used elsewhere in the manual. You can find out much more about building an index in the section on indexing in Chapter 14.

Lists of Figures and Tables

You may think that lists of figures and lists of tables exist in software manuals as a matter of convention. In fact, they make up a main element in the usability of a document.

Notational conventions

This section discusses notational conventions used in this book.

`Bold monospace`

In command examples, text shown in **`bold monospace`** identifies user input that must be typed exactly as shown.

`Monospace`

In paragraph text, `monospace` identifies command names, system calls, and data structures and types.

In command examples, `monospace` identifies command output, including error messages.

In command syntax diagrams, text shown in `monospace` must be typed exactly as shown.

Italic

In paragraph text, *italic* identifies new and important terms and titles of documents.

In command syntax diagrams, *italic* identifies variables that must be supplied by the user.

`{ }`

In command syntax diagrams, text surrounded by curly brackets indicate a choice. The choices available are shown inside the curly brackets and separated by the pipe (|) sign.

The following command example indicates that you can enter either a or b:

`command {a | b}`

`[ ]`

In command syntax diagrams, square brackets indicate optional data.

The following command example indicates that the variable *output_file* is optional:

`command` *input_file* `[`*output_file*`]`

...

In command syntax, horizontal ellipsis shows repetition of the preceding item(s).

The following command example indicates you can optionally specify more than one *input_file* on the command line.

`command` *input_file* `[`*input_file ...*`]`

xvi   SPP-UX System Administration Guide

FIGURE 10.19   Notational Conventions
This example shows how a manual announces the cuing patterns in the front matter of a manual under the heading "Notational Conventions."

Lists of figures and tables allow users to see quickly if they can find an example of a screen in a figure (see Figure 10.20). Similarly, many users expect to find lists of commands, syntax conventions, and procedures in tabular form (for easy reading). The list of tables, in this case, can direct users to the right information.

Lists of Screens

In some cases you should include a list of screens in your manual as an access tool for users. This list works when you have a program with easily recognizable screens. For example, university administrators use programs to track and register students. These programs contain a "student information screen," a "transcript screen," an "enrollment history screen" and so forth. Computer operators routinely call up these screens using acronyms and numbers, such as "SIS" for the student information screen. They relate to the system through these screens, as in "That program has 120 screens; dang!" Most of an operator's time gets spent calling up screens and filling out or updating fields. Where screens make up a prominent feature of a program's interface, users expect to scan a list to find a screen and information about it.

The list of screens should appear early in a manual or in the primary index screen of a help system. Format it like the table of contents or other lists, including the chapter and page numbers.

Interrelated Examples

You use interrelated examples when you follow the same example from one procedure to another. Doing so builds continuity into your document design. In documents where you use a lot of procedures, the use of interrelated examples becomes increasingly important. You should always use examples in a manual or help system because examples allow the reader to view the use of the program in terms he or she can understand. When the examples work this way we say they "contextualize" the information of the program—put it in the user's work context.

The best way to uncover realistic examples is to look for them during the process of user analysis discussed in Chapter 5. The earlier you can find them, the more valuable they can be because they increase understanding. Using interrelated examples provides the following benefits to the user and to the writer.

- **Creates a *learning curve*.** When you use aspects of the same example throughout the document, you allow the user to build on previous learning. The user doesn't have to remember the details of the example each time.

- **Ties the document together.** Using interrelated examples helps give unity to your document. Depending on the kind of program you're working on, you may have a set of data for a business or work environment that you can use from one procedure or lesson to another. You can re-use images, forms, names, and so on, to give the same reference point to different parts of the document.

- **Makes the writer's job easier.** If you follow through with the same example in the whole document, you don't have to think up new examples each time you need one.

Solutions to the Design Problem for Online Documentation

With an online help system you face the same objective as you do with print: getting the user to the right information. But you don't face the same constraints as you do with a

FIGURES

FIGURE 10.20   A List of Figures Helps Users Find Information
This list of figures, from the TC8215 Sectra Management System for Windows
User's Guide, illustrates the use of detail in a figure list to help users find the
right information.

book. While topics bear some resemblance to pages, the system for delivering topics to
users differs significantly. Help systems provide a much larger array of features—tools
for use—that the user can employ for finding information. Whereas books provide

headers, tables of contents, and so on, help systems provide these plus jumps, hypertext links, history displays, and many more tools for the user. When you design a help system, you should familiarize yourself with the access tools it provides your user. Below are descriptions of some of these tools common to help systems and how help systems differ from print documents. You'll find a more elaborate explanation of these features in Chapters 2, 3, and 4. Figure 10.21 shows an example of a help topic.

Non-Scrolling Regions

Headings differ in online documentation because you can keep them in a region of the help page that doesn't scroll. These non-scrolling regions appear at the top of the screen and stay there while the user scrolls through the procedure or topic. This ability to stay in the user's view represents an advantage over print documents because the user doesn't lose sight of the topic and thus can keep a clearer focus on the task while performing it.

Expanded Text

Expanded text, sometimes called "stretch text" allows you to embed more detail into a topic so that the user can click on the expanded text link to view the detail. In Figure 10.21 you can see that the help system makes extensive use of expandable text in two ways. The first way is to show definitions by simply "stretching" the text and inserting more words (in a green color so the user can identify the added definition). The other way to show expanded text is to include a "show more" link that fills in details under a heading. Often the "show more" link is in blue and has an open and closed icon (the tiny triangle) to indicate the state of the display.

Keyword and Whole Text Searches

Keyword searches refer to the ability of a help system to electronically find topics that the user types into a keyword search box. The system looks through all the topics and

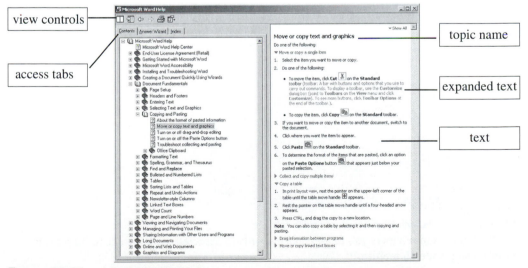

FIGURE 10.21   A Typical Help Topic

finds those that were pre-set in the system. Whole-text searches refer to the ability of a help system to find topics containing any word or combination of words the user types (Figure 10.22). You don't have to pre-set words with whole-text searches. The help driver (program) does the searching. This kind of search resembles the searching done in a library, where you indicate the subject or subject identifiers and the system coughs up the resulting topics.

How does this differ from a print index? Print indexes can contain the same information as keyword indexes, but the user has to turn pages to get from the locator in the index entry to the resulting page or procedure. Keyword searches have the advantage of ease of use over print indexes. As for whole-text searches, print documents have nothing parallel. But whole-text searches of online documents can pose

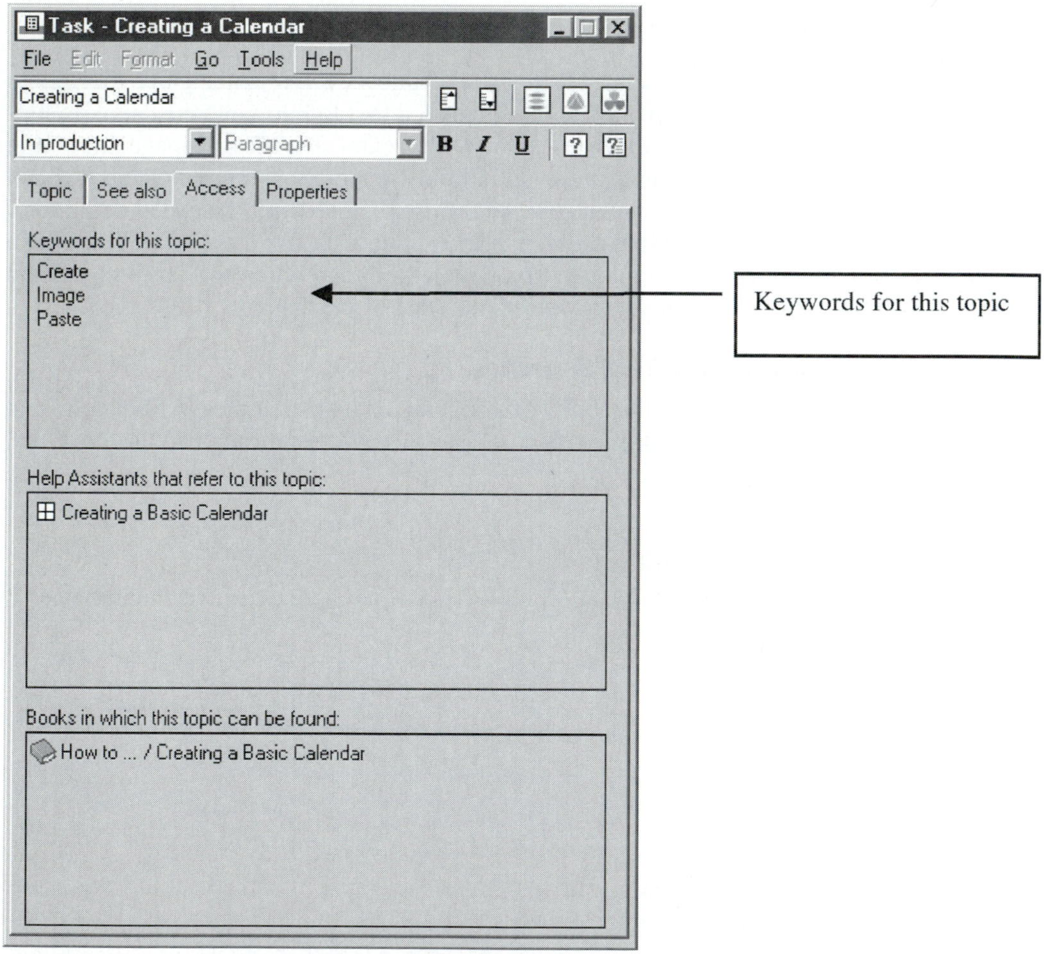

FIGURE 10.22   Associating Keywords With a Topic
An authoring program allows you to identify keywords to associate with a topic to help users access it using a search feature.

problems for users unfamiliar with doing them. If the user can't tailor the search in the right ways, the result may be a flood of irrelevant information.

Indexes

An index shows an alphabetical view of all the important topics and terminology used in a help system. Most help compilers will generate automatic indexes by creating a concordance of all the terms. The author has to identify some terms, such structural elements of sentences as articles and pronouns, for the index to ignore when presenting the list to the user.

Links and Jumps

Links and jumps in a help system allow users to go directly from one topic to a related topic. With a book you can make these references (called cross-references), but the user can only get there by turning pages. Here, online help systems have a definite advantage. Not only can you link topics as part of the design of a system, but you can allow users to go back and forth between topics.

Popups

Popups provide a way to handle glossaries in an online system. With a book you can highlight your glossary entries in the text with bolding or some other cuing technique. But to get to the glossary, the user has to turn to the page or to the glossary section in the book. In an online system the user just has to click on the term in the topic to see a window containing the definition. Another click of the mouse closes the popup window.

Context Sensitivity

With a print manual, the user goes through a chain of events between identifying a problem and finding the solution. He or she has to name the problem, then consult the table of contents or the index or thumb through the book to find the right procedure or explanation. With context sensitivity, the ability of a help system to present information based on the current state of the program, the user goes directly from a problem with a screen or a field to an appropriate help topic containing a solution. The user doesn't have to identify or name the problem because you, the writer, have already done it. You put the tags into the program (probably with the help of the programmer) that tell the program which topic to get when.

Of course, you have to put the tags into the program, using up more time in development. And after you recompile, test, edit, review, and revise, you can't guarantee that the user will get the right information after all. If the user has a question about format, and presses the help key while the print box is on the screen, the system will provide the wrong advice consistently. This example shows how the randomness of user questions limits the usefulness of context sensitivity as an online feature. Nevertheless, context sensitivity gives you a powerful and flexible tool to tailor information to the user's tasks.

Histories

History buttons allow users to trace their steps. They can pull down the history list and easily go back to a previous topic. They can even save histories and refer to them later. Users can't do this with a book, unless they make notes or leave bookmarks to retrace steps. And given that people solve problems by trial and error, using a path like this, and having the option to retrace steps, might encourage more exploration and enhance their problem solving.

Browse Sequences

When you identify a series of related topics, such as those relating to printing or to formatting, you can easily include the relationship in the form of browse sequences. When the user clicks on a topic that's part of a browse sequence, the system displays forward and backward arrows to facilitate moving from topic to topic. Unless you group your procedures or other print topics, you can't set up this kind of sequence easily with a book. Nor can you set up as many of them as you could with an online system.

Bookmarks/Annotation

With a book the user can easily mark a place in the book and then return to it—possibly using sticky notes, pieces of paper, dog-eared pages, and so forth. With early help systems you didn't have this advantage. You can't put a mark on a screen. However, newer help systems have elaborate ways to incorporate bookmarks. This is an example of online documents mimicking a feature of print documents.

The same goes for writing on pages. Books have a marvelous surface for marking on that screens don't. However, you can also do this with online systems, using annotation features. Not only can you annotate online systems, you can do other things, like collect your annotations, print them, revise and delete them, or share them with colleagues.

Glossary

authoring environment: a software program for writing online help files that provides functions for creating topics, linking related topics, assigning keywords, and other features of help.

advance organizer: a paragraph at the beginning of a section of a document that tells the user the structure and organization of the information to follow.

bleeds: illustrations or graphic elements (lines, shaded boxes) that extend to the edges of the page and remain slightly visible when the book is closed.

cognitive science: a branch of science that studies how people think and solve problems.

graphic overview: a drawing using symbols to represent the parts of a document or a program.

human factors: an area of research that studies the design of tools to fit human physical and psychological needs in the workplace.

hypertext: documents stored electronically with the capability to move from one topic to another automatically using pre-set links. The user of a hypertext document uses the mouse pointer to click on special areas of a document, which results in the system presenting another part of the document.

learning curve: the plot of progress of a user's learning, with time on the x-axis and skills on the y-axis.

media: the technology you choose to deliver instruction or other information to the user. Media usually includes print or online, but may include "multimedia," which uses sound, images, and animation on the computer screen to deliver instruction.

protocol method: a method of researching that records and then analyzes the actions a computer user makes when using a program or a manual. Sometimes the researcher asks the user to talk aloud while using the program, telling about decisions and reasons for actions. The record of these comments helps researchers understand user behavior.

rule: a horizontal or vertical line used to indicate column widths or to separate vertical columns.

☑ Checklist

Use the Document Design Planning Guide (Table 10.6) to plan and analyze your document design. In the column on the left, jot down important characteristics of your users, based on the user analysis. Then in the columns on the right, make notes to yourself about the kinds of design features you think would meet the users' special needs. The planning guide allows you to compare your users' characteristics, which you write in the left-most column, with the document design features discussed in this chapter, which are summarized for you in the three right-hand column headings. This planning guide groups the document design features into three categories: visual elements, textual elements, and structural elements. The example, in the first row, shows how you might use the planning guide for a grade calculation program. In this example, the program has two very distinct user groups (teacher users and teachers' aides). The designer has considered creating a special job performance aid to cover the teachers' aides' three basic tasks: opening the program, entering the daily grades, and closing the program. The example also shows that the user also plans an overview of how the program works to explain it to teachers unfamiliar with computerizing grade books.

Like all planning guides, this one aims to help you think. Use a pencil. Expect to revise it as you progress toward achieving just the right mix of features to tailor your document to the user's needs.

Practice/Problem Solving

1. Inventory Page and Screen Features

Identify a print or help document that you think is strong in it's content design and analyze it for the features described in this chapter. What techniques did the authors use to meet the users need to find information? What information needs do you think the user might have that the document does not meet? List both of these and then write a list of changes you might recommend for revising the document.

TABLE 10.6   Design Planning Guide

User Characteristics	Design Solutions		
	Visual Elements	**Textual Elements**	**Structural Elements**
	cues, rules, headings, labels, icons, illustrations, progress indicators, color,	layering, advance organizers, scenarios, parallel structures, hierarchies, document overviews, examples, explanations	job performance aids, organizational strategy, lists of figures, tables, screens, table of contents, index, links, browse sequences
User Groups:			
Learning, Motivational, Technical Problems:			
User's Informational Needs:			
Workplace Tasks:			
User Experience Categories:			
Users' Subject-Matter Knowledge:			
Workplace Characteristics:			
Learning Preferences:			
Usage Pattern:			

2. Explore Design Problems

Find a computer user who uses manuals daily, who knows a lot of programs, and seems to be a "computer person." Also, find a computer user who is just beginning or who isn't a "computer person." Discuss their use of computer documentation, and prepare a short report comparing how they differ. How do their different experiences affect the way they see and use documentation? What design problems/implications does this have for the software documenter?

CHAPTER 11

Laying Out Pages and Screens

In this chapter we examine the two main elements of document layout: the design of the screens and pages and the design of type. Seeing pages and screens from the designer's point of view, as *communication spaces,* means that you acknowledge the degrees of *modularity* and structure in pages and screens, and you learn how to look at their density, balance, and legibility. It also means that you follow a design process for creating pages that communicate the program interface information effectively. Common page formats—one-, two-, and three-column—use page design in different ways to achieve usability goals. We also see how to use many elements of design— the parts that make up the structure of a page—to help the user find information quickly. Practice using two designer's tools, the *page grid* and the *thumbnail sketch,* can help you become a good page and screen designer.

Designing the type for manuals and online help means determining the size, font, and style of the letters used to make words. The goals of the designer in using type consist of helping readers recognize words and building a pattern of information that allows readers to understand and navigate the document easily.

How to Read This Chapter

All readers will benefit from a study of the differences between the examples in this chapter and those in Chapter 10, "Designing for Task Orientation." These two chapters complement one another and contain important principles on which to build a successful task-oriented document. Guideline 4 gives specific advice for planners, and the Discussion section contains many examples on which to base further discussion or design decisions.

Examples

The examples in Figure 11.1, Figure 11.2, and Figure 11.3 offer you, in different ways, overviews of major trends in page and screen design. Figure 11.1 uses thumbnail sketches to outline five basic manual page layouts and one Windows-based help

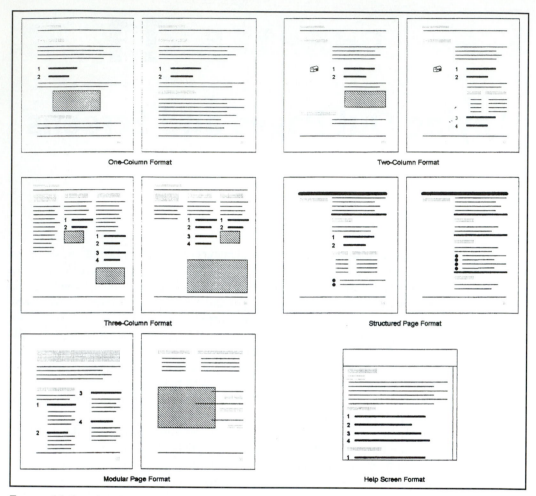

FIGURE 11.1   Thumbnail Sketches of Common Formats
These sketches illustrate the variety of formats available for page and screen designs.

screen format. Figure 11.2 focuses on an example of the two-column format, show-ing how the physical elements of the page combine to ease the reading experience and lead to workplace solutions with the software. You may want to compare this example to the one in Figure 10.1. Both figures show the same page, but the callouts in Figure 11.2 draw attention to elements of page design instead of document design. Figure 11.3 carries the concepts of page design into screen design, showing points of similarity and difference in basic guidance documentation.

Guidelines

Figure 11.4 lists the guidelines for designing pages and screens.

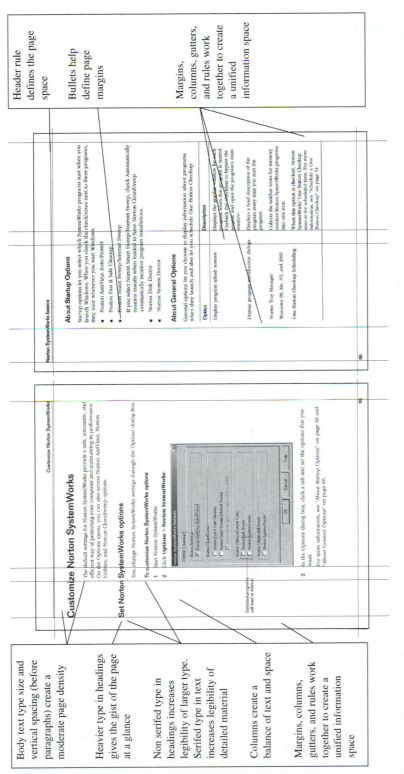

Customize Norton SystemWorks

The default settings for Norton SystemWorks provide a safe, automatic, and efficient way of protecting your computer and maintaining its performance. On the Options menu, you can also access Norton AntiVirus, Norton Utilities, and Norton CleanSweep options.

Set Norton SystemWorks options

You change Norton SystemWorks settings through the Options dialog box.

To customize Norton SystemWorks options

1 Start Norton SystemWorks.
2 Click **Options > Norton SystemWorks**.

Selected programs will load at startup

3 In the Options dialog box, click a tab and set the options that you want.

For more information, see "About Startup Options" on page 66 and "About General Options" on page 66.

65

About Startup Options

Startup options let you select which SystemWorks programs start when you launch Windows. When you check the checkboxes next to these programs, they load whenever you start Windows:

- Norton AntiVirus Auto-Protect
- Norton Fast & Safe Cleanup
- Norton Smart Sweep/Internet Sweep

If you select Norton Smart Sweep/Internet Sweep, check Automatically monitor installs when loaded to have Norton CleanSweep automatically monitor program installations.

- Norton Disk Doctor
- Norton System Doctor

About General Options

General options let you choose to display information about programs when they launch and also let you schedule One Button Checkup.

Option	Description
Display program splash screens	Displays the startup window for each program when that program is started. Uncheck this checkbox to bypass the splash and open the program's main window.
Display program introduction dialogs	Displays a brief description of the program every time you start the program.
Norton Tray Manager	Collects the taskbar icons for memory resident Norton SystemWorks programs into one icon.
Windows 98, Me, NT, and 2000	
One Button Checkup Scheduling	When this option is checked, Norton SystemWorks' One Button Checkup runs at the scheduled time. For more information, see "Schedule a One Button Checkup" on page 51.

66

Header rule defines the page space

Bullets help define page margins

Margins, columns, gutters, and rules work together to create a unified information space

Body text type size and vertical spacing (before paragraphs) create a moderate page density

Heavier type in headings gives the gist of the page at a glance

Non serifed type in headings increases legibility of larger type. Serifed type in text increases legibility of detailed material

Columns create a balance of text and space

Margins, columns, gutters, and rules work together to create a unified information space

FIGURE 11.2 A Well-Designed Page
This page from Norton System Works *User's Guide* shows how elements of page layout lead to a unified communication space.

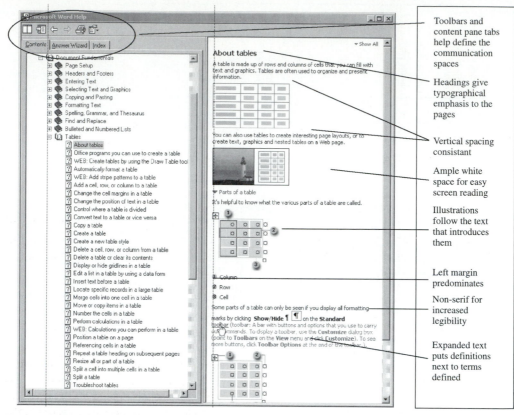

Toolbars and
content pane tabs
help define the
communication
spaces

Headings give
typographical
emphasis to the
pages

Vertical spacing
consistant

Ample white
space for easy
screen reading

Illustrations
follow the text
that introduces
them

Left margin
predominates

Non-serif for
increased
legibility

Expanded text
puts definitions
next to terms
defined

FIGURE 11.3   A Well-Designed Screen
This page from Microsoft Word Help shows how elements of screen layout lead to a unified
communication space.

1   Review the User Analysis

As you did your planning for your documentation project (See Chapter 6, "Planning and
Writing Your Documents,") you identified, based on your user analysis, what you thought
your pages would look like. You may have articulated these page designs in your documentation plan in terms of type size, fonts, page layout and so on. It may be that you are
rewriting an existing manual that already has an established page layout, or that you are
following guidelines for page design dictated by your department's style guidelines. In
these cases your contribution to the overall design of pages is going to be reduced because
you can't make sweeping changes. On the other hand, knowing how established layout
and format guidelines work (which also is part of this chapter) can help you understand
how the format you're using works. In either case, the material in this chapter is meant to
extend your ideas on page design and give you tools for further refinement of those ideas.

Document layout basically falls into two kinds of activities: designing page layout
and designing type. Designing page layout means determining the best arrangement of
words and images on a page or screen to achieve maximum usability. Designing type
means determining the proper font, size, and style of characters as well as determining

1. Review the user analysis.

2. Create page grids.

3. Define the page grid using styles.

4. Draw thumbnail sketches.

5. Set up pages and styles in your work processor.

6. Determine the layout of help documents.

FIGURE 11.4   Guidelines for Designing Pages and Screens

the format for tables, lists, and paragraphs. As in the overall design of the document discussed in the previous chapter, constraints complicate the design process: constraints of user types and experience, of the user's problem-solving techniques, and of the documenter's resources. You must account for these constraints in any intelligent design of page and screen layout to achieve your design goals.

The Goals of Page and Screen Layout

The goals of layout, again, resemble those of document design in general. For one thing, the layout should allow the user to overcome the design problem; the layout should meet dynamic needs with a static document. (And even though a help document seems less static than a page, it differs from the manual only in degree of being static.) The document should support task orientation by helping the user perform information-related tasks efficiently and productively. The well-designed page or screen should lead users to the right kinds of information. This doesn't just mean helping them not get lost; it means making it easy for them to move from a problem in their work to a solution in the manual or help.

For its second goal, then, good page and screen layout should support overall task orientation. In some manuals, for instance, this means that the task becomes the primary unit of information, as opposed to the older style of making the function or command (a system-oriented element) the main unit of information. In addition, the well-dressed page or screen should bear some consistency with the interface of the program itself. This means using screens and *icons* to help the user make the connection between page or screen and program.

Finally, the layout should accommodate the visual needs of the user, the need to learn and do through images rather than words. Our knowledge of how people perform real tasks tells us that speed increases when the reader can see the steps and see the results of steps, instead of reading about them. While words insure accuracy, more and more the look of a page or a screen—the overall impression created by the size and placement of text and images—determines its effectiveness.

2 Create Page Grids

The user analysis creates a kind of model, telling the characteristics of the users in a way that allows you to design documents (using the software program as content) to help the user perform meaningful, automated work. The overall content design process is covered in Chapter 10. The model of your page design first emerges as a page grid. Page grids define communication space by drawing invisible "fences" around the areas of a page. An example of these invisible lines is shown in Figure 11.2 and Figure 11.3. In these figures the lines that define the page grid have been penciled in so that you can see how they work. A page grid also acts like a *scaffold* or *framework* onto which you put text and graphics. The grid resembles a rack, a kind of shoe tree for words and pictures, or a word and picture parking lot with the spaces falling within the invisible grid lines. Figure 11.1 shows you a number of page grids that you can start out with for your pages.

To design a page well you need to know about grid lines and the other parts of page grid:

- **Grid lines.** Lines drawn where the page and column margins would fall
- **Margins.** Areas of actual space between the text and paper's edge
- **Columns.** Spaces between the grid lines marking columns
- **Gutters.** Space between columns
- **White space.** Space, inside the margins, where no text or pictures may go
- **Baseline.** Grid line at the bottom of the text and graphics area that defines the bottom margin

Making a page grid forces you to see the page in abstract terms. Page grids show us columns, text areas, graphics areas, margins, and heading areas—but only in a general way. When you compare one page grid design to another, you can see how some are better suited for highly structured pages, others not. When you create a page grid for a manual, you see the arrangement of the page more clearly, and in advance. If you want to get good at seeing the underlying page grids of all pages, practice drawing thumbnail sketches (see Guideline 4).

3   Define the Page Grid Using Styles

When you have identified a grid for your pages and screens, you've identified the basic pattern you will follow throughout your manuals and online help systems. Some pages will look different from the grid pages: Tables of contents pages and index pages, for instance, may not stick to the columns you define for other pages. But they will stay the same size and use the same header and footer scheme you identify in your page grid. Remember that a page grid provides the road map for almost all your pages.

Once you have decided on the grid for your pages, you should identify the styles you want to use to set up the pages. As a rule, your styles for page and screen components should include all the components listed in Table 11.1.

Notice that for screen presentations you don't identify margins in a one-column format because the margins (usually about 1/8") get set by the variable-sized window in which the help screen appears. Notice also that the font for the body text is a serifed font (Times Roman) for printed documentation and a non-serifed font (Arial) for screen presentation, because the non-serifed font is more easily read on a screen. The rest of this chapter discusses these and other concerns in the difference between print and online text.

TABLE 11.1   Styles for Page and Screen Components

Component	Unit(s) of Measurement	Example: RoboHELP 2.0 *User's Manual* and *Help*
Page:		
Top and bottom margins	Inches from the edge of the page	Top 1", bottom 3.5"
Left and right margins	Inches from the edge of the page	Left .5", right 2"
Column margins	Width in inches	Left .75", right 4.75"
Gutters between columns	Width in inches	.25"
Line spacing above	Points	6 points
Line spacing below	Points	None
Icon caption	Font, size, style, capitalization	Arial, 10pt, italics, title case
Page number	Font, size, style, justification	Arial, 12pt, bold, right justified
Tabs	Inches	.25 inch
Body text	Font, size, style, justification	Times Roman, 10pt, plain, left justified
Headings	Font, size, style, capitalization	Main: Arial, 14pt, bold, upcaps Second: Arial, 12pt, bold, upcaps Third: Arial, 10pt, bold/underlined, sentence caps
Rules	Color, point size, length	Black, 1.5 pt, 5.5"
Bullets	Symbol, point size, indentation	Circled numbers, 12pt, .25" indent
Numbers	Style, point size, punctuation	Arabic, 12pt, followed by a period
Step	Font, size, style, punctuation	Times Roman, 10pt, plain, sentence caps
Screen:		
Columns	Width in inches	One-column, variable
Headings	Font, size, style, capitalization	Main: Arial, 14pt, bold, upcaps Second: Arial, 14pt, plain, sentence caps
Step	Font, size, style, punctuation, justification	Arial, 12pt, plain, sentence caps, flush left
Body text	Font, size, style, justification	Arial, 12pt, plain, left justified
Bullets	Symbol, point size, indentation	Solid, 12pt, .5" indentation
Line spacing above	Points	6 points

■4■ Draw Thumbnail Sketches

Like a diagram of a building or a football field, a thumbnail sketch uses lines and spaces to show how pages get organized. You would make thumbnail sketches as part of your planning effort for your publications department or to experiment with different plans for your company's style manual. As a student of software documentation, making thumbnail sketches helps you sharpen your eye for effective page designs that encourage usability. The following section will guide you through the process of making your first sketch.

Pick a page that you think exhibits good page design. Does it have just the right balance for you, or just the right degree of formality you like? Did you find it in a manual that clearly succeeds in helping readers use a system? Once you have chosen a representative page to draw, follow the steps below to draw a sketch of it. At the end, compare your page to the original to see if it has the same balance and look as your sketch. These instructions focus on drawing on a section of a regular page, but you can use a draw program for this exercise if you want.

1. Fold a piece of paper in half, then into quarters.

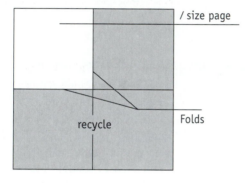

Keep the page proportions the same when you reduce the page size. Reducing the page proportionally forces you to draw a more economical composition. Smaller also works better for doing thumbnail sketches, because the well-done sketch abstracts and shrinks the original. You should try to spot the general elements of page design and represent them in your diagram. So look for the forest, not the trees.

2. Draw lightly around one quarter to mark the edges of the quarter-size page.

You actually have four smaller pages. Pick any one you want to draw in. If you make a mess out of the first square, start again in a different one.

3. Based on the original page you chose as your model, draw grid lines for the margins.

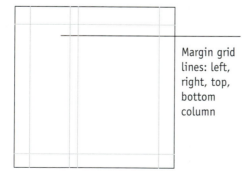

Grid lines should be accurate but light, so hold your pencil a little away from the tip and don't bear down. Keep the lines straight. Draw them all the way to the edge of the paper, so that they cross one another. The space defined by the top and bottom margins should be the same proportion as the original.

4. Using the original as a guide, draw grid lines for columns.

Again, draw these lines lightly but accurately. They too go all the way to the edge of the paper. If you have a two-column format, one column would be the graphics column, another the text column.

5. Sketch in the page features.

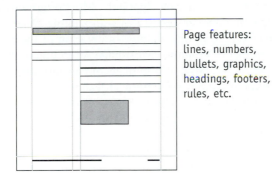

Page features: lines, numbers, bullets, graphics, headings, footers, rules, etc.

As you sketch in the page features, you should try to dither them. Dithering means replacing each and every page element with some symbol: a line or a box, or a shaded area. A landscape seen at a distance by a near-sighted person will get dithered by the eye's inability to focus. If you imagine seeing your page from, say, twenty feet away, the length of a room, you can imagine what it would look like dithered. Try to make your thumbnail sketch look like the page seen from twenty feet away.

Sketch the page features using the guidelines below:

- **Text.** Draw straight lines for all body text lines. Use a ruler if you want, but if you don't have one, learn to draw lines *sketch-straight:* Put your pencil on the starting point, look at the ending point and, using short, straight pencil movements (about 1/2 inch each), work your pencil across the page to the ending point. The resulting line will look straight and will not have a tendency to drift up or down. This technique takes practice.

- **Graphics.** Draw all the graphics using shadows, abstract sketches, and circles and lines. Draw tables and lists using lines to represent text. Give the scribbles in your thumbnail graphics the same relative darkness or lightness that they have in the original.

- **Headings.** Headings usually appear in larger fonts than body text, so draw them as shaded rectangles, again using their relative size in the original as your guide.

- *Rules,* **boxes, other features.** In your sketch, you may need to draw rectangles around rules to give them the same value on your sketch as they have on the page. Make sure you include all the graphics and text in the headers and footers.

TIPS FOR DRAWING THUMBNAIL SKETCHES

1. Drawing a thumbnail sketch may take from ten to twenty minutes. Slow down, and make it accurate.

2. Keep the page items in proportion while trying to include everything that is in the original.

3. Keep the values of darkness, density, lightness, and spaciousness the same in your sketch as in the original.

Drawing thumbnail sketches will help you identify the layout of your pages and screens. You should identify grids for all the kinds of pages you want to use in your document.

5 Set Up Pages and Styles in Your Word Processor

Once you have identified your styles and written them down, you can set them up in your word processor or desktop publishing program. You handle the *specifications* for pages in two ways: styles for the text and page setup. Depending on your word processor, you will do these either together or separately. Most programs do the layout of the pages with one function and set text styles with another. When laying out pages, you tell the program the margin dimensions, orientation of the pages (landscape or portrait), and the number of columns. Most page setup functions allow you to arrange your pages in a *mirrored page* or two-page spread fashion, each pair of pages having a right-hand and left-hand page. Figure 11.5 shows the page layout dialog box used in the Microsoft Publisher program.

You should set up your page as you did in your planning with Guideline 2, with the columns and margins identified and set on the page as you want them in your manual. Usually you will use a different word processing function to set up the styles that define your text. Find whatever function in your word processor that defines styles and use it. A style defines the format you specify for a page or screen element (such as body text, headings, etc.), stores your choices, and applies them, when you want, to any text you type. You might ask yourself, "Why should I go to all this trouble and not just set the styles as I type in the words?" Good question. Basically, setting styles saves you time in the long run for two reasons.

- You can change the styles later, and you don't have to change each instance of a certain text. Thus, if you get into the project and decide that you want all your steps to print in boldface, you only have to change the definition of the step style. Voila: all the steps you have formatted using the step style change to boldface.

- Setting up styles insures consistency in your document. Unless you have a perfect memory, you will end up with a patchwork of formats by the time you get to the end of the document. I learned (repeatedly) the hard way by setting different formats throughout a document because I forgot how I did it in earlier pages. The result costs money later in editing for consistency, and making a lot of changes to get things straight.

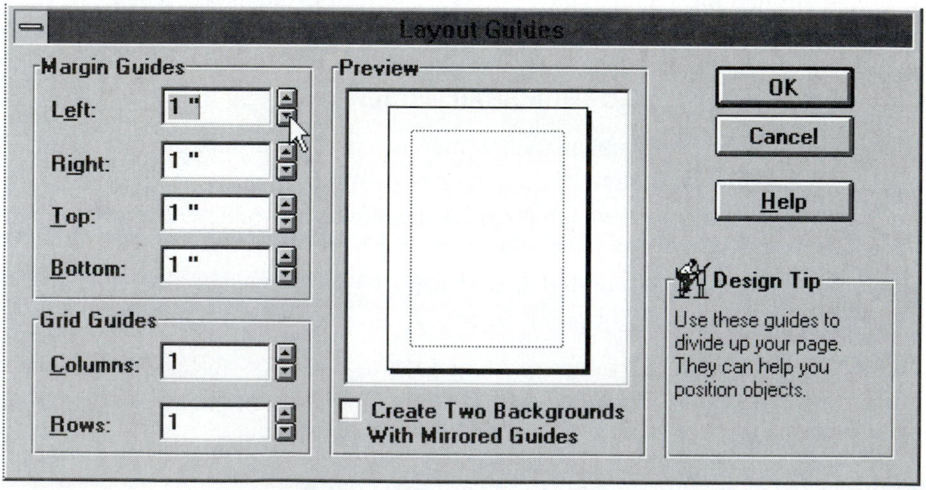

FIGURE 11.5   A Page Layout Dialog Box
Use a dialog box like this one from Microsoft Publisher to set up your page grid.

Figure 11.6 shows a list of styles I defined once for a manual for a database program. I used WordPerfect as the word processor. Each style contained the point size, margin dimensions, and other information to set the pages up just as I wanted them. Once you've got your styles set up in your word processor, you should use them to write up a few procedures and have them reviewed by the development team or your client. These pages will show the layout and allow the other team members to suggest changes or improvements. Make these changes at this stage to avoid having to make them later.

6 Determine the Layout of Help Documents

Styles in screens differ significantly from pages, in that the screen resolution is usually much less than that of printed pages. For this reason the number of fonts that work on a screen is much smaller than those that will work on a page. Also, the screen size is variable so styles are more simplified to accommodate screen adjustments. Finally, in terms of layout, screens offer an array of features that you can use to create a usable and intuitive design. Among these features are: frames, narration strategies, hypertext

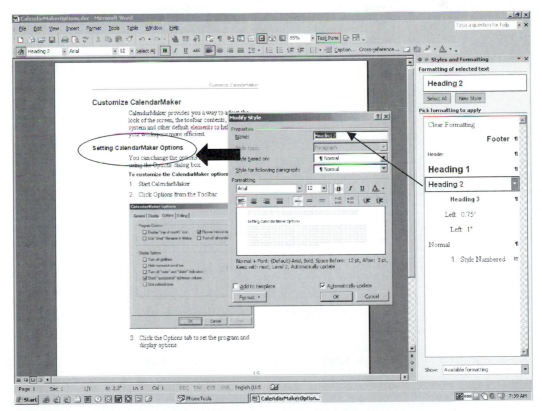

FIGURE 11.6   An Example of Using Styles to Format a Manual
This figure shows the styles in a task pane on the right, and the Modify Style dialog box used to change the settings for a particular style.

links, and image maps (images with regions that link to other documents). Because of these different features, users approach screens in ways they don't with pages.

User behavior in searching for information on a screen is very different from that of page readers. The main difference is the user's tendency to browse. Software users looking through a web help document or in a help window are looking with both their eyes and their mouse pointers. These tools work together as users are drawn using their mouse to link text (often in blue for the web or green for help documents) and with their eyes to headings and images. As they "hunt" for information they will often follow a trail of hypertext links until they reach the procedure they're looking for. So the layout of screens needs to accommodate browsing.

The constraints of pages and screens result from the various sizes of pages and print capabilities, the user's type of display, and other technical elements. So you have to examine these limitations to determine the differences between putting the same steps in different formats. To help you overcome some of these limitations, consider the list of layout elements for pages and screens shown in Table 11.2.

TABLE 11.2   Differences in Layout Elements for Pages versus Screens

Element	Page	Screen
Headings	Horizontal spacing by line proportions	Horizontal spacing by blank lines
Fonts: general	More flexibility in using fancy fonts with wide stroke variations	Less flexibility: simple and bold work best
Body type size	Smallest type size: 8 point	Smallest type size: 10 point
Body text	Unlimited range of fonts	Font range limited by the user's computer
Page length	Limited to the edges of the paper	Unlimited because of scrolling ability
White space	Use to balance text, create "soft" boundaries	Use to decrease clutter
Indentation	More because of larger communication space: more hierarchies	Less and for smaller increments because of the smaller communication space: fewer hierarchies
Cuing	Greater detail in small cuing graphics like icons and dingbats	Detail limited by the graininess and decreased legibility on the screen
Lists their	Use as much as possible	Use as much as possible, but limit length
Graphics	Greater detail	Less detail
Icons	Greater detail	Less detail
Line length	Between 4" and 4.5"	Variable by re-sizing the help window
Rules	Wide variety possible	Less variety possible
Margins	Larger, up to 1" all around	Smaller, usually ¼"

Discussion

Designing Communication Spaces

The documenter needs to decide two important things: the degree of modularity pages need, and the degree of structure they need. These two elements will determine the overall look of the communication space, regardless of the pattern of columns and words (discussed below) you choose. Follow these important principles because they bear directly on how well one puts a task orientation into layout.

Degree of Modularity

Modularity means breaking information into chunks of text and graphic units and fitting them onto a one- or two-page spread. Using a modular format, the writer would follow the one-task-per-page idea, being sure to include at least one image per task. Usually the graphic consists of a result screen (showing what happens when the user follows the steps). Chapters and sections do little to make a page or screen modular. Modules, stand-alone elements of information, each contain just the information needed to complete the task, no more no less.

MODULARITY AND PAGES.   You can gauge the degree of modularity of a page or screen by asking this question: Does this communication space contain all the information the user needs to perform the task and understand the concepts in the task? For instance, a task that presents steps on one page but refers to screen captures on another page (or in a list of figures) does not give all the information needed at the time. Keeping tasks self-contained increases modularity. Often this means doing the following.

- Repeat background information where necessary.
- Repeat screens when necessary.
- Include orienting information about the relationship of a task to other tasks.
- Keep all relevant steps on the same page.
- Minimize cross-references.

Edmond Weiss, in his book *How to Write Usable User Documentation,* champions the cause of modularity.[1] He points out that modules are functional (based on tasks the user performs), independent (don't rely on the reader having read previous modules), and small (fit on a page and are easy for the user to hold in short-term memory). Weiss uses compelling logic in promoting the modular concept. He sees the following advantages of the modular-format idea:

1. Text is kept with images so the user avoids having to turn the page to see the screen that relates to what is being read.
2. The act of designing modules helps the writer structure the writing process, from the research phase to the editing phase.
3. The modular concept is task-oriented and thus helps the user in the work environment.

You can easily see why modular design would help address the documentation problem. Greater ease of access unlocks the information. Theoretically, modularity accommodates both the experienced and advanced user because each may select only those modules that solve a particular problem. Similarly, different user role types—managers, system administrators, users—can easily choose the modules they need. But beyond these advantages, the modular document provides the writer with the ideal format for working with elements of the task analysis. Whether the writer creates documentation to teach, to guide, or to support the reader, the tasks identified in the task analysis seem naturally to fall into the modular format.

The tradeoffs of modularity lie in the costs associated with producing modular documents. Because some modules can take less than two pages, the resulting manual often has empty spaces; the next task starts on the next page. To publications managers or project sponsors the bottom line is often the total page count and the cost per page. Their main concern is keeping the manual production costs within budget. A writer working under these constraints may need to modify the modular design, keeping the concept of task orientation, but filling pages as much as possible.

MODULARITY AND SCREENS. You should also note that modularity has less and less to do with online help systems. Because of the physical constraints of a page, you have to put all the necessary information in one space; otherwise the user has to go to another page. With a help system, however, you can overcome this problem with things like popup windows and *expandable text.* For example, in a *page* telling how to put interactive fields into a form you would have to define the term field for the user. In a *help topic,* you would simply include a link to the glossary entry or put the definition in a *roll over,* and the user who didn't know the term could simply click or position the mouse pointer to get the definition. Because of online help's ability to link information, help documents can segregate information more clearly. For this reason you see help systems with many different parts. The parts simply represent information categories, not physical locations as with pages.

Degree of Structure

Structure in page design means that we place the information on the page according to patterns, with certain kinds of information only in certain places. This process reserves certain areas of the page or screen for certain kinds of information. As an example, imagine a section of farm land: Some parts are used for grazing cattle, some for growing wheat, some for the farmhouse and barn, some left alone for woods and a pond. Similarly, the structured page has certain areas for headings, certain areas for overviews, others for screens. Highly structured pages use gardens of *bulleted* lists, tables of commands, and indented margins for steps, cautions, warnings. Highly structured pages also use fence-like vertical and horizontal lines, called *rules,* to separate and help the reader keep track of information on the page.

When you structure writing on a page, you must develop the knack of breaking down information into types. Often called *chunking,* this technique helps the reader identify what kinds of information the page presents. Helped by headings the user (ideally) can quickly learn where to look on a page to get what. This idea makes sense from the point of view of information processing. Researchers have determined that readers locate information in computer manuals (in fact in all documents) by remembering the physical location of information on the page, rather than the more abstract terms of chapter or section numbers. Additionally, the use of lists and tables gives the user the option to look over a list to select the appropriate function. In this way, the dynamic needs of the user may mesh more often with the static organization of a manual or help.

How much structure you build into your pages and screens depends on the degree of clutter you will allow and on the amount of white space you will need in order to balance the text items. Indeed, as Figure 11.7 shows, the structured method packs a lot of usable information into a small space, because all the format conventions help keep things in order. As Figure 11.7 also shows, a less-structured page, illustrated on the left, also shows efficiencies for its purpose. Uninterrupted paragraphs make a suitable format for reading background information. Elements that contribute to structure in your pages include the following:

- **Rules.** Various lengths and thicknesses help the user tell the reading area from the heading or scanning area.
- **White space** (or quiet space in screens). Helps the user focus on page elements such as graphics without having to process impinging information. Helps create a balanced page or screen.
- **Bullets.** Help the user identify the kind and organization of information at a glance
- **Chunks.** Help the user identify reading information in overviews and elaborations.

How to Look at Pages and Screens

To learn how to design pages for a software manual, you must first learn how to look at pages. You should always give a manual the flip test and register an impression of the overall layout of the book. In your study of page layout you should make a point to browse through computer manuals found at computer stores or on your shelves. The best place to get a good idea of different page layouts is a software company because often these companies purchase and use a wide variety of software. Many businesses that use software keep shelves of program packages that you can learn from. In studying layout, try to develop an eye for the following elements of page design.

- **Page density.** Comparing the pages of one manual to another, which seems darker, or heavier, or more crowded with text? Which seems lighter, using white space as a soft barrier between kinds of information?
- **Balance.** Compare how one manual balances white space and text space with one another. Do some pages seem top-heavy, bottom-heavy, right- or left-heavy? Well-balanced pages have a unified effect.
- **Legibility.** Compare the ease of reading of the type font and style among manuals which you think read clearly. Some combinations of fonts seem to work well together; others don't.

Experience will help you develop a feel for pages; you should also try to develop a sense of how the values listed above—density, balance, and legibility—get determined by and controlled by page elements like margins, columns, and images (screens, rules, and icons). The next section will discuss how these elements contribute to the overall design of pages.

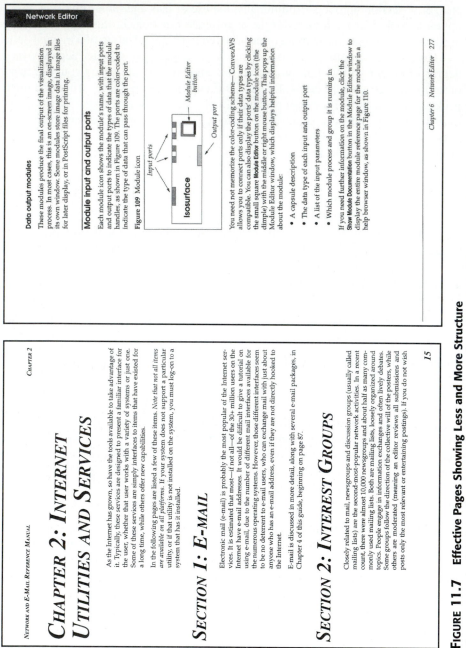

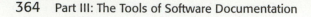

Network Editor

Data output modules

These modules produce the final output of the visualization process. In most cases, this is an on-screen image, displayed in its own window. Some modules store image data in image files for later display, or in PostScript files for printing.

Module input and output ports

Each module icon shows the module's name, with input ports and output ports to indicate the types of data that the module handles, as shown in Figure 109. The ports are color-coded to indicate the type of data that can pass through the port.

Figure 109 Module icon

Input ports
Isosurface
Output port
Module Editor button

You need not memorize the color-coding scheme—ConvexAVS allows you to connect ports only if their data types are compatible. You can also display the ports' data types by clicking the small square Module Editor button on the module icon (the dimple) with the middle or right mouse button. This pops up the Module Editor window, which displays helpful information about the module:

• A capsule description
• The data type of each input and output port
• A list of the input parameters
• Which module process and group it is running in

If you need further information on the module, click the Show Module Documentation button in the Module Editor window to display the entire module reference page for the module in a help browser window, as shown in Figure 110.

Chapter 6   Network Editor   277

NETWORK AND E-MAIL REFERENCE MANUAL   CHAPTER 2

CHAPTER 2: INTERNET UTILITIES AND SERVICES

As the Internet has grown, so have the tools available to take advantage of it. Typically, these services are designed to present a familiar interface for the user, whether that user works with a variety of systems or just one. Some of these services are simply interfaces to items that have existed for a long time, while others offer new capabilities.

In the following pages are listed a few of these items. *Note that not all items are available on all platforms.* If your system does not support a particular utility, or if that utility is not installed on the system, you must log-on to a system that has it installed.

SECTION 1: E-MAIL

Electronic mail (e-mail) is probably the most popular of the Internet services. It is estimated that most—if not all—of the 30+ million users on the Internet have e-mail addresses. It would be difficult to give a tutorial on using e-mail, due to the number of different mail interfaces available for the numerous operating systems. However, those different interfaces seem to be no deterrent to e-mail users, who can exchange mail with just about anyone who has an e-mail address, even if they are not directly hooked to the Internet.

E-mail is discussed in more detail, along with several e-mail packages, in Chapter 4 of this guide, beginning on page 87.

SECTION 2: INTEREST GROUPS

Closely related to mail, newsgroups and discussion groups (usually called mailing lists) are the second-most-popular network activities. In a recent count, there were almost 10,000 newsgroups and about half as many commonly used mailing lists. Both are mailing lists, loosely organized around topics. People engage in information exchanges and often lively debates. Some groups follow the direction of the collective will of the posters, while others are moderated (meaning an editor reviews all submissions and posts only the most relevant or entertaining postings). If you do not wish

15

FIGURE 11.7   Effective Pages Showing Less and More Structure

The page on the left from the Texas Tech University *Network and E-Mail Reference Guide* has less structure, to accommodate reading of background information. The page on the right from the Convex AVS Operating System *User's Guide* has more structure, to accommodate scanning and selective reading.

Common Page Designs

Many designs used in software manuals incorporate the concepts of modularity and structure to varying degrees. Often we discuss pages in terms of the two-page spread: the left, or even, page and the right, or odd page. Talking about and viewing pages in this way helps the designer get a feel for the binding, which joins the pages in the middle. All designs leave about one-half inch on the binding edge to accommodate spirals and loose-leaf rings. Page layout also has to do with the number of columns.

Screen design needs to accommodate many of the same elements as pages: columns, headings, text, and graphics. But screens also need to arrange space for the non-scrolling region of the screen: the part that contains the topic name. The page designs described below by no means make up all the designs you can have for the page. But they do represent some very popular ones, and studying them can get you started on the design of pages for your projects.

Two-Column Format

Most software manuals today display a two-column format, shown in Figure 11.8. Most two-column formats have a graphics column (an area reserved for screens, icons, diagrams, and headings) and a text column (an area reserved for explanations, steps, notes, cautions, tables, and screens). Most patterns follow this basic design. Many such pages contain rules that either separate the graphics from the text column, or separate task names or other headings in the text column.

The graphics column contains icons and headings: the signposts for the user to the explanation of the task he or she needs to perform. The text column contains the explanation, in the form of steps, lists, definitions, explanations, and lower-order headings. The writer arranges the words in bulleted lists or numbers (for steps) and keeps the text width to about 4 1/2 inches (or one and a half alphabets). Some variations of this format have the graphics column very narrow, and others have it wider in order to include features such as tips or illustrations.

ADVANTAGES AND DISADVANTAGES OF THE TWO-COLUMN FORMAT. The two-column format allows the reader to distinguish easily between guidance information and support information. Guidance consists of those elements on a more general level—icons, headings—with which to navigate the document. At points where the user needs more precise information—steps for actual performance, explanations of commands—he or she may stop browsing and read carefully in the text column. The impatient reader can easily pick up the pattern of general to specific—the most common and preferred pattern of all information.

The two-column format works best with guidance-level documentation: procedures, step-by-step, installation, getting started. It does a good job where readers read selectively, when they read to do. But on a space-to-unit of information ratio, the two-column format uses more space per information unit than dense one-column format. For sustained reading, say of background information, use a one- column format.

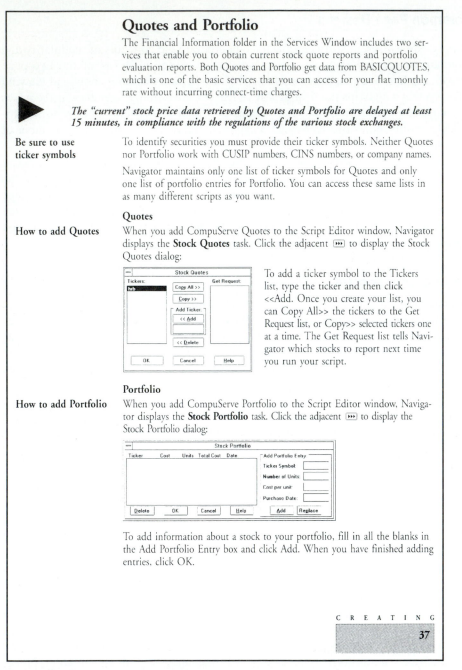

Quotes and Portfolio

The Financial Information folder in the Services Window includes two services that enable you to obtain current stock quote reports and portfolio evaluation reports. Both Quotes and Portfolio get data from BASICQUOTES, which is one of the basic services that you can access for your flat monthly rate without incurring connect-time charges.

▶ *The "current" stock price data retrieved by Quotes and Portfolio are delayed at least 15 minutes, in compliance with the regulations of the various stock exchanges.*

Be sure to use ticker symbols

To identify securities you must provide their ticker symbols. Neither Quotes nor Portfolio work with CUSIP numbers, CINS numbers, or company names.

Navigator maintains only one list of ticker symbols for Quotes and only one list of portfolio entries for Portfolio. You can access these same lists in as many different scripts as you want.

Quotes

How to add Quotes

When you add CompuServe Quotes to the Script Editor window, Navigator displays the **Stock Quotes** task. Click the adjacent ▸▸ to display the Stock Quotes dialog:

To add a ticker symbol to the Tickers list, type the ticker and then click <<Add. Once you create your list, you can Copy All>> the tickers to the Get Request list, or Copy>> selected tickers one at a time. The Get Request list tells Navigator which stocks to report next time you run your script.

Portfolio

How to add Portfolio

When you add CompuServe Portfolio to the Script Editor window, Navigator displays the **Stock Portfolio** task. Click the adjacent ▸▸ to display the Stock Portfolio dialog:

To add information about a stock to your portfolio, fill in all the blanks in the Add Portfolio Entry box and click Add. When you have finished adding entries, click OK.

C R E A T I N G
37

FIGURE 11.8   A Two-Column Format

The two-column format, illustrated here in the CompuServe Navigator Windows Version *User's Guide,* allows users to navigate by topics and icons in the left column and then read details in the right column. Note also the use of prose-style instructions, discussed in Chapter 13.

One-Column Format

The one-column formatted page, shown in Figure 11.9, arranges both graphics and text in the middle of the page, in effect filling the page. Smaller page size on manuals using this design allows for the text lines to be kept at around the 4 1/2-inch optimum for reader comprehension (optimum line length is discussed below). This format can help pack plenty of information into smaller pages of reference manuals for large systems, but it might impede the progress of the user trying to learn material for the first time.

In the one-column format, all graphics, icons, headings, and text obey a strong vertical left margin. Some text items—like notes, cautions, syntax examples, examples of commands to type in, and messages from the program—get indented. Because this page layout fills up the main portion of the page with text, it has a less dynamic look than pages with space reserved for cartoons and icons on which to rest the eyes.

ADVANTAGES AND DISADVANTAGES OF THE ONE-COLUMN FORMAT. In some ways the one-column format helps a writer modularize a document because it makes it easy to keep task information together in a linear form. The task or module just keeps on going until the next one starts, sometimes spanning pages. But because it does not have the large guidance column presented by the two-column format, the one-column format forces the reader to rely more heavily on other guidance elements, such as main headings that start new pages or tasks, and section, chapter, or topic notices contained in headers and footers. With a one-page format, I always find myself consulting page numbers to keep track of my progress.

In this regard, writers have used the one-column, full-page format for tutorials—which tend to have longer passages of prose—because they think that they can get away with cramming information on the page. Uncorrected, this mistake leads to the tutorial section not being read at all, or read complainingly. The one-column format works best for procedures and reference manuals and in documents where the margins are large and numerous graphics and other visual aids help the reader follow along.

The Elements of Page Design

We have seen how page design can contribute to overcoming the design problem. The writer's decisions about the degree of modularity and of structure, and choice of one- or two-column format, will determine how well a given audience can use the manual. Pages consist of the arrangement of many complex elements. But by mastering a few of these, you will begin to understand the building blocks of pages. The following paragraphs describe these elements.

The Left Margin

Text and graphics (the edges of boxed graphics, screens, or the imaginary "soft" edges of icons surrounded by white space) align according to various margins, but most important, they align according to the left margin. The indent margin, the right, top, and bottom margins all play their part, but the left margin rules the page, so to speak. This makes sense, because most of the items on the page use the left margin as a starting place and are defined relative to it. In a two-column format the left margin

USING A BOOT ROM WITH THE DOS ODI DRIVER

If your adapter uses a boot ROM, you must install the program RPLODI.COM (written by Novell) before you install the Thomas-Conrad driver. RPLODI.COM is available from the same sources as LSL.COM.

To install RPLODI.COM, install it between LSL.COM and TCNSW.COM. For example, enter the following commands from the DOS command line, or put them in the AUTOEXEC.BAT file:

LSL
RPLODI
TCNSW

CREATE NET.CFG IF NECESSARY

The NET.CFG file includes configuration information about the network adapters in your workstation. If you are not using any of the options listed in this section, you do not have to create NET.CFG.

❑   You will need a NET.CFG file,

❑   If you do not use IRQ2.

❑   If you do not use memory address D0000.

❑   If you want to include additional shell commands.

❑   If you want to add additional protocol stacks.

If either of these conditions applies to your workstation, take the following steps. Otherwise, you are finished installing the workstation driver.

STEP	ACTION
1	Create the NET.CFG file.

NET.CFG resides in the root directory of the workstation's boot drive. If NET.CFG does not already exist, use your text editor or the **COPY CON** command to create the file.

Insert the following statement into the NET.CFG:

LINK DRIVER TCNSW

Page 6-4   DOS ODI Workstation Driver

FIGURE 11.9   A One-Column Format
Toe one-column format, illustrated here in the Thomas-Conrad Corporation's award-winning manual TC3045 TCNS ADAPTER/AT *Installation Guide,* shows how you can present a lot of information in a small space and still allow for easy use. Note the innovative style for introducing steps that reinforces a sequence–action pattern.

has the extra importance of dividing two kinds of information: images and headings for tracking, and text for close reading.

Columns

Most software will set up columns for you easily. When your desktop system or word processing system does this, it reveals them to you in the form of grid lines (imaginary lines where the margins lie). Grid lines help you see the page or screen before you fill it with words and pictures, and they show the spaces between the columns, called gutters. Columns can work in two ways: as newspaper columns and table columns.

Newspaper columns fill the page by snaking text from the bottom of one column to the top of the next. The effect is to fill one column before the next one gets any text. Table columns, on the other hand, treat the columns on the page as discrete items, and will move text lines from the bottom of one column onto the top of that column on the next page. For most manual projects, table columns work best, because they give you the option of not filling up the bottom of the column before the next one gets filled. For two-column formats, this helps, because each column does indeed contain different information: one with graphics, one with text. Table columns make this arrangement easier to manage.

Headers and Footers

No manual page will work unless it has headers and footers to help the reader keep track of sections, topics, and page numbers. Additionally, these text areas also contain product names, version numbers, company or program identification icons, and rules (lines) to help define the page space. For taking up such a small part of the page, they do a big job. How much stress you place on your headers and footers depends on the kinds of user support your specific pages need to give and the kinds of page design you choose. As we saw earlier, the one-column format places a heavy burden on headers and footers to help the reader navigate the information. The two-column pages shift some of the navigational burden onto the text and graphic elements of the page.

Icons and Diagrams

Increasingly, the easy-to-use page or screen contains many visual elements. These include icons, screens, diagrams, pictures, and rules. Subsequent chapters will treat these elements in detail; here we will see how they contribute in a general way to overall page design. Icons and diagrams function mostly to help the reader identify the needed information, to move from one section to another. They support the cuing system used in a manual. If the page designer has done the job well, they will fit, have adequate white space around them, and occur often in the document.

Screen Shots

No documenter can avoid including screen shots in a manual and to do it well you need to know their types. These include the following:

- Full screens (showing the entire computer display)
- Partial screens (showing usually half of the display or an important part such as command lines)

- Menus showing just the pull-down type menu, or the objects (buttons, sliders, check boxes) that a user can choose from

Because the user needs to refer to these screens during careful reading, design them so that they have accurate details in large-enough type. You will often need to compromise legibility of these screens with the size needed. Bigger usually works; better might mean smaller, to save space. Also, don't forget that screens need captions and labels, which increase the amount of space the screens take up.

Rules

Rules consist of lines of varying width and length that you place on a page to help line up columns or to distinguish types of information. Rules come in thick or thin, tones of gray, double, hairline, color, and other types of border artwork. As a visual element, they present a wide variety of options for the software manual writer. More often than not, rules get used to separate levels and thus help the user keep track of the depth of information. The user will seldom read highly technical material unless he or she knows that it contains the right information.

No matter how many columns on the page, rules used along with headings (rules that are often the same length as the heading) help signify levels of detail. In highly structured documents, rules play a crucial role in keeping page areas distinct and defining columns. When used in headers and/or footers, rules help the reader maintain a sense of the size of the page and thus aid in recall of information locations.

Pagination

While it may seem like a simple concept, how you paginate a manual can greatly affect its usability. If you paginate a document from front to back, starting with page one, then the manual has sequential pagination. If you paginate a document chapter by chapter, with each chapter starting again at page one, then the manual has modular pagination.

Sequential pagination works best with documentation sets where different user's needs are met in different books. When you have a supervisor's guide, a user's guide, and an installer's guide, then use sequential pagination in each one to help give each a sense of unity. When your tutorial reference section fills its own book, you can get away with sequential numbering because it encourages the reader to keep reading.

Modular pagination has the advantage of making the reader more aware of chapters and their contents. It works well in user's guides that contain tutorial and reference documents all under the same cover. Because each chapter starts with page one again, you can easily maintain this pagination scheme. When you add to a chapter, simply reprint that chapter, not the whole book.

Like the modular scheme, sequential pagination works with other user navigational elements like headers and footers. Headers and footers contain names and numbers for sections, parts, and chapters. Working with your pagination method, these add to the degree of structuring each page has. You should also use care in designing pagination for appendices, indexes, and front matter pages (usually they just have lower-case roman numerals). Make it clear that these sections of the manual differ from the body portion; also make clear what these sections contain.

Common Screen Designs

You will find a lot of variation in help screens among software programs, especially in interactive tables of contents screens and index screens. The following paragraphs discuss some of the common screen formats.

Windows Screen Format

The windows screen format, shown in Figure 11.10, contains the usual system features of a window: It contains a non-scrolling region (or regions) that can help the user keep track of his or her position in the levels of information. It usually uses a one-column design with hot areas that take users to related topics. In some variations of this screen design you can divide the screen into two columns, with the index entries in the left column and the selected topic in the right column. In this way the user can move among the help topics on the left, select from the list, and see the topic on the right. This design has the advantage of always letting the user see the table of contents—to navigate the system with it always in view.

You can also set up screens using this format to illustrate the functions of a program as they relate to the interface. In Figure 11.11 the dialog box contains hot areas. When the user clicks on a hot area, a popup window explains that function. Another click closes the popup window. You can use this format for illustrating screens, toolbars, menus, and other interface elements. You can also include links to procedural documentation on this topic. In this example the layout has the title in the non-

FIGURE 11.10   A Procedure in Windows Screen Format
This format from the Microsoft Works *Online Reference* shows a one-column screen with the task name and other information in the non-scrolling region at the top.

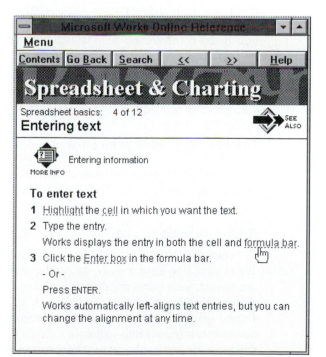

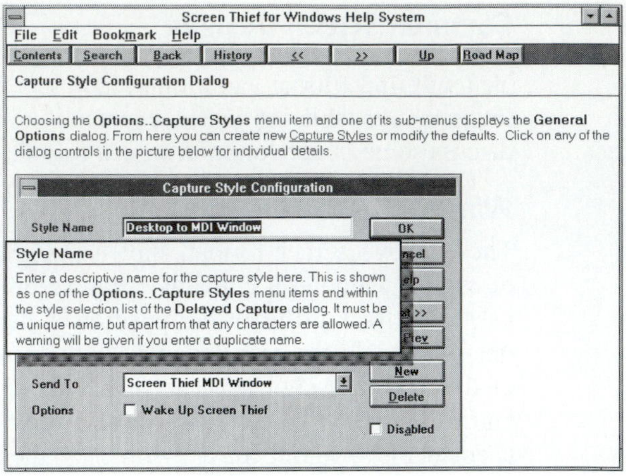

FIGURE 11.11   A Screen Topic
Showing Parts of a Dialog Box
This screen from the Screen Thief *Help*
shows a common format: a dialog box
or other interface item with pop-up
explanations.

scrolling region, the definition of the dialog box, and then a bitmap of the dialog box
with the hot areas pre-defined by the system designer.

Man Pages Format

Many users will recognize the man pages (short for "manual pages") format if they
have any experience with operating systems like VMS or UNIX. It basically consists
of a handy format for dumping print documentation online. It has no left margin or
other niceties of a windows-like bitmapped screen.

The man pages format, shown in Figure 11.12, consists of three parts: a heading or
title area, a text area, and a navigation area. The title area usually falls at the top, and in
it you find the title of the topic or command. The text area falls in the middle of the
screen, and it contains definitions, examples, and references to associated commands.
The navigation area at the bottom of the screen contains a prompt. Variations of this format can contain other help system commands such as to print or search the help system.

The Elements of Screen Design

Designing screens can get very complex because screens get easier to read as new
technology emerges. The following paragraphs discuss some of the design challenges
you face and offer advice for writing highly usable screens.

A Changeable Space

The idea of a communication space changes when you compare pages and screens.
With pages you have a static space, the same with each page; with screens the user
can re-size, scroll, and in other ways change the communication space. How can you
design information effectively for this space? As a starting place, I find it useful to
see the space as a frame, one of many in front of the user at any given time. The document you're presenting to the user shows through the frame. So design for the document, so that when the frame changes, the user can still read it effectively. Consider
following these guidelines to design for help screen frames:

FIGURE 11.12   A Topic in Man Pages Format
This format, common to mainframe operating systems and other large systems, presents information in an easy-to-use format compatible with just about any screen.

```
HELP

    Enables you to obtain information about the Mail utility.

    To obtain information about all of the Mail commands,
    enter the following command:

        MAIL>HELP

    To obtain information about individual commands or topics, enter
    the HELP command followed by the command or topic name.

    Format

        HELP [topic]

    Additional information available:

    ANSWER    ATTACH    BACK    COMPRESS  COPY    CURRENT  DEFINE
    DELETE    DIRECTORY  EDIT    ERASE  EXIT    EXTRACT  FILE
    Files   FIRST   Folders   FORWARD  Getting_Started  HELP
    Keypad   LAST   MAIL   MAIL_Commands  MARK  MOVE
    NEXT    Overview  PRINT  PURGE  QUIT  READ  REMOVE
    REPLY   SEARCH   SELECT   SEND  SET-SHOW  SHOW   SPAWN
    Usage_Summary

    Topic?
```

- **Forget line length.** Don't bother adjusting the line length as you might in a column of a page because the user can bring the right screen border in and change the shape. In most cases you only have to design for the left and other indent margins.

- **Avoid lots of scrolling.** You can expect the user to do a degree of scrolling to get to the end of a long procedure of explanation. But too much scrolling can disorient just about anyone, especially if the text doesn't contain cues like numbers or characters of the alphabet (as in the case of an index) to give a sense of location. You should make your topics no longer than two screens, 3 1/2 inches, in length. Break up longer topics.

- **Indicate the extent of the topic.** If the topic contains more than one screen and you move screen by screen, indicate so with a "More" button, or a "1 of 3" progress indicator.

Multiple Window Management

You have to determine how and whether users can manage information in help windows. The help system program that you use will provide you with a number of tools for doing this. When making decisions about how to manage the help windows, you might want to follow these guidelines.

- **Don't obliterate the user's work.** Some help systems splash over the entire screen when you call them up. But this can disorient users because suddenly their existing application vanishes. Allowing the help screen to cover only part of the screen reassures users that their application didn't go anywhere.

- **Avoid window clutter.** Try not to let too many windows clutter the screen. A window within a window within a window can confuse users because they loose a sense of the levels of topics. Of course, with some systems you can't avoid the clutter the user might create. With WinHelp you can implement an "On Top" button that allows users to keep a topic (usually a procedure) on the screen to read while performing the procedure. You can't stop users from leaving a number of these topics open, but you can try to minimize the number of windows the system itself presents.

- **Give the user control over frames.** Some windows will close automatically when the user presses a key or clicks with the mouse. Others can close automatically in a pre-set time. But you should make sure that when it's up to the user to close the window, you make it clear how to do it.

Color

Use color to cue important elements. Color provides you with a powerful tool for helping users manage the help system. You can use it to indicate sections of the screen and to help identify important kinds of information like navigational tools, tips, interface tools, and so forth.

Graphics

Use simple graphics. On most systems graphics appear grainy and may not allow you the kind of detail you get when you print them on a page. Overall, you find fewer graphics in help, for this reason, and because of the space they take up and the limitations of some systems in reading and displaying them. In general, use simplified images that don't rely on details to communicate: cartoons, simple icons, screens.

Screen Grids

As discussed above, the grid you use for the screen differs from the one you use for a page. Most of the differences result from the differences in the lower legibility of screens and the flexibility of the screen frame. Consider these differences when setting up screen grids:

- **Use narrower margins.** On a page you would use a 1" left margin, whereas on a screen you only need about 1/8" to 1/4" margin. You don't need the passive white space of a margin to define the communication space; the plastic edge of the computer monitor does that for you.

- **Use less indentation.** Indent only about 1/4" rather than the standard 1/2" you would indent on a page. For one thing, you don't want to waste the space, and for another, you don't need as much indentation with the larger fonts used on screens.

- **Define the grid for single pages only.** It may seem obvious, but you don't have left-hand and right-hand pages online. Similarly, you don't need to plan for a binding offset.

- **Use rules sparingly.** With screens you tend to define topics as whole units and don't need as many rules to segregate information and establish hierarchies.

Line Spacing

In general, when you're setting up screens, you put your lines closer together or at least almost always single spaced. And because of the overall reduced legibility of screens, you use larger type in headings to make them more visible.

Designing Type

The second major component of document layout, type design, challenges you to use the complex writing tools at your disposal to overcome the design problem: getting users linked up with information. Years ago, small-budget documentation projects had few tools to work with. Now, powerful desktop publishing systems, and even low-cost PCs with page design software, allow writers many possibilities for designing type.

A study of type design should begin with what we know about how users recognize words, from a researcher's point of view. Then we will apply some of that knowledge to the specific job of building task orientation into software manuals.

Helping People Recognize Words

Much of what we already know about users becomes important when we consider ways to help them use the text portions of the pages to their maximum capacity. We know that the user's experience with computers, with other programs, and with the subject matter of the program will play great parts in our determination of type size, font, and style. Designing type means accounting for as many idiosyncrasies as possible. But beyond that, type designers need to understand some fundamental principles of how readers get meaning from words and how readers process words on a page into ideas.

One of the things we have learned about the way people recognize words reinforces the idea we have discussed earlier: that people prefer a visual orientation. Manuals that support a visual orientation consistently test higher in usability evaluations. For this reason, we should focus our attention on the shape of words, researchers tell us, as well as the fact that they consist of letters.

Software documenters need also to acknowledge the shape of words. For evidence of this, remember the times you have heard someone read a sentence like this: "Don't forget to store the disk accessory in the startup folder," substituting the incorrect word disk for the correct, look-alike word desk. Words that look like one another confuse us at times, because we read by shape. For simple words like *the* and *and* we may not even see more than the vague outline or part of the shape before we have already understood the word and gone on to fit it into the overall meaning of the sentence. Long, tall, short, mixed: Words speak to us by their shape, as well as their letters.

The serifs on type characters add to the shape of the word. Serifed type has little strokes at the ends of the letters that almost connect the letter to its neighbor. Sans serif fonts—Helvetica, the "modern" fonts—don't have these little connectors. (The sans part of the phrase comes from French—the word for without.) The difference between serif fonts and sans serif fonts resembles that between hand-printing and cursive, or script. Script letters, similar to serif characters, have lines that connect one letter to the next. Hand-printed characters, like sans serif characters, consist only of straight-ended strokes.

The distinction between serif and sans-serif type means a lot to the software documenter. Because close reading by users occurs so infrequently, the design of the type should support scanning, make it as easy as possible. In addition, reading should take place in as small a space as possible. Serifed fonts work well for this because those little strokes on the ends of the letters contribute to the shape of the word and contribute to legibility. The serifs help the eye move quickly across the page. Non-serif fonts, on the other hand, work best when the font size makes up for any loss of legibility. The smaller the type, the more important the serifs.

Design Advice

- **Choose a type face the user knows well.** Reserve exotic type faces like Albuquerque or NASA for the cover or title page.

- **Choose a font with definite ascenders and descenders for small-sized type.** Ascenders are the part above the line in "b" and "d"; descenders are the part below the line in "p" and "y."

- **Choose a type style familiar to the user.** These include Times Roman and Courier (for input information).

- **Avoid long passages in italics or all caps.** Both italic and all caps make the shape of a word more difficult to recognize. Using all caps, because all letters stand at the same height, blurs the distinction between one word and the next and slows down comprehension.

- **Keep headings short.** Use the implied "you" as in "Installing the Program" instead of "Here's How You Install the Program."

- **Use serif for body text, sans serif for headings.** Headings speak most boldly when they stand out and stay short.

Building Patterns with Type

As we discussed earlier, the main way you help the reader overcome the design problem consists of giving your manual a recognizable pattern (the cognitive scientists call it a schema). Just as you would use a schema for finding your car in a parking lot ("big antenna, rack on top") to help you distinguish your car from the others that resemble it closely, so you want to show your user how to use a schema for recognizing important information in your manual. In our study of software document design so far we have seen a number of ways to do this: using cuing techniques, page structures, headings, rules, and other layout elements.

Type design also contributes to the pattern of a manual because you can use it to signal kinds of information for the reader and to assist the reader in getting the needed information from closely read passages. Table 11.3 shows the differences between these three elements. As a designer of type you face three important questions: what size, what style, and what font.

Design Advice

Keep it simple. If you have doubts as to which font, size, or style to use, change them in this order: first the style, then the size, then the font. Thus, if you decided that you wanted to give, say, glossary terms a special look, make them bold, or italics, first. If this does not make them recognizable to the user, then, keeping the same font, make the type bigger by two or four points. After that, as a last resort, change the font to make them stand out better.

TABLE 11.3   Type: Style, Size, and Font

Type Characteristic	Examples	Used to ...
Style	plain **bold** *italics* <u>underline</u>	Add special emphasis, distinguish foreign terms, identify titles, and make headings stand out.
Size	very small small medium large	label callouts, present warranty information, identify body text, identify headings, and identify document title pages.
Font	Times Courier Helvetica *Caliban*	distinguish body text from headings, represent "typed" text, and add emphasis.

- **Arrange for no more than three levels of information, preferably two.** If you limit the number of major hierarchies on the page, you limit the number of times you might have to change size, style, and font.

- **Use a sensible cuing pattern.** Don't overdo your pages with three or four kinds of cuing devices (usually based on size and style of type) because it makes the page look cluttered and adds to the user's cognitive load.

- **Use type design consistently.** Throughout the manual and help system, use the same design in associated documents.

- **Think in terms of styles.** Decide what you want your paragraphs to look like and specify your type style, font, and size, along with your indentation and margin specifications. Most word processors work in this way: They allow you to set the way you want paragraphs to look (their layout and type) so that you can determine the look of the text by evoking the style from a saved style guide. When you work like this, you help ensure consistency in the whole document.

The Idea of Body Text

When you select type style, font, and size for the main text portion of your document, you make one of the most important choices in document design: You determine body text. Body text fills the pages and screens of your documents. Explanations, introductions, advance organizers, glossary definitions, index entries—all these convey their information to the user through variations on the type style, font, and size that you selected for body text. You can modify body text in size and style (but not font) to achieve different effects without introducing significantly different elements onto the page or screen. While the general consensus on body text dictates that readers like a plain style, serifed font, between 10 and 12 points in size, many constraints affect the actual choice of body text style. Your main constraints follow, in descending order of importance:

- **Page size.** Generally, the smaller the page, the smaller the size and the less dense the font. On smaller pages you need a smaller size in order to economize. When you reduce the size of the type from 12 to 10 points, you gain an additional 20 percent of space. This gain can allow you to modularize more effectively and to keep manual size to a minimum. As for a less dense font, such as Times (less dense) instead of Schoolbook (more dense), the smaller page looks less crowded with a font with slimmer lines. And the reader has less ink to process.

- **Media.** Pages allow you to use much smaller, more detailed fonts for body text, whereas screens allow a more limited range of fonts. Some help programs come with their own fonts, recommended by the help developer. Usually these consist of sans serif fonts of slightly larger size, say 12 or 14 points. You want a font that defines the character clearly (and from the distance between eye and screen) but that doesn't include fine details (like serifs) to confuse the recognition. Sometimes a combination font (such as Chicago or Geneva) works best online.

- **User expectations.** The document designer should pay attention to what kinds of type users see regularly in software documents. For the most part, you can predict what the type design for body text will look like before you open a new manual. However, in some manuals you will find sans serif fonts (Helvetica, etc.) used for body text, and you will also find it used as body text in other documents people read regularly: books, newsletters, journals. If you decide to set body text in one of these fonts, make sure you have a good printer and the manuals get reproduced on high-quality copiers. This way you offset any loss of legibility due to sans serif fonts.

Non-Body Text

A manual or help system consists of more than just body text. Once you determine the style, font, and size of body text, you next have to deal with the variations of body text that will make up the rest of your pages and screens. Readers need to have information other than explanations and introductions set apart—given a distinctive look—so they can remember them, look forward to them, and relearn them. What other kinds of information require such fancy treatment? Many of them found their way into the discussion of basic document design above: headings, cautions, warnings, notes, input instructions, and computer messages. Below you will find discussions of these elements and suggestions for selecting type for them.

Headings

Headings function to help the reader locate important information, and they do their job best when the reader can easily distinguish them from body text. For this reason we put them on separate lines, in special columns, and in larger, attention-grabbing type. Try to make the different levels of headings (chapter titles, section titles, tasks, etc.) consistent in type design.

Conventional wisdom dictates that they should look larger, use a sans serif font, and use a bold style. This choice makes sense when we consider the lookability of type. The size catches the user's eye; the size also makes up for the lack of word-shape cues in the sans serif font; and the bold style gives the heading a distinctive density, apart from that of the body text.

Other possibilities for heading type design include keeping the same font as the body text but varying the size and style of different levels of headings proportionately. So you would end up with small, medium, and large versions of basically the same font. The users of the document get their navigational needs met, and the writer's task stays a little simpler.

In choosing the type for headings, designers should also remember that such elements as the use of rules can significantly affect one's decisions. Again, the manual's type design should conform to an overall vision of the manual as a system. All the parts need to work together.

Hints, Notes, and Cautions

Like the other informational elements of manuals, hints, notes, and cautions give the user extra, or special, information. Some writers call these "asides." They, too, must read easily and use visual cues to catch the reader's eye. Thus, you will format them using headings, indentation, boxes, icons, and color to make sure the user gets the message.

With all these other visual cues helping the user, the type design decision often amounts to this: Should the writer adjust the type size, font, and style, too? Often these messages occur in the type used for body text because, unlike some features discussed below, they consist of information from the writer to the user. You decide to include one of them as a way of saying, "Pay attention to this information" or "Note the variation possible here." You may, thus, choose to leave them in body text type because, after all, you want the reader to see them as body text, but body text of a special kind.

If anything, change the type characteristics of the heading or cuing word for the note or caution. You will see many of these elements set in larger, sans serif font, bolded italics style. This way the word itself catches the reader's eye and identifies the information. The rest of the processing event gets managed by indentation and rules or boxes.

User Input, Computer Output

Unlike the features discussed above, software documentation also consists of information that the user has to type in, and that the computer program displays. The writer does not write this information; the interface of the program dictates it. Commands and displays make up the substance of computer interaction, whether the user gives commands with a keyboard or mouse, or whether the computer displays, by means of the screen, the speaker or the printer.

For these reasons, then, writers usually change the font of input and output messages from that of body text. For input, most users expect the Courier font, because it resembles typewriter text, and the analogy of the keyboard to the typewriter helps them understand that they should now use the keys to enter a command or a file name, or whatever. For output, warnings and other messages may catch the reader's eye better in a small-sized sans serif font like Helvetica. A font like this allows the message to resemble body text but still look distinctive.

Tables and Lists

As with hints and notes, discussed above, your main decision with tables and lists will revolve around how to do the headings: The substance of the tables and lists—words—works best set in the type size, font, and style used for body text. Don't change type elements just for tables; instead make the tables differ in indentation and column layout. That way, you emphasize that they resemble text but have been reformatted and condensed for greater clarity. You can also call attention to tables and lists by altering the style of headings and adding rules.

Glossary

bullet: a heavy dot, filled square, or other graphic device that calls attention to important points.

communication spaces: in document design, any area within both print and online media containing text and images designed for viewing and communicating information. The communication space of print usually consists of pages in a relatively fixed one- or two-page spread and presents the best space ever invented to communicate highly detailed information. The communication space of online media consists of the screen, a relatively flexible and dynamic space presenting one of the worst spaces to communicate highly detailed information.

gutter: space between columns; also, the page margin on the inside (binding) edge.

icon: a graphic element that cues the reader to a function of the program, such as a wastebasket to signify file deletion.

mirrored pages: pages making up a two-page layout, with page numbers located on the outside edge of each facing page.

modularity: a page design in which the page contains all the information required about a particular topic so that understanding the topic or task doesn't depend on the information that appears in previous topics or tasks.

page grid: a device for designing pages that uses lines and boxes to identify page margins and to mark the main communication spaces of a page and used as a pattern.

rule: a horizontal or vertical line used to indicate column widths or to separate vertical columns.

sketch-straight: a line made by short pencil movements that looks straight but doesn't require a ruler to draw.

specifications: statements of the requirements of page design that tell the dimensions of pages and columns, size, font, and style of text and other details. You write up specifications for page layout in the documentation plan.

thumbnail sketch: a miniature sketch of a page showing the placement of text and graphics, used in page design.

☑ Checklist

Use the following checklist as a way to keep you on track when making the many decisions involved with page and screen layout.

Document Layout Checklist
Overall

❑ Has the client, sponsor, and/or user reviewed your selection of page or screen design?

❑ Do all of the elements of your manual or online help set contain parallel information for consistency and efficiency?

Layout Goals

❑ Include elements to help the user overcome the documentation problem.
❑ Include support for overall task orientation.
❑ Accommodate the user's visual needs.

Modularity

❑ Repeat background information where necessary
❑ Repeat screens when necessary.
❑ Include orientation information about the relationship of a task to other tasks.
❑ Keep all relevant steps on the same page.
❑ Minimize cross references.

Structure

❑ Rules   ❑ Bullets
❑ White/quiet space   ❑ Chunks

Page and Screen Designs

Page: **Screen:**
❑ Two-column format   ❑ Windows help format
❑ One-column format   ❑ Man-page format
❑ Multiple-column format

Determine the Following Layout Elements

Page: **Screen:**
❑ Left margin   ❑ Sizing and scrolling
❑ Columns   ❑ Multiple window management
❑ Headers and footers   ❑ Cuing and highlighting
❑ Icons and diagrams   ❑ Icons and buttons
❑ Screens   ❑ Margins and columns
❑ Rules   ❑ Line spacing
❑ Pagination

Determine the Following Text Elements

	Size	Style	Font
Body Text			
Headings			
Hints			
Notes			
Cautions			
Warnings			
User input			
Computer output			
Tables			
Lists			

Practice/Problem Solving

1. Learn from the Competition

Elements of page layout should accommodate different kinds of documentation. Analyze the design of two or three manuals for the following elements, explaining how you think the writer did or did not adjust them for the appropriate:

- Page density
- Balance
- Legibility

2. Practice Drawing

Draw thumbnail sketches of three different page designs in computer manuals that address user's needs at the same level (e.g., three different designs for procedures, three different designs for tutorials, etc.). Submit these sketches, along with a brief, one-page report analyzing how each design met the user's task needs. Which one do you think worked best, and why? How much of the success or failure of a document can you attribute to its design as revealed in the thumbnail sketch? What elements of page composition do you think affect the document the most? What would an ideal page design look like?

3. Analyze Flyer Pages for New Ideas

Often software packages come with flyers, telling the main features of the program and showing sample screens. Often these documents contain support information for decision-making: "Should I buy this program or not?" Examine some of these flyers— available at a software or computer store—for ideas about how to construct such a document. Create a list of page layout options and styles that you find. Discuss how you can take some of the basic ideas in this text and apply them to the variety of layouts you find in software flyers.

4. Use Thumbnail Sketches in the Design Process

Prepare a series of thumbnail sketches for the program for which you are currently writing documentation. Get these designs approved by your sponsor or client. Revise the sketches and create a final, mock-up version. You may want to do this with your word processor or page-design program.

5. Design for the Page and the Screen

Study the task illustrated in Figure 11.2 or Figure 11.3. How would the layout of the information differ if you developed it 1) to teach a clerk the basics of the word processing program, or 2) to guide a marketing executive who knows the word processing program well enough to use the online help when writing letters to clients? You might include more elaboration for one user, and include the procedure as a task in a help system for the other. Create page designs for the layout and text for two different pages or screens, and list differences in the two screens as they pertain to the two users.

CHAPTER 12

Getting the Language Right

In this chapter we study the role language plays in helping the manual and help system attain the goal of supporting information-oriented work. You have to think about the users of the software and say things in a way that has value to them. Following the guidelines of *performance-oriented,* structured writing can help ensure that your language supports the overall task-orientation of the manual or help system. Our study of language and style needs to start with what research tells us about how we process language and how we remember and learn through words. We process language by providing a task context for the words, by bringing meaning to the words and not getting meaning from them. We learn and remember easily when we have patterns and structures in language and when the language in manuals and help does not violate the structures we expect. We will also see that many problems in the language of software documentation all revolve around two central difficulties: failure to write so that the user can perform the task easily, and failure to write as if we were speaking to real human beings.

How to Read This Chapter

You should study the Example and read the Discussion section first, and then the Guidelines, because in this case you can see a clear logical connection between the psychology of language processing—even though only the basic concepts are presented—and guidelines like "Focus on Functionality Rather Than Functions."

- If your project or the work of other writers places you in the role of editor, you may appreciate the discussion of "Style Problems in Software Documentation."

- For those readers reading to understand, the entire chapter is focused on problems specific to software documentation and not other forms of writing in the computer industry.

Example

The example in Figure 12.1 comes from an award-winning software manual that uses style to its best advantage. Written for scientists, this passage embodies important

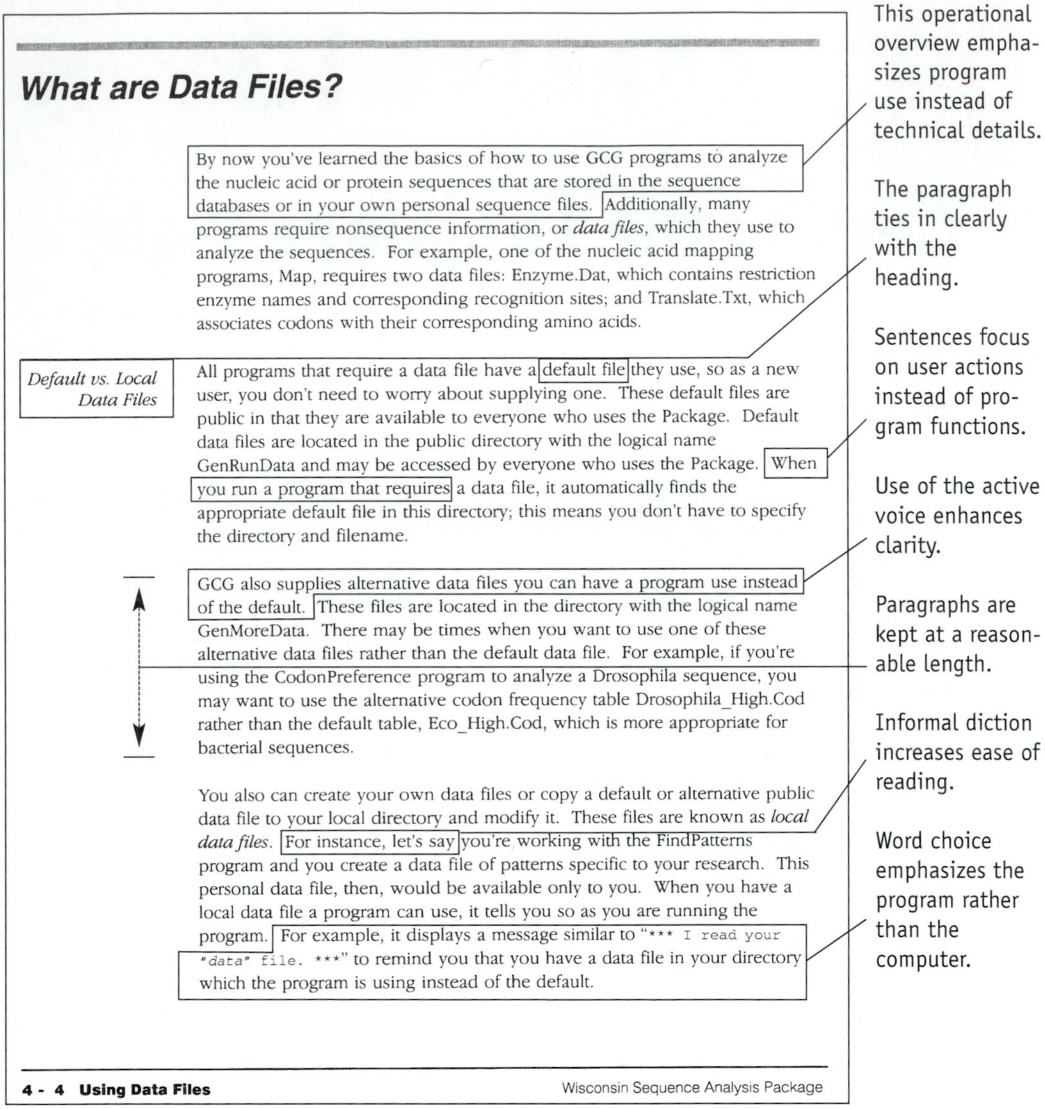

What are Data Files?

By now you've learned the basics of how to use GCG programs to analyze the nucleic acid or protein sequences that are stored in the sequence databases or in your own personal sequence files. Additionally, many programs require nonsequence information, or *data files*, which they use to analyze the sequences. For example, one of the nucleic acid mapping programs, Map, requires two data files: Enzyme.Dat, which contains restriction enzyme names and corresponding recognition sites; and Translate.Txt, which associates codons with their corresponding amino acids.

Default vs. Local Data Files

All programs that require a data file have a default file they use, so as a new user, you don't need to worry about supplying one. These default files are public in that they are available to everyone who uses the Package. Default data files are located in the public directory with the logical name GenRunData and may be accessed by everyone who uses the Package. When you run a program that requires a data file, it automatically finds the appropriate default file in this directory; this means you don't have to specify the directory and filename.

GCG also supplies alternative data files you can have a program use instead of the default. These files are located in the directory with the logical name GenMoreData. There may be times when you want to use one of these alternative data files rather than the default data file. For example, if you're using the CodonPreference program to analyze a Drosophila sequence, you may want to use the alternative codon frequency table Drosophila_High.Cod rather than the default table, Eco_High.Cod, which is more appropriate for bacterial sequences.

You also can create your own data files or copy a default or alternative public data file to your local directory and modify it. These files are known as *local data files*. For instance, let's say you're working with the FindPatterns program and you create a data file of patterns specific to your research. This personal data file, then, would be available only to you. When you have a local data file a program can use, it tells you so as you are running the program. For example, it displays a message similar to "*** I read your "data" file. ***" to remind you that you have a data file in your directory which the program is using instead of the default.

4 - 4 **Using Data Files**   Wisconsin Sequence Analysis Package

Annotations (right margin):

This operational overview emphasizes program use instead of technical details.

The paragraph ties in clearly with the heading.

Sentences focus on user actions instead of program functions.

Use of the active voice enhances clarity.

Paragraphs are kept at a reasonable length.

Informal diction increases ease of reading.

Word choice emphasizes the program rather than the computer.

FIGURE 12.1   Effective Writing Style
This example from the Wisconsin Sequence Analysis Package *User's Guide* shows some of the features of good writing that help users gain a sense of control over the program and understand how it will work for them.

background information and leans toward the reading-to-understand user—hence the lack of a lot of page structure in order to maximize the number of words on the page. But when you write in this compact way, the style needs enough strength and clarity to encourage the user to read and understand.

Guidelines

Figure 12.2 lists guidelines for ensuring that your style is oriented to users' needs.

1   Write About Actions Rather than Functions

One of the main differences between system-oriented, "default" manuals and help and task-oriented manuals and help lies in their language. With every menu item you describe you have the opportunity to tell users the name of the function and what results from its use (system orientation) or to tell users the name of the function and what workplace task it will perform (task orientation). Consider the following statement:

> **Hyperlink:** *Inserts a hyperlink or edits the selected hyperlink.*

This statement focuses on the function of the hyperlink button in Microsoft Word. It clearly and succinctly defines the button. (I got to this definition by using the "What's this?" function under the Help Menu.) As a definition it works fine. But the user has to remember it in order for it to work because it only does one thing: define the button. It doesn't answer the implicit question in the user's mind, which is "what can I do with this button?" or "Why would I want to insert a hyperlink?" Now consider this revised statement:

> **Hyperlink:** *This button allows you to link selected text to other text or documents or edit the properties of selected hyperlinks.*

The second statement does two things: it identifies the object as a "button" and it tells what the user can *do* with the button. It puts the object into a class of "tools" for the user by grouping it with other "buttons." But more importantly, it acknowledges that the user inserts a hyperlink for some reason (to achieve coherence in a document, refer the reader to another part of a document, or whatever).

The first kind of statement puts the users at odds with the program—as something apart from them that they will have to learn to use; the second, and preferred, kind of statement, puts the program on the users' side—as a set of tools designed to

FIGURE 12.2   Guidelines for a User-Oriented Style

1. Write about actions rather than functions.
2. Revise for the active voice.
3. Revise to keep writing simple.
4. Rvvies to build parallel structures.
5. Add operational overviews.

help them do their work. The more you focus on usefulness rather than on functions, the greater the likelihood that your users will recognize that usefulness, which is the point your manual or help.

To write effective performance-based sentences you should create direct sentences that focus on user actions, such as those in Table 12.1. Tags like "You can . . ." or "X function helps you to . . ." work well and should be used frequently.

2   Revise for the Active Voice

The active voice puts a subject at the beginning of each sentence, a verb in the middle, and a receiver of the action at the end. "You can use the File menu to . . ." or "The File menu displays . . ." uses the active voice. "The File menu is used to . . ." uses the passive voice. The negative results of the passive voice abound: Your sentences will fill up with noun words ("functionality," "usefulness," "motion") instead of verbs ("function," "use," "move"). Nouns clutter things up; verbs get things done.

Using the active voice in writing requires you to understand the grammatical constructions that make up the passive voice (the "to be" verb and the past participle, plus the word "by" to include the subject) and it requires you to know how to revise your sentences to eliminate this awkward way of saying things.

One way to identify and start to understand passive constructions is to read a passage made up entirely of them. For example:

> *The Toolbar is accessed and the Preferences button is pushed. The preferences for File Names is set and the Save Options box is checked. The Automatic*

TABLE 12.1   Examples of Statements that Emphasize Actions

Focus on Functions	Focus on Actions (Preferred)
"The Apple Menu has four main functions."	"The Apple Menu indicates the four main tasks you can perform."
"The tutorial is organized in a step-by-step manner so that the user will gain a basic understanding of the program and be able to create readme files."	"The tutorial is organized in a step-by-step manner, so that you can understand the program and create readme files."
"The greatest speed in List Box scrolling is achieved from the keyboard by typing one character to get the part of the list where the entries begin with that character."	"You can scroll quickly in the List Box by using the keyboard to type one character to get the part of the list where the entries begin with that character."
"The main function of the Calendar menu is to make changes or set-ups to the existing calendar."	"The main function of the Calendar menu is to help you with any changes or set-ups you want to make to the existing calendar."
	or
	"You can use the Calendar menu to change the set-ups for the existing calendar."

Sharing is set to "on" and the directory for saving shared files is typed into the directory field. The Okay button is clicked by the user and the Preferences panel is closed by the program.

The same passage using the active voice would read as follows:

On the Toolbar press the Preferences button. Select the File Names and Check the Options box. Set Automatic Sharing to "on" and type the shared files directory into the directory field. Click the Okay button. The Preferences panel closes.

Can you see the difference? The first passage is impersonal; it can make the reader feel manipulated uncaringly. The second passage is more personal and direct and makes the reader feel empowered. You can see that while the active voice is, at the grammatical level, a way to structure sentences, it is also, at the tone level a way to sound impersonal and distant.

3 Revise to Keep Writing Simple

Simplicity helps every aspect of software manual writing. You should strive for simplicity in each sentence. Often you will find sentences that suffer from the reading difficulty of the following one:

"Warning: Do not install other desk accessories in either the System file *in the* Server Folder *on the startup volume, or in the* System file *in the* System Folder *on the* Server *disk." (From a computer manual)*

The preceding sentence makes perfectly good sense. But what price must the reader pay to get at that sense? For one thing, the writer has crammed it with look-alike words (*System file, Server Folder, startup volume*). Consider the burden these words put on the reader trying to get the point the first time. Without *first-reading* clarity the reader must waste twice the time rereading and pondering. Not good. Break sentences like the one above into more than one sentence. Separate out or find acceptable substitute phrasing so sound-alike words don't bombard the reader. Consider something like the following: "Warning: Do not install other desk accessories in the System file in the Server Folder on the startup volume. Do not install other desk accessories in the System file on the AppleShare Server Installer disk."

The next example shows how separating the subject from the verb creates a sentence which forces the user to put understanding on hold while the sentence presents related data.

subject

*"*An example *of a program that was the first to implement a Windows capability and has demonstrated a high rate of speed* is *Netscape." (From a help system)*

verb

In this case, you should revise the sentence to keep the subject and verb together. Something like "Netscape implemented a high-speed Windows interface long before any other program" might work.

4 Revise to Build Parallel Structures

Of all the solutions to the problem of using the user's short term memory effectively, none works as well as *parallelism.* Parallel items acknowledge the similarities between concepts and express that similarity in matching grammatical structures. Headings that all end in *ing* follow the principle of parallelism. So do steps that all begin with a command verb and so do sequences of results described from the user's perspective. Parallelism helps readers remember even though they may not recognize the pattern.

Such subtle patterns should abound in your writing. In Table 12.2 you will find some examples of ways to use them in all aspects of manuals and help.

5 Add Operational Overviews

The way you present conceptual information makes a huge difference in documentation because of the highly abstract nature of software work. Often users react negatively to manuals and help because they do not have the necessary theoretical and technical background to understand the vocabulary and ideas. Primarily, users want

TABLE 12.2   Types of Parallelism in Software Documentation

Type of Parallelism	Example of Use
. . . ing	Task Names
	Opening a file
	Saving a file
	Closing a file
Noun First	Lists of options
	Options for setting the display
	Options for setting the transfer directory
	Options for specifying file types
Parallel Sentences	Giving overviews of commands, or suggesting alternatives:
	To get a closer view of the drawing object, use the Zoom+ command; to get a farther view of the drawing object, use the Zoom-command.
Imperative Voice	Steps
	Select File from the Draw menu.
	Type in the filename.
	Click on the OK button.

to use a program and thus care mostly about how it operates and how it can make them efficient. Your manual is like a bathroom key: People want to get their hands on it not because of its intrinsic properties but because it lets them do what they need to do. That motivation to do can lead users into reading carefully for meaning or it may lead users into scanning the text before reading, depending on how important the program is to them.

Because users often read for meaning, you should provide prose passages (paragraphs) containing clear overviews of concepts as well as straight procedures (steps); users appreciate learning the conceptual model of the program and how the program does its processing. Of course, your well-crafted examples and accurate metaphors should point out the benefit of the model to the user. Notice how the following passage points out the benefit to the user of storing the location of project files.

> *Aside from the fundamental organizational benefits of keeping things structured beneath the Project Directory, RoboHELP provides some additional affordances for this organization. As you might guess, RoboHELP needs to keep track of the files which make up your Help Project. RoboHELP stores the location of these files (in HAPPY.HPJ and HAPPY.RBH) as 'relative' paths based on the Project Directory. The main benefit of this technique is that it allows you to move the entire Help Project to another drive, computer or directory (or all three), without confusing RoboHELP or the Microsoft Help Compiler. For instance, if you've maintained this organizational hierarchy beneath the Project Directory and you're suddenly called to move your project onto a laptop computer in order that you might complete the project while sipping piña colada under the Caribbean sun, you are free to do so. Once the files are transferred to the laptop, you can immediately open the Project Document in your word processor and at once be productive."* (From an online help topic: RoboHELP's Model of a Help Project)

You can choose how to emphasize your explanations of abstract concepts, but writers primarily use three: the theoretical (emphasizing the theories behind the working of a program), the technical (emphasizing the technical functioning of the program), and the operational (emphasizing the application of the program.) Overall, for end users interested in productivity, the operational emphasis works best. Consider the examples of these three kinds of overviews in Table 12.3.

Discussion

In the previous chapter we looked at ways the physical layout of pages can help the manual overcome the design problem. In this chapter we examine the part language plays. We look at how word choice can affect how a user applies the program to information work. First, we have to concern ourselves with how language creates meaning and how the skillful writer can craft sentences that convey useful ideas. Most important, the practical advice you will find in this part of the chapter bears directly on the writing of software manuals. As in previous chapters, the primary

TABLE 12.3   Ways to Present Conceptual Information

Emphasis	Example	Use
Operational	"Word allows you to write documents easily, and also lets you communicate numbers using graphs and ideas in pictures and drawings."	Novice and experienced end users in business and other professional fields
Theoretical	"Word attempts to combine word processing with other communication domains, such as presentation of numerical data and visual symbols."	Experienced and expert users; linguists; computer professionals
Technical	"Word is a DOS-based text editor and page design program using a point and click interface with embedded draw and graphics functions."	Expert users in the computer industry with a high degree of technical background

focus of using language effectively will fall on building task orientation into the document.

In software documentation, you will often use different writing styles and different tones to support different user tasks. Tutorial documentation, for example, involves a much closer relationship between the writer and the user and thus uses more words and contains more familiar diction, even humor. Reference documentation, on the other hand, involves a more distant relationship between the writer and the user, requiring a more formal, businesslike tone. Other levels of user support require stylistic adjustments that are treated in Chapter 10, "Designing for Task-Orientation." Here you will find general advice that you can apply to most writing in software documents.

How Do We Process Language?

To set goals of language use to support task orientation, first look at how readers process language. For most people, the words in the document take a second place in their experience. They might have a problem using a computer, head for the manual, look up the task in the manual, return to work, and try it out. This simple scenario, however, gets more complex the more you examine it. The "looking up" part requires that the user scan the index, scan the chapter, pick a task description that sounds good, read one or two sentences of the introduction, scan the steps quickly to get the gist of the procedure, and return to work. This act of reading requires that the concept in the user's mind bond with words on the page and that the words register as significant. Significance is meaning. And you don't have it unless the word connects with the user. So in this way, we say that the words don't convey meaning to the reader, but the reader brings meaning to the words.

We should also factor in other variables, such as the fact that the motivation to consult the manual in the first place comes from subtle elements of the task context. The person using the computer might want to effect change in his or her organization. He or she might want to improve job performance, fulfill personal goals, increase job visibility, generate or express power. Moreover, the person has specific searching behavior (which some information scientists study) or problem-solving strategies (which cognitive scientists study). These elements of the job context give complexity to the created meaning of the words. The language of the document should help appropriate the energy the person puts into the task. It should fit into the dynamic psychological pattern of effort and reward, exertion and satisfaction.

The examples above are simple models of the act of processing language. While these models lack the complexity of some of those constructed by linguists to interpret how users make meaning, you probably get the point. Meaning, according to these researchers, depends on the context of the whole reading event, not just the context of the sentence. Instead of our words containing meaning, they say, we should think of words as getting meaning from our readers. Not only does meaning come from the user and the user's job context, the amount of meaning a text has—something finally impossible to measure—depends on the demands of the user's current task. The more clearly writers can anticipate that task—in layout as we saw earlier, or in language, as we see in this chapter—the more task-oriented the manual.

Performance-Oriented Language

Clearly the user's guide or manual could benefit from clear explanations of how to perform using the software system. In software documentation I call this making the language "performance oriented." The bet that the reader will slog through extended passages for simple information has not paid off for many documenters. Using the active voice, using the *you* pronoun, and using the imperative verb add to the performance orientation of the style. In a larger sense, too, the skillful documenter will write as if writing to a real human being. So often computer-related language sounds cold and alien to the novice user and probably bores the experienced user too. Written as if to an imaginary person, software language does not have to be stuffy, formal, or robot-like.

For example, consider this passage from the TC8215 SECTR Management System for Windows *User's Guide.* While most of the manual contains very technical information, this section, called a "Guided Tour of Sectr," uses relaxed diction to emphasize performance.

> *An address search can find a lot of objects, but you can stay active while the search is underway. For example, you can create objects by clicking on them and then clicking the Create Object(s) button. Another good move would then be to click on the Hide Dialog button so you don't have the Address Search dialog box in your face while you do other things.* (TC8215 SECTR Management System for Windows User's Guide, p. 2–5)

Notice the diction: "a lot of," "another good move," and "in your face." We would speak to real people using words like this, so why shouldn't we use "real" words like this in a manual? Of course, this kind of writing would likely flop in a reference section, where users don't expect to dally on ideas. But it does work in a tutorial section, where readers temporarily give control to the guiding voice of the manual writer.

How Do We Remember and Learn?

So, good writing in manuals and online help leverages the users' need to do work. But what happens, exactly, when users finally come across an idea they can employ: a procedure, a command, a technique? They have to remember the idea until they get back to work. How do they remember and learn?

When you read technical information—documents, guides, manuals—you have to rely on both short-term and long-term memory. Your ability to understand depends on words triggering memories you have of stored information. Some writing requires you to search extensively through your memory databases to find something to connect with the words on the page. Some writing seems to trigger these memories much more easily, so you don't have to put out so much effort in understanding. As a rule of thumb, the writing that requires a lot of effort in connecting words to meaning will not work as well. You should strive for writing that makes the connecting seem effortless.

Structured Sentences

The requirement of putting as few demands on the user's memory as possible means that the software documenter needs a thorough knowledge of how to structure information clearly in sentences. For one thing, the degree of structure in sentences will not necessarily show up as fewer words. More likely, it will show up as patterned, highly parallel sentences, balanced sentences, and sentences that end in three rhythmic clauses like this one. Well-structured sentences means that sentences are simple (using easy-to-understand language) and short (so as not to overburden the user's short-term memory). Below, we will explore many ways to build structures into the language of software manuals, with the intention of controlling the amount of structure in the style and thus adapting to the job needs of the user.

Style Problems in Software Documentation

Problems with style in software documentation relate directly to your overall goal of making software easy to use. Like it or not, users actually read a passage once all the access features of the document or help system have gotten them to the correct nugget of information. And that information, as you know, can take many forms: not just steps, but explanations, encouragements, overviews, interpretations, and other forms of sophisticated and well-designed language.

The next section details some of the problems that befall instruction writing in software documentation. It may help you to study the overview of some of the ways style hinders the use of software for productive work.

TABLE 12.4   Problems with Style

Ways Style Goes Wrong	Explanation
High Level of Abstraction Users confronted with terms that don't connect directly and immediately to daily work situations.	Acronyms: shortened versions of words: *DOS, MAC, UNIX*
	Synonyms: words that mean the same thing: *hit, press, push*
	System orientation: an emphasis on using commands rather than doing work: *"Using the File command."*
	No analogies/poor transfer leaning: poorly developed mental model of the system: *"The program runs efficiently in batch mode."*
No Performance Support Users confronted with the need to do work, but a language that emphasizes passivity and acceptance.	Passive voice: cluttered language based on helping verbs and nouns, *"The file was displayed by the program."*
	Definitional emphasis: language that tries to define rather than instruct, *"F1 is a function key used to open the context-sensitive help module."*
	Ambiguous task/step names: terms that name commands rather than giving steps, *"Printer."*
	Formal tone: language that ignores the way other users talk in the workplace and that discourages help from the user community, *"Entering data into the data-entry window of BossPro is quite intuitive."*
Overly Complex Syntax Users confronted with complex sentences that diverge greatly from the simple meaning.	Faulty parallelism: nouns, . . . ings, and verb phrases not in grammatical alignment: *"Saving a File, Open a file, . . ."*
	Lack of fluency: paragraphs don't develop the intended subject in the heading (*under the heading*) **Overview of the Program** *"Printing is very easy to do with BossPro."*
	Noun addiction: using nouns in front of words to define them, causing intense cognitive overload in the user: *"The advanced processor playback allocation file stabilizing fault-anticipator bit should be set to 'on'."*
	Contorted syntax: overly complex sentences that block first-reading clarity, *"The Enter key, which causes the data-entry screen to disappear entirely without saving changes, should be hit."*
	Confusing sequences: instructions that don't clarity steps properly, *"Before you open the file you should first have set up the file directory."*
	Sentences/paragraphs too long: unnecessary length where shorter, well-focused paragraphs would work better, *a "brief overview" that takes more than one or two short paragraphs.*

Table 12.4 gives you some idea of just how the very words you use and the way you arrange them in a document can affect the key goals of your manual. The following discussions of these and other maladies in language and style can help you target areas where your style can benefit by careful adjustment to achieve maximum ease of use.

TABLE 12.5   Acronyms and Abbreviations Abound in Manuals and Help

Type of Acronym or Abbreviation	Example
Abbreviation	Win, demo, doc, exec
Acronym	I/O, TEDIUM, BASIC
Three Letter Acronyms (TLA's)	IBM, VAX, DOS, RAM
Contractive Acronym	dcmnt = document
	dbms = database management system
	txt = text
	hlp = help
Command Contractions	autonum = automatic numbering
	fldsrch = field search

Acronyms

Most computer users have learned the more familiar *acronyms* used in programs: **DOS** for disk operating system, **RAM** for random access memory and so on. They have learned these even though they may not know their meaning completely. Anyone beyond a novice knows these terms, but that does not guarantee that their use leads to task orientation. Users often can understand common acronyms like these, but the difficulty comes when some of the terms constructed as part of the computer program or system appear in the document. This results in writing that emphasizes the system and its construction, rather than the user and the user's work.

Usually these uncommon acronyms consist of *contractive acronyms,* like "txt" for text, and "dcmnt" for document. The difficulty increases when these contractive acronyms refer to parts of programs and don't make intuitive sense, as with INSCOMP or PROC. Throw in some numbers and the problem gets out of control—as in the following example:

> *The AFC17 is a flexible, high performance, multi-channel analog-to-digital (A/D) converter. Under program control the AFC17 performs a 13-bit A/D conversion at a rate of 200 channels per second.*

Consider the kinds of acronyms and *abbreviations* illustrated in Table 12.5. You can't avoid them, but try to use only a few of these in a sentence at a time.

You can easily see that these acronyms need to appear in documents because software systems contain hundreds of them. They function very well for programmers because they allow for easy naming of program parts (called routines) that get generated during program development. Acronyms save time. But the trade-off for the saved space costs the reader in lost usability. Every acronym that you use should be shadowed by its meaning, either in parentheses or in the context of the sentence.

Synonyms

Some of the most common tasks in software use go by some of the most diverse names. What terms can we use to convey the idea of doing something as simple as

TABLE 12.6   Confusing Synonyms

User Question	Synonyms
"I want to start my program."	Open, boot, start, run, call up, load
"I want my program to work."	Install, setup, configure
"I want to stop my program."	Quit, exit, terminate, kill, abort
"I want to use a key."	Hit, press, depress, mash, punch, push, strike
"What kind of program do I have?"	Software, shareware, netware, freeware, dareware, beggareware, bread-on-the-waters ware

pressing a key? A partial list would include the following: *hit, depress, punch, push, touch, and strike.* (I like *mash,* myself, as in "Mash on the Enter key.") Some of these terms seem to come from slang usage, such as those for starting programs: *open, boot, start, run, call up, load, bootstrap.* Some of these terms have special meanings that distinguish them from one another, but in the common vocabulary of the computer user, some of whom have no computer vocabulary at all, they all mean about the same thing. Some terms for computer tasks even evoke disgusting metaphors, as in purge or depress or dump. Others sound violent: *kill, terminate, abort.* Some terms evoke paradoxes. My favorites include *"Press Enter to Exit"* and *"Press Return to Continue Quitting."*

Along with these synonyms, you also find terms that change meaning from program to program. For example, take the concept of saving a document to a disk file. Some call the process *saving* and see the document as saved. Others call the process *transferring* and see the document as transferred (to the disk). Some programs put numbers in *boxes,* others put numbers in *cells,* or *fields.* What should you do when faced with a program that uses a variation of one of these terms (some of which you see illustrated in Table 12.6) and the user knows other programs that use different terms?

Usually these terms have developed as ways to describe overall tasks, but you should always use them consistently, and as accurately as possible. You should anticipate those terms that your readers most likely will recognize and build in crossover techniques, such as using synonyms in parentheses, in the index, in tables, or just in text. Some have suggested that the prudent writer will index terms and even commands from competing products.

Paragraphs and Sentences Too Long

Another typical problem we find in software manuals stems from overly long paragraphs and sentences. Some writers see their mission in life as initiating the user into all the intricacies of computers in one manual. The length of their paragraphs and sentences attests to this. Ideally, paragraphs should focus on explanations, not performance, and not on steps telling the reader what to do. Paragraphs work best when they support a simple concept, such as definitions for terms, or distinguishing between functions or elements of the program. They help explain what happens after a step, and, because the user will not usually tackle paragraphs unless he or she must, they should read as quickly and easily as possible. It would probably be good for the documenter to forget about paragraphs altogether and just think in terms of lists and chunks of no more than three sentences.

Emphasizing the Computer Instead of the Program

When you think about computerized work, you realize that it involves basically three components: the user, the program, and the computer. The computer makes the work happen by taking the input from the user and processing it into output. But from the standpoint of the software manual, everything that happens results from what the program does, not what the computer does. We interact with the program, not the computer.

In conversation we often speak as if the computer does the interacting. Clerks enter data "into the computer," or "computers make the cartoon images on TV," or "computer simulations show how molecules attach." In reality, all these computer effects result from specific programs, but the computer gets all the credit. But the result of this attribution is that the user can start to see the computer as usurping control, as making decisions beyond the influence of the individual user.

To return control of software to the user, the manual writer should try very carefully not to fall into the trap of ascribing the actions of the program to the computer. For sure, the computer does a lot of things: It starts up, it shuts down. But for the most part, it follows the instructions from the program. Besides, the documenter often wants to promote the program by name to help the reader identify the functions of the program. Besides clarifying who does what, keeping this focus helps the user with learning, which translates into more doing. The following sentences emphasize the program, not the computer:

> *"The **Edit program** will store your document in the folder you specify."*

> *"When you press Save for the first time, **Microsoft Word** uses the first line of your document as a suggested filename."*

No Connection between the Heading and the Topic

Often you will read a passage from a manual or *readme* file that contains headings followed by sometimes long paragraphs explaining a program. They tell how it works, what computers it runs on, and other stuff. The problem with these passages is that often the paragraph doesn't contain the information announced in the heading. Instead, the writer has included one or two relevant details about the program, which get presented in highly technical language. Details like this require topic sentences, elaborations, examples, and other information elements to help the reader put the particular piece of data into some meaningful—work-related—context. An example:

How BeyondHelp Operates

Special care has been taken to adjust the BeyondHelp graphics filtering file format so that it processes files in half the time it did in previous versions. This change allows BeyondHelp to handle much larger files.

If you get right down to specific details, you miss the point of the heading. The reader of this example probably expected an explanation organized from general to specific that explained the basic principle of the program. The paragraph should have followed the heading with specific explanations of how the program operates, as in these two examples: "AccountMaster uses Excel spreadsheets to maintain bookkeeping information for small businesses" or "PagePro translates page formats among a large number of word processing and desktop publishing programs."

Too Formal Tone

Software documentation that you write for real people using the program in their work should sound conversational, not too formal. This example shows that the writer needed to relax: "A blank disk has been provided for you. It would be to your benefit to make a back-up copy so that you may always have an extra copy of the program if one should fail." Part of the formality here ("It would be to your benefit to . . .") comes from the writer not realizing that sentences must communicate with another real person. So often, the writer only thinks about getting the idea out.

But the idea really only counts if it makes contact with the motivation of the user, which usually means getting some job done. A well-intentioned writer might offer this hint: "Now that you know how to add an occasion to the file, may I suggest that you add some occasions of your own, such as your birthday." In a less formal tone, that sentence might read as follows: "Now that you know how to add an occasion to the file, try adding some occasions of your own, like your birthday." Speaking in an informal tone—without being overly familiar or presumptuous—makes contact more quickly and evokes the user's desire to do well on the job. You can identify an informal tone by incorporating the following characteristics into your style:

- **Use of contractions.** "Once you've set the parameters, you can use Alt+M to return to the Main menu."
- **Reference to other users.** "Most people use the Home Page as their base of operations."
- **Humorous aside.** "Note: Only the left mouse button has function. (Even on the rare fifteen-button mouse found only in the Australian outback.)"

When to Use Humor

The topic of humor relates to that of tone in software documentation. In manual writing, humor sparks as much controversy as it does good feeling. Glaringly inappropriate when it fails, bad humor can cause the user to reject a manual. Common sense tells us that humor will not work in all kinds of documentation, especially in support sections (reference, appendices, etc.) where the user simply looks up information and expects it to come sliding across the countertop without a lot of extra posturing on the manual's or writer's part. The psychology of the reference situation complicates the problem of humor because often the urge to insert a wry comment occurs when you describe errors or the results of errors. If the joke does not work, then the user will in no way appreciate the fun at his or her expense.

It makes sense to write reference documentation in a cooler tone. The relationship between the writer and the user takes on a more distant feel, not one conducive to back patting or rib tickling. Also, experienced users, for whom reference sections function best, tend to value accuracy over an open, more intimate style. Finally, not every writer has the talent to see the little quirks of computer interactions that give rise to humorous remarks. Even writers who do have that talent should consider what their manual will sound like in later editions when updated by a new writer who may not have a talent for humor.

When humor works, it works because it breaks some rule of formality. An early word processing manual showed a picture of a little granny being escorted away by two huge, uniformed police officers; the picture had the caption: "What happens if you press the wrong key." Humor takes a risk that many writers do not want to take, fearing that the comment will fall flat. Another difficulty lies in the exclusiveness of humor. Humor often leaves someone out—a powerful argument for keeping a businesslike tone.

So when should you use humor? Never in reference documentation, seldom in procedures, from time to time in tutorials and background information when a lighter, more familiar tone seems desirable. Never overdo it, and always edit carefully so you don't offend anyone. Always test drive passages that have asides or puns in them, to gauge users' reactions.

Ambiguous Task Names

Part of the problem of weak language in documentation comes from some writers' tendency to refer to tasks vaguely. In task-oriented documentation you should name tasks clearly, with a sense of planning for the user's new vocabulary. A task name like *Remove Applications* does not convey the idea clearly. Remove applications from what? Or does it mean applications that remove? *Removing Applications* or *How to Remove Applications* does a better job. You should try to make task names into headings or short sentences that predicate the user's action. Table 10.8 gives examples of headings.

The name of the task, in performance-oriented phrasing, should appear frequently in the text of the manual. Task names help steady the user on the right task, and they help focus on usability by allowing you to keep track of just what needs explanation. Because task names occur in headings, you may hesitate to repeat the task names in the text, fearing redundancy. Not repeating task names, however, leads to an overuse of the relative pronoun this, as in:

> *Click the Speedo Icon to start the program. This will cause the icon to flicker and the program will display the main screen.*

or as in

> *Whetstone: This classic benchmark is primarily a test of floating point math with a heavy emphasis on transcendental functions.*

The problem here relates to vague pronoun reference, but it also signifies a lost opportunity for the writer to reinforce the task orientation of the manual. Revised, the second example above would read "Whetstone: Using the Whetstone test (see page xx) allows you to examine . . ." The result makes better sense as good syntax and reinforces the idea that the whetstone test can help the user.

Step Not a Step

Very often the beginning documenter will not clearly understand the nature of a step in software documentation. As we saw in Chapter 5, the *step* constitutes the basic element of human-computer interaction. As such, you should articulate the action element of a step very carefully. Where does the following step fail? "Step 2: You will be prompted with a dialog box that will ask you the name of the file." This step fails because it does not express an action. Actions come in the form of a command, so the step should read: "Step 2: Type in the name of the file when the dialog box appears." To avoid this problem you should always examine your steps to make sure they contain a clearly stated action, often using an imperative form of the verb.

Omitted Articles

It's very common in documentation to fall into the telegram style of writing. Telegram style refers to the time when people paid per word to send messages, and so they left out articles like the and an to save a buck. Telegram style writing sounds like this:

> "*Use doclst only when existing doctor or referring doctor treats patient at other hospital or not at office site.*"

In fact, we don't have to pay per word anymore for our messages, and such attempts to sound official fall flat. Why not communicate with real people, using something like this:

> "*Use the doclst function only when an existing doctor or referring doctor treats the patient at another hospital or not at the office site.*"

Glossary

abbreviation: a shortened form of a word, used to save space and avoid repeating long names. Example: "sys" for "system."

contractive acronym: an acronym made by removing vowels or other redundant parts of words. Example: "rnw" for "renew," "lst" for "list."

acronym: a shortened form of a word or group of words that uses the first initial of each word in the phrase.

first-reading clarity: a kind of clarity in writing that allows the user to understand on the first reading.

parallelism: the technique in writing characterized by repeated structures of grammar, syntax, or other sentence elements for the sake of emphasis and easy learning by repetition.

performance orientation: a way of using language based on the premise that the user will more readily understand if explanations are put in terms of performing useful workplace actions.

☑ Checklist

Use the following checklist as a way to analyze your writing style for usable task orientation.

Style Checklist

❑ Do your sentences emphasize actions the user can take rather than functions of the program?

❑ Do your sentences use the active voice predominantly?

❑ Have you reviewed your sentences to make sure they have first-reading clarity?

❑ Have you avoided overly long and complicated sentences?

❑ Do your sentences contain easy-to-use parallel sentences?

❑ Have you included clarifying operational overviews in your documents?

❑ Do your sentences reflect a performance orientation?

Analyze your style to make sure you avoid the following problems:

❑ Overused and unexplained acronyms and abbreviations

❑ Confusing synonyms

❑ Overlong paragraphs and sentences

❑ Emphasis on the computer rather than the program

❑ Lack of connection between headings and topics

❑ Too formal tone

❑ Inappropriate humor

❑ Ambiguous task names

❑ Confusing statements of steps

❑ Omitted articles

Practice/Problem Solving

1. Analyze Writing Style

Analyze the style in two comparable manuals or help systems according to the principles expressed in this chapter. Given the differences you find between them, which do you think contributes more to the task orientation of the document?

2. Learn from Others' Mistakes

Find an example of weak language and style in a procedural manual (a user's guide or other document designed to lead the user through steps). You may want to review examples of in-house manuals written by persons not adequately trained in performance-based writing. Determine whether the language reflects a system or task orientation. Use specific sentences and phrases from the manual that, to you, indicate opportunities for improvement. Practice by rewriting the passage so that you think it reflects a task orientation.

3. Revise for an Emphasis on Actions

Revise the following sentences to shift them from a focus on functions to a focus on actions.

a. Zoom. Enter a magnification between 10 and 400 percent.

b. The following section will explain the use of the List menu.

c. In order for these instructions to be used effectively, the user must have a working knowledge of WordPerfect.
d. The Reporting sub program provides a number of features, including Print Setup, Multiple Report, Fax Printing, and Color.

4. Revise for the Active Voice

Revise the following sentences putting them in the active voice.

a. The following examples are to be used as a reference guide.
b. Check to see if the data are entered correctly.
c. The prompt will be shown as a dollar sign.
d. Events do not need to be all entered at once and can be modified accordingly.

5. Revise for Correct Parallelism

Revise the following sets of lists and put them in grammatically parallel format.

Example 1

- Custom Background and Font Colors
- Insert Your own Logo
- Custom Headers and Footers
- Choice of Time Zone, and Date Format
- Choose Your Banner Type
- Thread Format Choice
- Convert Smile Symbols to Images

Example 2

- Receive a Discount
- Brand as Your Own
- Your Own Price Charged
- Increase Your Recurring Revenue & Sales of Core Products and Services
- Improve Customer Retention and Satisfaction
- Ease of use
- No Installation, Completely Browser Based
- All Inclusive Self-service Customer Sign-up and Development
- Interface provides complete e-commerce resources
- Retain Your Customer for All Their Needs

6. Revise to Simplify

Revise the following sentence to make the grammar and style as simple as the idea they express.

a. Before you start using ModemMaster you should have set up your computer by plugging it into a power source, attaching the keyboard and mouse, and connecting any peripheral devices, such as printers, which are described in Section 4: Printer Setup.

b. For each event you are asked to enter planned, revised, and actual event start dates and event end dates.

c. If at any point during the entry of the maximization function and the constraints the data are entered incorrectly the following may be done at any point during the program to edit the data.

7. Write Operational Overviews

Rewrite the following paragraphs shifting the focus from the technical to the operational.

a. Sending mail is controlled with QuickMail's message windows, toolbars, and icons. Message composition occurs offline, reducing the amount of connect time. Composing offline represents a significantly cost-effective approach for users.

b. In the first place, there is only one official version of FlagMaster and changes to the basic program are discouraged. A "file change" mechanism allows the program to be compiled differently as dictated by the hardware and system software constraints that exist at any particular computer installation, but these changes should not modify FlagMaster's formatting capabilities.

8. Relax the Diction

Rewrite the following passages in a more conversational style that encourages performance among users.

a. Before you can begin using the Party Planner program, you should have set up and configured your computer. Setting up involves plugging it into a power source, attaching the keyboard and mouse, and connecting any peripheral devices, such as printer or fax machines. Instructions for setting up the computer are in the owner's guide.

b. If the user is printing a large database, the command prompt "Press any key to abort print" appears. Pressing a key will eventually result in the "printing process abort confirmation" prompt.

c. As the user progresses in the comprehension of the software application program, facility in executing and customizing the utility of the Analyze features will increase substantially.

9. Make the Diction Less Abstract

Examine the following sentences that demonstrate the difficulty of understanding abstract terms. Rewrite them to make them more concrete.

a. As soon as the chart data form appears on your screen, press F9/Options and select Option #6, Bullet Options. Use the SPACE bar to cycle through the selections until the word "None" appears (or use the F3/Choices function key to display a complete listing of bullet types.)

b. If you hit auto-configure on boot you may set up the wrong file table, so you should press Esc to terminate the load.

10. Eliminate the Clutter

The following sentences don't support workplace tasks because of their wordiness. Revise them using fewer words and making clearer sentences.

a. You are prompted to enter a chapter name when you use the Create Chapter function to create a new chapter. You may also change the current chapter name at any time while running the program by selecting the Rename Chapter function under the Chapters menu.

b. The following page shows a photocopy of the model that is produced by this set of instructions. This copy may be used as a reference to ensure that the program has been correctly completed.

c. Installing the Race Sheet program on a hard drive requires that the software program is on a non-copy protected program distribution disk. The user of the program must insert the program distribution disk into the floppy disk drive of the computer.

11. Reduce the Number of Acronyms and Abbreviations

Rewrite the following passages to reduce the use of acronyms and abbreviations.

a. There are several reasons to name BNN files, not only to use them with the RETRF and RENAM functions, but to avoid confusing with DOS or UNIX filenames.

b. To round out our discussion of \def, we will note that there exists a \let primitive that is somewhat analogous to \def but differs from \def in timing. The difference can be explained by noting that \let\b=\a sets \b to \a at the time when \let\b=\a is read by MegaForm.
(Hint: \b and \a are "values" and \let\b=\a is an "expression.")

c. Use .ip, .ep, .sp, and .lp to set VARS for paragraph and chunk sizing in SETUP or REDO modes when running in DOS.

12. Create Real Steps

Rewrite the following sequence of steps, eliminating the nonsteps, and organizing the information in an action/result pattern.

1. Enter the maximization function at the colon prompt:
   :MAX 3x+2y
2. Press Enter
3. Note: At this point, a ? prompt will appear.
4. Type the following at the ? prompt: Note: This command stands for "subject to" which signals the computer that the next set of data entered will be the constraints.
   ?S.T.
5. The program will display the ? prompt.
6. Press , the program will display the ? prompt.
7. Type ?END at the ? prompt. The program will respond by asking you to enter a disk in drive A:

13. Add Articles

Revise the following sentences to include articles and make them sound more human.

a. If insurance company found in list, then type in number of carrier. Press Return.
b. Turn on computer and monitor.
c. If doctor is not in history file or has changed diagnosis data but not updated patient record, then do 1-main menu, 1-update patient record to update patient record.
d. Find data and resource files in cabinet next to computer.

14. Add Humor

Imagine you're working with a scientific calculation program that displays some irregularities in its overall operation. This program, DigiDoc, displays the odd feature of requiring the user to input the same information (a date parameter) twice in the same sequence of steps, once at the beginning of the sequence and again at the end. If the user doesn't enter the date in the exact same format as the original date, the program assumes the user has given a "nul" date variable and stops the operation, returning the user to the system prompt. Other than that, the program runs pretty straightforwardly.

Your task: write a paragraph or so about this behavior, informing and warning the user. The description will go in the introduction to a chapter called Inputting Calculation Variables. Use the technical description above as the starting place, but put the experience in terms of the user's point of view.

15. Shift the Focus from the Computer to the Program

In the following examples, revise the sentences to emphasize the program and its functions rather than the computer.

a. When you ask for an inquiry, the computer searches the hard disk for the file containing the account records for that vendor.
b. The computerized version of the document will appear on the screen, and you may edit it using the draw functions.
c. To choose a control code, click in the box next to the number one (the first column) and type in a code. The screen shows the code you entered.
d. When you press Tab the computer automatically records the default data for that field and moves to the next field.

CHAPTER 13

Using Graphics Effectively

This chapter covers the use of illustrations in manuals and online help. Examples of graphics in manuals reinforce the key role images play in informing the user. The chapter offers guidelines for selecting and designing types of graphics that are essential to software documentation. It discusses the functions of graphics and presents an overview of the types of graphics and graphic elements.

How to Read This Chapter

You should read this chapter in conjunction with the previous chapter. This chapter on visual elements complements Chapter 12, "Getting the Language Right," just as text and images complement one another on the page and screen. Both the project-oriented reader and the reader for study and understanding should study the examples and eventually read the entire chapter.

- Those planning a project or involved in a project will want to consult the section on "Types of Graphics in Software Manuals and Help" for a quick overview of the main options available in these media. In creating graphics, the section on "Elements of Graphics," and the Checklist at the end of the chapter can aid in double-checking work.

- For the reader wishing to master the basics of graphic design in manuals and online help, the Discussion section provides a clear overview; you should probably read it, then the Guidelines.

Examples

The two figures provided in this Examples section show you two ways you can support task orientation by using graphics. Figure 13.1 illustrates how graphics can present printing options to the user in a way that makes selecting the correct printing paper easy and efficient. This example uses graphics to support user decision making. As you'll see from this chapter, supporting user decisions presents just one of many

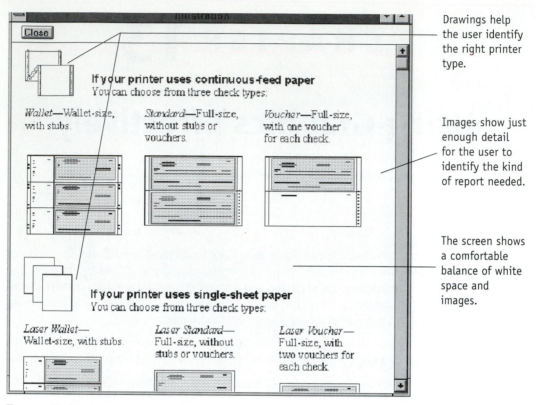

Drawings help the user identify the right printer type.

Images show just enough detail for the user to identify the kind of report needed.

The screen shows a comfortable balance of white space and images.

FIGURE 13.1   An Illustration to Support Decision Making
These graphics from the Microsoft Money Help system assist the user in making decisions about how to apply the program features to the workplace.

opportunities to use graphics. You can also use graphics to indicate the *interface elements* of a program's main screen, as shown in Figure 13.2. This reference-type documentation is a staple of manuals and online help, and one that you should prepare carefully, with close attention to the user's task needs.

Guidelines

Figure 13.3 lists guidelines for using graphics effectively.

1   Identify Needs for Graphics by Your Users

As with all elements of a well-constructed manual or online help, the graphics should support user questions: "How can I use the program easily?" and "How can I put the program to work?" To meet the first need, use graphics to help the user locate and act: to get to and operate the actual features of the program. To meet the

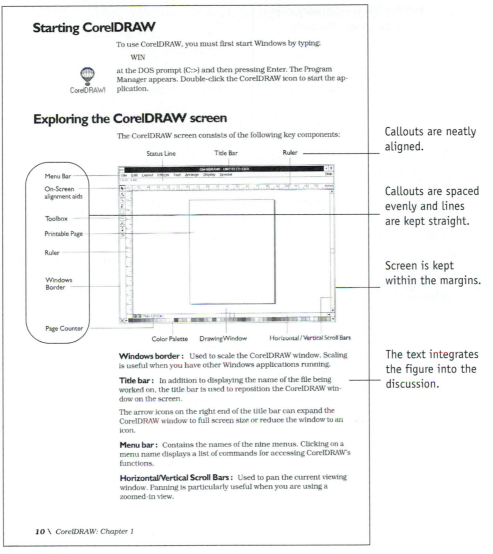

Starting CorelDRAW

To use CorelDRAW, you must first start Windows by typing:

WIN

at the DOS prompt (C:>) and then pressing Enter. The Program Manager appears. Double-click the CorelDRAW icon to start the application.

Exploring the CorelDRAW screen

The CorelDRAW screen consists of the following key components:

Callouts are neatly aligned.

Callouts are spaced evenly and lines are kept straight.

Screen is kept within the margins.

The text integrates the figure into the discussion.

Windows border : Used to scale the CorelDRAW window. Scaling is useful when you have other Windows applications running.

Title bar : In addition to displaying the name of the file being worked on, the title bar is used to reposition the CorelDRAW window on the screen.

The arrow icons on the right end of the title bar can expand the CorelDRAW window to full screen size or reduce the window to an icon.

Menu bar : Contains the names of the nine menus. Clicking on a menu name displays a list of commands for accessing CorelDRAW's functions.

Horizontal/Vertical Scroll Bars : Used to pan the current viewing window. Panning is particularly useful when you are using a zoomed-in view.

10 \ CorelDRAW: Chapter 1

FIGURE 13.2   A Fully Labeled Screen
This example from the CorelDRAW *User's Guide* shows how to do one of the basic illustrations in your manual: the screen shot.

second need, use graphics to help the user understand: provide images that encourage education, guidance, and support for workplace tasks. Thus, instead of using graphics for decoration, you use graphics strategically to locate and direct user actions in using the program, and you also use them to explain concepts and illustrate examples.

You can meet the user's needs by following an easy process of first identifying user questions and then choosing graphics that respond to those questions. Go back

FIGURE 13.3   Guidelines for Using Graphics Effectively

1. Identify needs for graphics by your users.
2. Set graphics styles
3. Revise and edit
4. Revise for typography

over your user analysis and sift through the observations it makes about specific difficulties your user has. Generally, use more cartoons and animation with novice to experienced users and technically-oriented charts and diagrams with experienced to expert users. Respond to the questions you think your user would have about a task by designing graphics to support those questions. Some typical user questions are discussed next.

Where Is It?

Use graphics to help the user locate buttons, rulers, sliders, check boxes, menu commands, and other interface elements on the screen. Table 13.1 indicates some of the things users need to find and suggests ways to signal locations.

Making things visible means three things:

- Show the user where to look to perform tasks (see Figure 13.4).
- Show concrete versions of abstract things.
- Make visuals clear.

Show the User Where to Look to Perform Tasks

Users of programs need to focus their intention to perform a task (set a format, delete a drawing object, run a program) to the right interface object (button, ruler, icon, link, etc.). By showing the user the button rather than naming the button, you automatically encourage use. Also, the picture of a button requires less thought than the word *button*.

In simple terms, let the page reflect the screen. Use capture and art programs to bring elements off the screen and into manuals and online help. Usually, capturing a button takes little time or space. When you describe tasks involving complicated or hard to find screen objects, show them. In Figure 13.5 the manual makes locating screen objects even easier by magnifying them.

Show Concrete Versions of Abstract Things

Making things visible means helping make the abstract concrete. We saw earlier how metaphors help explain abstract concepts. Figure 13.6 helps users see the abstract concept of a relational database. The user is probably the biggest abstraction in your

TABLE 13.1   Help Users Find Things

Elements User Needs to Locate	Ways to Signal Locations
Screens	Arrow
Buttons	
Menus	
Checkboxes	
Scroll bars	
Sliders	Circle
Toolbars	
Message/Status boxes	
Page/Section numbers	

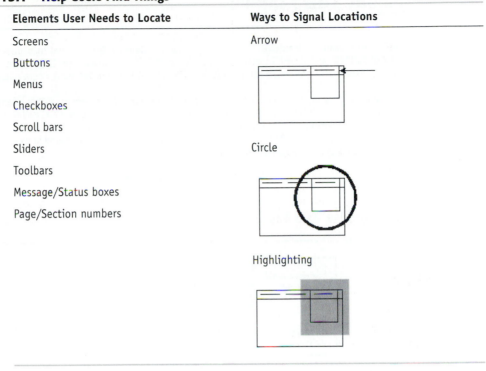

Highlighting

manual or help system. You should try to present images of your user in the manual to emphasize important points. Showing the user in images (as in Figure 13.14 explaining how the program works with a workgroup) provides a role model for the user. For novice users, such a role model can help them see the productive interaction of their work, themselves, and the program.

Make Visuals Clear

Make sure your users can see the things you want them to see. Avoid screen captures of buttons and keys that clutter the page or that you have had to reduce and squash to fit into the lines while sacrificing overall sharpness of the image. You will face the temptation to make your graphics smaller at times, to make them fit on the page, but you should remember that they do no good if the normal person can't distinguish details easily.

What Is It?

Use graphics to help define concepts unfamiliar to the user. These needs often fall into the area of subject-matter knowledge or background information the user needs about a program or idea. User testing and interviews can help you identify those concepts that you need to define. Graphics that are used to reinforce concept understanding fall into two types: examples and metaphors.

About This Manual

This manual contains instructions for installing and configuring the National Instruments AT-GPIB/TNT or AT-GPIB interface board and the NI-488.2 software for DOS. The NI-488.2 software is intended for use with MS-DOS version 3.0 or higher (or equivalent).

This manual uses the term *AT-GPIB/TNT* to refer to a National Instruments GPIB board for the ISA (PC AT) bus equipped with the TNT4882C ASIC. The term *AT-GPIB* refers to a National Instruments GPIB board for the ISA (PC AT) bus equipped with the NAT4882 and Turbo488 ASICs. This manual also uses the term *GPIB board* in cases where the material can apply to either board.

This manual assumes that you are already familiar with DOS.

How to Use The Manual Set

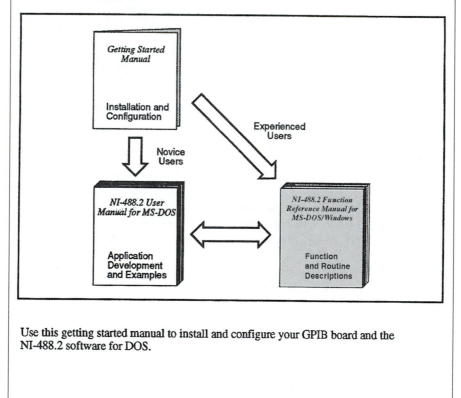

Use this getting started manual to install and configure your GPIB board and the NI-488.2 software for DOS.

© National Instruments Corp.   *ix*   *AT-GPIB/TNT for MS-DOS*

FIGURE 13.4   Graphics Show the Document's Structure
In this example from a manual called *Getting Started with Your AT-GPIB/TNT and the NI-488.2 Software for DOS,* the images convey what information users need from what documents. Users can identify their use class and react appropriately.

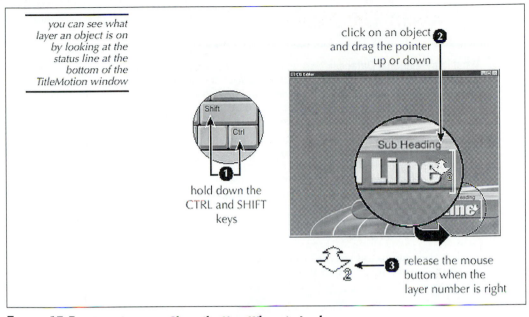

FIGURE 13.5   Area Captures Show the User Where to Look
These captures help users find the right place to click.

Examples show documents, reports, printouts. Examples also, as in the example in Figure 13.7, show sample data, helping the user identify the various aspects of a procedure. Mind you, inserting an example doesn't make it a graphic in the sense that it communicates visually. But examples benefit greatly when you explain them and label them as in Figure 13.7. You should use examples liberally, especially in highly technical, *command-driven* programs. Such programs include compilers, scientific graphing and analysis software, and operating systems.

Metaphors show the basic nature of an idea by relating it to something the user already knows. Common metaphors that abound in software manuals and online help are shown in Table 13.2. Metaphors allow the user to know something without having to learn it from scratch. Users can rely on their previous experience in the world to do some of the explaining for them. Metaphors are workarounds for the general abstractness of computer terminology. They compare two things: the abstract or hard-to-understand technical concept, and a concrete object or idea known to the user but in another domain. For example, novices can understand the abstract concept of the workspace of a computer by relating it to a concrete workspace they know well: the desktop. Metaphors abound in the kinds of interface elements in programs: radio buttons, sliders, pull-down menus. To help us understand the program's capabilities, designers have portrayed them using pictures of familiar objects. This familiarity increases the likelihood of successful interaction with the processing capabilities of a program. Figure 13.8 illustrates the power that graphics have to make the abstract concrete through metaphors.

You could argue that metaphors belong under the subject of language because we usually use words to convey them: "The program has three levels" or "Take

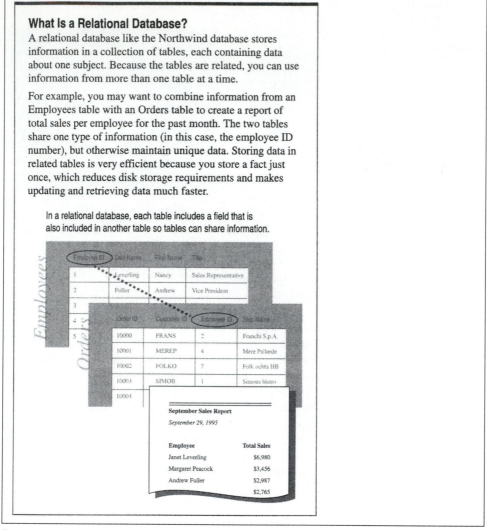

Use Your First Microsoft Access Database

What Is a Relational Database?

A relational database like the Northwind database stores information in a collection of tables, each containing data about one subject. Because the tables are related, you can use information from more than one table at a time.

For example, you may want to combine information from an Employees table with an Orders table to create a report of total sales per employee for the past month. The two tables share one type of information (in this case, the employee ID number), but otherwise maintain unique data. Storing data in related tables is very efficient because you store a fact just once, which reduces disk storage requirements and makes updating and retrieving data much faster.

In a relational database, each table includes a field that is also included in another table so tables can share information.

FIGURE 13.6   Make the Abstract Visible
Graphic examples such as this one from the Microsoft *Getting Results* guide enable the user to associate something visible with something abstract.

the MS-DOS 5-minute workout." But such metaphors of language, where we compare two things (a program and a level, a lesson and a workout), gain strength when we support them with the actual images suggested by the words. Thus, the MS-DOS five-minute workout booklet contains an abundance of runners, rope skippers, and images of tennis shoes and gym equipment to help create and extend the metaphor.

FIGURE 13.7   Graphic Showing an Example
This topic from Microsoft Excel Help shows an example and how to copy the example for use in the program.

How Do I Do It?

Users wanting to know how to perform a procedure will welcome graphics that demonstrate and support sequential actions. These kind of images use drawings to introduce the step-by-step procedural work done in most computer work. Graphics that answer the question "How do I do it?" give overviews of procedures as a way of introducing the user to the steps that follow. In their simplest form overviews of procedures can consist of words inside boxes connected by arrows.

Because they use flowcharts in designing computer programs, many people working in the computer industry see these kinds of process charts frequently. Overviews, like the one in Figure 13.9, help the user build a ***mental model*** of the process before performing it.

TABLE 13.2   Graphics Help Users See Metaphors

Concept	Metaphor
Typing	Courier font (typewriter): `Type the filename.`
Work area on a screen	Desktop:
Fill areas in draw programs	Paint bucket:
Email	Mailbox:

Where Am I?

It's a better idea to show users their location in a manual or online help system than to tell them. Called *access indicators*, graphics like this tell users where the information they have before them fits into the organized whole. As indicated in Figure 13.10, you can reproduce a small-sized map of the document structure in the margin of your chapter title pages, and at other points, to keep your user from feeling lost. Access indicators consist of diagrams showing current location, history maps, and header and footer icons.

Where pacing counts, as in tutorial documentation, you want to make sure the user can easily see progress through a number of pages or lessons. For this reason you might use a *progress indicator* in the header or footer to keep track of lessons, as shown in Figure 13.11.

What's the Big Picture?

People new to computing, or new to a program, need to know the structure of the program—how it's all assembled. Beginners need this because they can't draw on a store of knowledge of computer systems to help them operate efficiently. Experienced users need reminders of how programs work. And expert users need this because they may never have worked with the program before, but their knowledge of computer systems in general leads them to expect that such a structure exists.

Using a good printer and publication program, you can produce a very high quality overview graphic for your manual or online help system. I think one of the tricks involves keeping things as small as possible when you're drawing an overview graphic. You have the capability to make elements very small in these drawings so that the clarity of line available on laser printers can pay off by allowing you to present a fair amount of legible information. You can use icons from the program to help users make

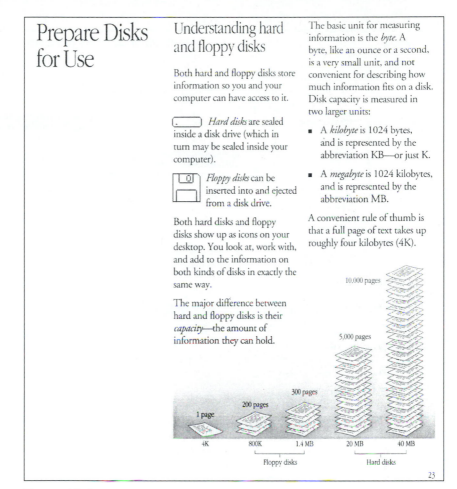

FIGURE 13.8   Metaphor Helps Define an Abstract Concept
In this page from the Macintosh *User's Guide,* the metaphor compares the capacity of disk drives to stacks of paper. The *familiarity* of the concept of stacks of paper to measure storage capacity helps the user understand the *unfamiliarity* of the concept of bytes to measure storage capacity.

the connections between the things they see on the screen and their workplace demands to learn, process, communicate, and store information. You can also use color, to indicate levels and to separate elements of the program for easy recognition.

The following section illustrates some of the graphic elements you can use to give overviews.

OVERALL PROGRAM DIAGRAMS. Overall program diagrams illustrate program system components so the user can see the flow of information. These comprise some of the most important kinds of information you can present and will result in greater dependability of user performance. Figure 13.12 shows an example.

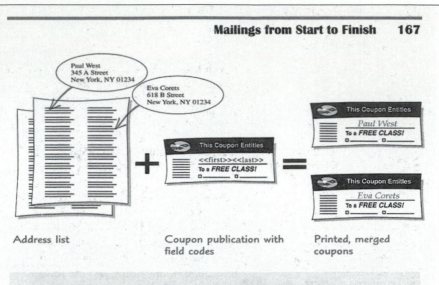

Address list

Coupon publication with field codes

Printed, merged coupons

Cost-effective direct mailing

A recent study of 3500 households to determine the response to direct, non-personal mail found that 70 percent of all households read their advertising mail. Of course, the fact that the mail was read doesn't necessarily mean that the reading led to customer responses or purchases!

The percentage of response to a mailing can vary widely, depending on factors such as effective design of the publication and the appropriate targeting of the mailing audience. One study showed a 7-percent return, which is much better than average for direct-mail advertising. Seven percent might not sound significant, but if you compare the cost of 200–300 pieces of mail to the cost of an ad in a newspaper or trade journal, or to the cost of a 30-second radio spot, you might find direct mailing to be a much more cost-effective method of generating new or repeat customer activity for a small office or home business. With newspaper, magazine, or radio advertising, there's no guarantee that someone interested in your product will read or listen to your ad. But a direct mail piece is delivered directly into the hands of potential customers. Of course, your mailing will be most effective if it's targeted to people who will be interested in and likely to purchase your products or services.

FIGURE 13.9   Diagrams of Processes Help Users Understand Steps
These diagrams give an overview of complicated software processes.

MENU MAPS. Menu maps abound in software documentation. Essentially they consist of program menus arranged on the page in the same structure as they appear in the program interface. As reference documentation, they help the user maintain a

FIGURE 13.10   Structure Diagram Showing the "Current Location"
You can use graphics, especially color, to orient the reader to a position in the overall structure of the information.

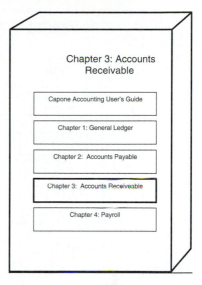

FIGURE 13.11   Progress Indicator Showing Lesson 3 of 4
In this example, each lesson shows as an increment in the graphic in the header to help the user stay oriented.

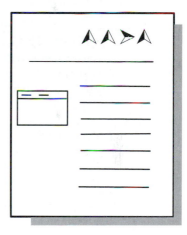

sense of organization of the program's features. The user can relate any specific function to the whole picture of functions in the program. Figure 13.13 shows a menu map.

CONCEPTUAL OVERVIEWS. Unlike program structure overviews, conceptual overviews reinforce the ideas of how to use a program. These essential and economical kinds of graphics help users see how to put the program to meaningful work, often, as in Figure 13.14 through the sharing of information and resources. Graphics of this type strengthen the emphasis on using programs in the workplace.

Follow these guidelines in designing conceptual overviews:

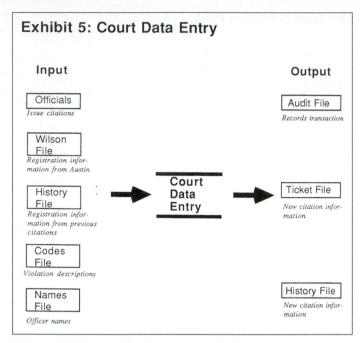

FIGURE 13.12   Overall Program Diagram
This diagram illustrates what occurs during processing. The diagram helps the user anticipate the process and fit unexpected events into the overall picture.

- **Use generic figures.** In all representations of the user, mix races, ages, and genders where appropriate (e.g., don't represent young persons in images of patients in nursing homes.) Reflect the users sampled in the user analysis. Find images that relate to the user's workplace. Use easy-to-identify images.

- **Embody the mental model of use to predict successful user actions.** Think of the mental model you want the user to have: working with a single file, storing in compressed format, dragging and dropping, filtering formats. Use a graphic that can be associated with the mental model.

- **Simplify use concepts, but make the images visually interesting.** Graphics don't work well for overly complicated ideas, and users like to rest their eyes on visually pleasing images. The more pleasing the image, the more the user pauses and thinks.

HOW TO USE THE MANUAL. Often you have the opportunity to use graphics to reinforce the idea of the big picture of your manual and online help system. In such cases you can route users to the right section in your book.

To help you decide on the kinds of graphics you need to respond to user questions use the table for Responding to User Questions in the Checklist section below. This planning tool summarizes the options described in the previous section.

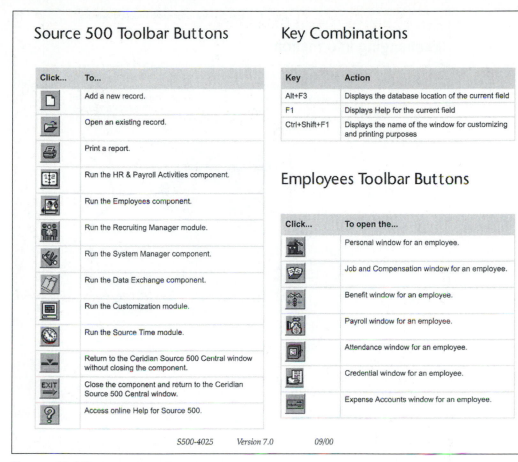

FIGURE 13.13   Menu Maps Help Users See Features
Menu maps like this one foster a sense of control over the program because they make all the features visible. This one is conveniently placed on the back cover of the manual.

2   Set Graphics Styles

You should use the same types and fonts, the same arrow styles, and box and frame styles throughout your document. Establish these standards for yourself early in the project, communicate them to your team members, and record them in the documentation plan. Update the standards as you encounter constraints of equipment or time and refine them as you go to your next project. Table 13.3 lists the elements of graphical styles.

Keeping styles consistent also has to do with the degree of realism you impart to your graphics. Depending on how much you want your images to reflect realistic details, you can gain some advantages by deleting details from your drawings and using more of a cartoon or line drawing style. This not only helps you focus the user's attention on specific elements of a drawing, such as report shapes or dialog box

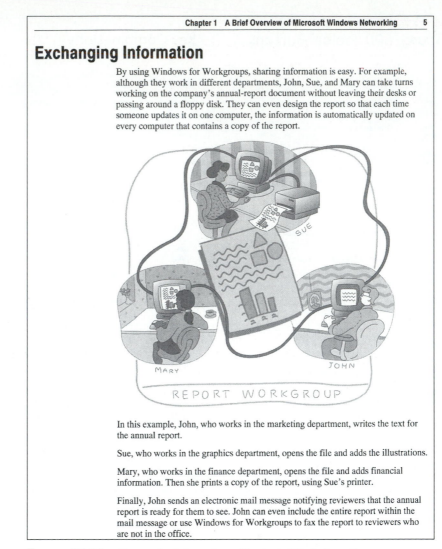

Chapter 1   A Brief Overview of Microsoft Windows Networking   5

Exchanging Information

By using Windows for Workgroups, sharing information is easy. For example, although they work in different departments, John, Sue, and Mary can take turns working on the company's annual-report document without leaving their desks or passing around a floppy disk. They can even design the report so that each time someone updates it on one computer, the information is automatically updated on every computer that contains a copy of the report.

SUE

MARY   JOHN

REPORT WORKGROUP

In this example, John, who works in the marketing department, writes the text for the annual report.

Sue, who works in the graphics department, opens the file and adds the illustrations.

Mary, who works in the finance department, opens the file and adds financial information. Then she prints a copy of the report, using Sue's printer.

Finally, John sends an electronic mail message notifying reviewers that the annual report is ready for them to see. John can even include the entire report within the mail message or use Windows for Workgroups to fax the report to reviewers who are not in the office.

FIGURE 13.14   Conceptual Overviews Illustrate Workplace Use
You can use graphics like this from the Microsoft *Windows for Workgroups User's Guide* to portray many kinds of concepts related to efficient and effective work.

choices, but it keeps the size of your drawings down. Photographic realism costs in hard disk or CD storage space.

Part of setting styles for graphics entails knowing what tools you have available for executing them. As you saw in Table 13.1, graphical elements often include screens and some form of label or focusing drawings (lines, circles, etc.) to indicate the relevant part of the screen. How you can label or otherwise annotate the screen depends on the tools you have available. Figure 13.15 shows some of the tools avail-

TABLE 13.3   Elements of Graphical Styles

Elements	Examples and Uses
Lines	Thin for callouts
	Thicker for highlighting choices
Fonts	Elaborate for titles
	Bold for tasks
Arrow Styles	Match sizes with uses, don't cover up part of an image with an oversized arrow
Box Styles	Use simple boxes for structure charts
	Use shadowed boxes for examples and reports
Frame Styles	Use simple shapes for error messages
	Use ornate shapes for tips and anecdotes

able in Microsoft Word. Graphics programs such as PhotoShop offer similar tools. When you are describing styles for graphics these tool selections can help you specify how you want your screens to appear.

You can read the values for lines or find examples of callout shapes to include in your documentation plan by finding the tools your word processor or authoring system provides. Give some thought to planning and setting standards for how you want graphics to appear. You have to make them fit in the overall page design, as one of the elements that has its own area or rules that the user can count on. And the sooner you can decide on the major issues (size, type) the better.

As you decide on the styles of your illustrations, record them in a table, as shown in Table 13.4. You will find this form below in the Checklist section of this chapter.

3 Revise and Edit

Once you have identified the kinds of graphics you want in your document, and you have a draft to work with, you can revise based on those standards, and for overall correctness and consistency. Here your eye for small spaces and alignment, your sensitivity to detail will serve you as an editor. For example, too much crowding of images and text creates an unattractive space, one the user has to work extra hard to get information from. Consider the following guidelines for revising your graphics.

Graphics, by their visual nature, present the user with the thing itself rather than the word for the thing. Because of this visual nature, they work best for recognition and retention in memory rather than for understanding. For this reason you should give your graphics a clear purpose—to support a task, to remind of an idea, to show use of a tool—so the visual element doesn't confuse the user. Software gives you the capability to make just about everything in a manual or online help system graphic. But should you use graphics as much as possible? Should you associate each task with an image?

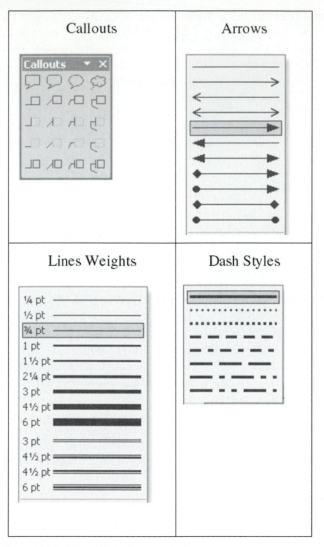

FIGURE 13.15   Tools for Labeling Screens
These lines and shapes provide a variety of tools for
implementing the styles of graphics in manuals and help.

TABLE 13.4   An Example of Planning for Graphics Styles

Graphic Element	Planned Styles
❏ Lines	4-point grey for subheads
❏ Fonts for titles	Arial 10-point
❏ Arrow styles	

I would answer, "No." Users have a craving for names: for the name of the thing (a word) rather than the thing itself (a picture). Words complete the user's sense of control over what the program does, so you should strive for a balance of both.

The following section offers additional advice for the care of your graphic elements. Under each of the headings you will find practical advice that, coupled with common sense and a developed eye for images, will help you design a scheme of graphics and text that will enhance the task orientation of your document set.

Titles

Not all images or screens require titles. As a rule of thumb, the more complex, the more the need for a title. Also, if the user can't easily see the relationship between the image and the text, then you should use a title to clarify the function of the image. Follow the guidelines below in creating titles for your graphics.

- Number the titles sequentially.
- List numbers and titles in the front of the manual.
- Don't use the same graphic over and over again because it's easy to use. Users will catch on.
- Use boldface titles, sometimes enlarged, in body-text style.

Labels

As with other graphic elements, you don't need labels on all your images. Often called *callouts,* labels point out the salient elements of a picture or drawing and direct the user to the correct and informative parts. If you label your graphics, follow these guidelines:

- Explain most (or all) figures. A figure without explanation sometimes works: icons and other spot graphics that work as special eye catchers for users don't need explanation; they explain themselves. Other figures presented as examples require explanation.
- Place callouts outside the image or screen. Callouts should go in a line at the left or the right of the image but should not cover part of the object they identify. Break this rule when you don't have space outside the image to put the callout.
- Use consistent capitalization. In general, the shorter your headings and labels, the more you should use all caps. In principle, identify a style and stick to it.
- Label components of screens used for presenting overviews and screen objects.
- Keep captions brief (don't try to say in a caption what you should say in the text).
- Make captions terminology consistent with the text.

Placement

Placement relates to where you put images on the page or screen. Follow these simple rules for placing graphics appropriately:

- Position table and figure titles consistently. Usually you put table titles above the table and figure titles below the figures.
- When you can, obey the text margins; or set margins for illustrations and stick with them.
- Set aside a region for graphics (like a column) and always put them there.
- Always place graphics as close to or following the text they relate to.

Rules and Lines

Rules and lines help define the communication space and give your page structure. Use them consistently to help the user navigate your manuals and online help. Follow these guidelines in using rules and lines:

- Keep lines straight and of the same value. Double-check to make sure you have the same size box for each item.
- Use large enough arrows. Make sure the user can see them but that they don't hide essential details.
- Make rules and lines straight and neat.
- Make rules conform to the style of headers and other cues.
- Use rules to indicate hierarchies of information in your text.
- Use grayscale rules when you don't want to waste ink and you need to save disk space.
- Use rules sparingly in help screens, usually only thin ones across the top.

Size

Images that the user can't read clearly don't do you any good in a manual. You should not make them too large either, or else they take over the page. Follow these guidelines for sizing your visuals:

- Use screens and icons liberally, but restrict yourself to one conceptual image per chapter (often on the chapter title page) and one cartoon per procedure.
- Try to keep your illustrations on one page (at least for larger ones).
- Flip oversized illustrations 90 degrees.
- Keep it within the margins. When you enlarge a screen to take up two columns, consider that its new margins consist of the left margin of column 1 and the right margin of column 2. Always put these limits on your images so they fit into the manual or online help system.
- Crop pictures for maximum impact.
- Design a hierarchy of sizes of illustrations and stick to it.
- Make it large enough to show up. You can make screens and figures wider than one column. Make screens about 3 to 5 inches wide.
- Give it enough white space/soft boundaries. Usually you should have about 1/2 inch white space around figures. You have less space on screens: 1/8 to 1/4 inch.

Colors

Color can add to the appeal and impact of your manuals, especially your online help documents (where color is "cheap"). But these elements should relate clearly to the scheme of information you have designed. In particular, color can help you identify kinds of tasks, information, reference, and other elements of documents to help unify the document. Follow these guidelines in using color:

- Relate color schemes to patterns of information.
- Keep elements the same tones of gray or the same families of intensity: pastels, primaries, earth tones.
- Use a single color for bars along the paper edge for cuing.
- Avoid "reserved" colors: red for danger, yellow for caution.

4 Revise for Typography

Using *typography* means giving an arrangement to the images based on a logic. In words we use headings, margins, lists, and other devices to shape the information to the user's needs. When using images, you should arrange and design them so that they convey the structure of the meaning you intend.

Follow these guidelines to give the kind of emphasis on meaning, information, action, and productivity that a successful manual or online help system should have.

- **Make important things larger.** Sometimes you can adjust the scale of surrounding objects in a drawing to make one object larger than the other. When you have the opportunity, reserve the largest category for your most important element.

- **Make important things darker.** Darker elements on an otherwise homogeneous page will attract the viewer's attention. Often manuals use this typographical technique to show the screen in a lighter value (say, 40 percent) and the menu the user should select in full black. The contrasting effect works because it shows the user where to find the menu.

- **Make important things central.** The viewer's eye naturally gravitates to the center of a page or screen, so you have a better chance that the person will see what you want if you put it in this critical location. Of course, this doesn't apply to lists with graphic bullets, where the central focus falls on the text, with the graphics serving as access reminders.

- **Make important things sharper.** The viewer's eye will more likely settle on a sharp image than a fuzzy image. Using picture and photograph processing programs, you can adjust the focus of surrounding details to guide the reader's eye to the important part of the image.

- **Align related things.** Because of our preoccupation with lines of type as a major structure of communication, we tend to ascribe importance to things that are aligned. In structure charts, we naturally associate things adjacent to other things. For this reason you should use alignment to help convey your information structures.

- **Put first things left, later things right.** As we do with other patterned responses to visual and verbal information, we tend to associate chronology with left and right, at least in the western hemisphere. Many manuals use a left-right progress from step to step in designing procedures spanning two pages. Because we read left to right, we ascribe firstness with leftness; the reverse—putting the first step in a procedure at the right—just wouldn't feel right for many people in our culture.

Discussion

To operate a computer program, a user must take some kind of action. Press a key or click a button, and the program performs a calculation or prints a character on the screen. If the documenter has done the job well, the user will not just press the correct key or button but do so in the context of meaningful work. To convey such *operations* and make them meaningful, the manual should focus beyond operation itself to the result of the operation in terms of meaningful workplace activities and actions. For this reason we say that the key to usability lies in describing operations in such a way that they're meaningful to users. The discussion that follows will show you how graphics can make clear connections between software operations and workplace actions.

Showing How Tools Apply to the Workplace

One of the most common kinds of graphics in software documentation, showing the use of tools, relates directly to the concept of task orientation. At the point where you tell the user to press a key or use any of the many interface elements of the software, you instruct in the simplest sense. All these keys make up what we call interface objects that the user can manipulate to do work.

The interface of a computer contains many tools, called interface elements. They include keys and key combinations, the mouse and its buttons, and the screen tools: scroll regions, radio buttons, data entry fields, check boxes, and so on. At times the devices for manipulating the software get complex, as in sophisticated control widgets and image sizers used on three-dimensional image processing computers. Using graphics in the documentation enables the user to see exactly which buttons and keys to press.

You can support the operation of these tools—keystrokes, mouse clicks, and interface objects—in two ways: by using images of the actions taking place, or by using tables showing the commands, the objects, and their definitions. Support for keystrokes and other interface elements usually falls into getting started sections of manuals, or guided introductions to tasks. In special cases, where the user must manipulate complex interface items, you would put support for tools in reference guides. Figure 13.16 illustrates how to show tool use.

Show Results of Software Operations

Nine out of ten images you will find in a software manual show the results of operations. Results of operations can come in various forms, but most commonly they

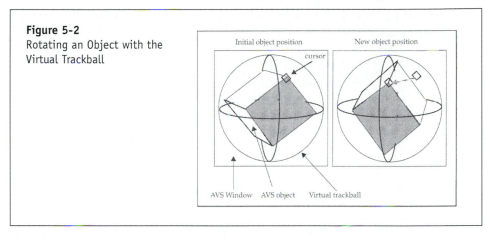

Figure 5-2
Rotating an Object with the Virtual Trackball

FIGURE 13.16   **Use Graphics to Show Tool Use**
This illustration from the Convex AVS *User's Manual* helps expert users manage a particularly difficult interface.

show the screen that appears after the keystroke. Typically, showing results occurs on all three levels of task orientation—teaching, guidance, and reference. At the teaching level, it accompanies the illustration of the tool in use: In fact, the two work together to support the operation/result movement found in tutorials. At the guidance level, it usually accompanies the most important step of a procedure, often the culmination or the goal state of the task. At the support level, it shows specific screens or other results the reader should see when the function is used. Figure 13.17 is a graphic that shows the result of an operation.

Present Overviews to Integrate Software with Workplace Activities

In all task-oriented documentation you will find introductions or operational overviews (described in Chapter 10) that help the user fit the procedures into his or her existing mental framework. Often images can do this job better than words (see Figure 13.18). The images we use to present overviews often consist of cartoons or drawings of various elements with process arrows (the arrows used to show "cause" between one image and another) to show how things fit together. They work particularly well with installation sections, showing the big picture of how the software on the disks in the box will end up as the system on the hard disk.

You will find that help systems tend to have overviews built in because of the interactive nature of the screen. The first screen of a help document tends to have icons showing the different elements of the help system that the user can choose from. But with manuals you need to explain the access elements more explicitly, so users will understand how you have designed the manual system. In large programs manual systems may require a number of books; in smaller programs manuals may consist of chapters or special sections of a single book. Graphics also help explain how users should read pages: what the headings look like, where to

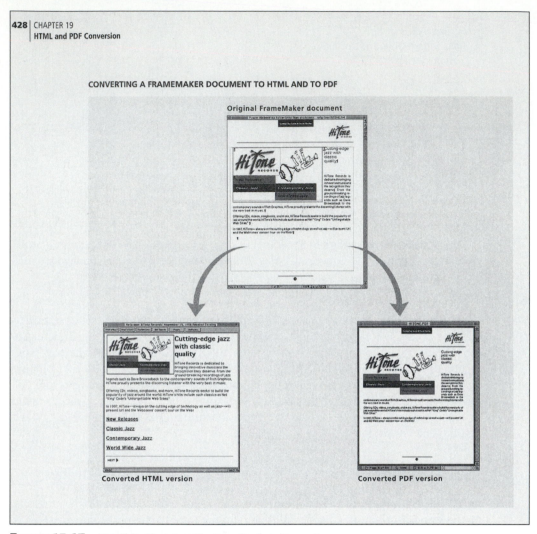

CONVERTING A FRAMEMAKER DOCUMENT TO HTML AND TO PDF

Original FrameMaker document

Converted HTML version

Converted PDF version

FIGURE 13.17   Graphic Showing the Result of an Operation
This illustration from the Adobe FrameMaker *User Guide* shows the results of a conversion operation.

find the step-by-step sections. In particular, graphic overviews of procedures can greatly increase the usability of manuals. Used wisely—consistently with other task descriptions and well-designed—graphic overviews can make the difference between a usable manual and a default manual.

Suggest Functions and Uses

Software manuals often contain graphics to illustrate problem-solving situations or work environments. These images fall into the category of graphics to suggest func-

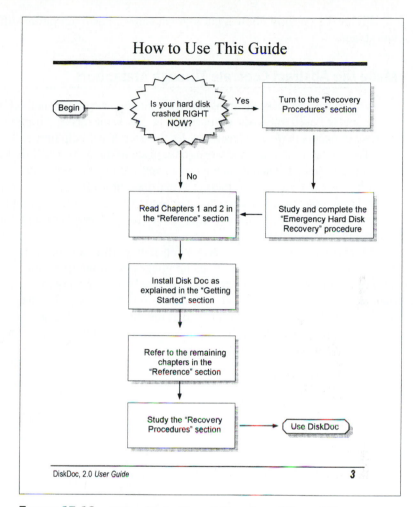

FIGURE 13.18   A Graphic to Give an Overview of Manual Use
This example shows how the graphic style (flowchart) fits the user—in this case, experienced users familiar with programming symbols.

tions and uses. These images include pictures of desktops or pictures of kinds of reports a program will produce. They often highlight the strongest, most dramatic feats a program can perform, like drawing 3-D graphs or plotting fancy maps. Documenters often use photographs, stylized drawings, or screens here. Photographs may show users at the keyboard. Usually these graphics are found in "Getting Started" sections that operate at the guidance level of task support.

In a more specific way, though, you can key suggestions of functions and uses to the task analysis by designing them so that they capture a *typical-use scenario* or workplace activity. The typical-use scenario outlines the most common use of the program—the actions and activities the program is designed to support.

Figure 13.19 shows how a photograph can suggest integrating software into the workplace.

Make the Abstract Concrete Through Metaphors

Users often have to make the transition from analog or by-hand methods of doing tasks to computer-based methods of performing the same tasks. Graphics that convey abstract concepts perform a basic task for the user: They help him or her see the invisible. They work well alongside words or in place of them. You will find them in places where the writer of manuals needs to convey a concept, especially an abstract one. Scrolling a text on a screen, merging two files into a third, connecting over phone lines: These concepts do not come easy to many users without graphics to help out. Figure 13.20 shows a graphic depicting a very basic abstract concept: that of the desktop in a PC environment.

Often images conveying abstract concepts portray a central metaphor of a program: the idea of communication using images of phone lines, the idea of capturing a screen using images of butterflies and nets. However, they most often show up in manuals for programs that produce graphic output: image processing software, desktop publishing programs, clip art programs, drawing programs. In these manuals, images of abstract concepts function as examples of the output of the program and often get put in special reference sections.

FIGURE 13.19   A Photograph Suggesting Uses
This photograph from *Macintosh IIci Owner's Guide* illustrates how to suggest the uses of software and computers to users. Showing the user's environment like this can provide a powerful image in documentation because of its direct relationship to productive work.

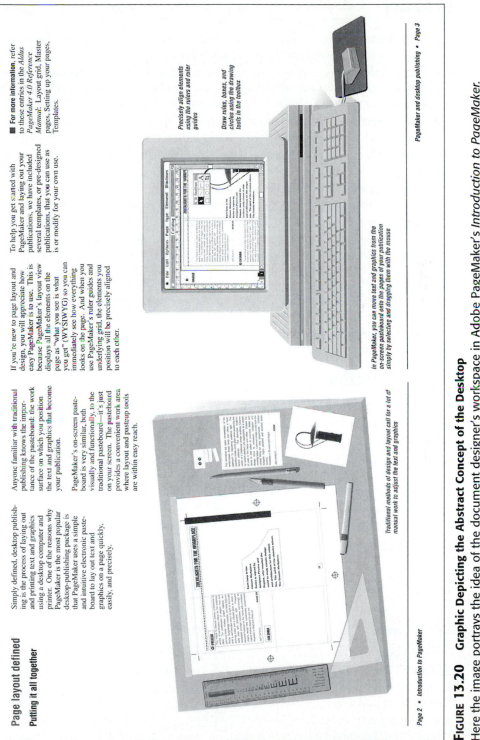

Page layout defined

Putting it all together

Simply defined, desktop publishing is the process of laying out and printing text and graphics using a desktop computer and printer. One of the reasons why PageMaker is the most popular desktop-publishing package is that PageMaker uses a simple and intuitive electronic pasteboard to lay out text and graphics on a page quickly, easily, and precisely.

Anyone familiar with traditional publishing knows the importance of the pasteboard: the work surface on which you position the text and graphics that become your publication.

PageMaker's on-screen pasteboard is very similar, both visually and functionally, to the traditional pasteboard—it's just on your screen. The pasteboard provides a convenient work area where layout and pasteup tools are within easy reach.

If you're new to page layout and design, you will appreciate how easy PageMaker is to use. This is because PageMaker's layout view displays all the elements on the page as "what you see is what you get" (WYSIWYG) so you can immediately see how everything looks on the page. And when you use PageMaker's ruler guides and underlying grid, the elements you position will be precisely aligned to each other.

To help you get started with PageMaker and laying out your publications, we have included several templates, or pre-designed publications, that you can use as is or modify for your own use.

■ **For more information**, refer to these entries in the *Aldus PageMaker 4.0 Reference Manual*: Layout grid, Master pages, Setting up your pages, Templates.

Traditional methods of design and layout call for a lot of manual work to adjust the text and graphics

Precisely align elements using the rulers and ruler guides

Draw rules, boxes, and circles using the drawing tools in the toolbox

In PageMaker, you can move text and graphics from the on-screen pasteboard onto the pages of your publication simply by selecting and dragging them with the mouse

FIGURE 13.20 Graphic Depicting the Abstract Concept of the Desktop
Here the image portrays the idea of the document designer's workspace in Adobe PageMaker's *Introduction to PageMaker*.

Glossary

access indicators: special types of graphics that present overviews of programs or manuals with the intention of orienting the user to his or her place in the system.

command-driven: a type of computer program that relies primarily on the user typing commands to perform functions. MS-DOS is an example of a command-driven program. Other types of programs include menu-driven programs where the user selects from a menu.

interface elements: parts of a computer program that the user can manipulate to perform work. Examples include buttons, rulers, sliders, check boxes, menus, and commands.

mental model: a conceptual structure of a task or activity that represents the kind of thinking needed to perform a task with a software program.

progress indicator: a kind of graphic that shows how far a user has come in a tutorial by showing the total lessons or pages, those already covered, and those still to come.

typography: the arrangement and appearance of images and text elements on a page.

typical-use scenario: a description in chronological format of the tasks that comprise the most frequent work done by users of a software program.

☑ Checklist

Use the following checklist as a way to evaluate your use of graphics on a project and to plan for incorporating graphics in your documents.

Graphics Checklist
Responding to User Questions

From your user analysis and program task list, find areas where users might have the following questions.

User Question	Option to Respond
❏ Where is it?	❏ Arrows
	❏ Circles
	❏ Highlighting
❏ What is it?	❏ Examples
	❏ Metaphors
❏ How do I do it?	❏ Process overviews
	❏ Diagrams
❏ Where am I?	❏ Access indicator
	❏ Progress indicator
❏ What's the big picture?	❏ Overall program diagrams
	❏ Menu maps
	❏ Conceptual overviews
	❏ How to use this manual

Making It Visible

❑ Do your screen shots clearly show the user where to look to perform tasks?
❑ Do you show concrete versions of abstract things?
❑ Have you made your illustrations clear and easy to interpret?

Keeping Graphic Styles Consistent

❑ In the area on the right, indicate what styles of graphics you plan for your manual or help project.

Graphic Element	Planned Styles
❑ Lines	4-point grey for subheads
❑ Fonts for titles	Arial 10-point
❑ Arrow styles	
❑ Box styles	
❑ Frame styles	
❑ Labels and callouts	

Typographic Techniques

❑ Identify elements that the user will see as important in the program or the workplace in your project that you can accommodate using typographic techniques. Indicate how you will make these elements clear to the user by selecting from among the following techniques.

Technique	Element
❑ Make it larger than surrounding text and graphics	Lassie (help) icon: 1/2-inch
❑ Make it darker than surrounding text or graphics	
❑ Make it central on the page or in the figure	
❑ Make it sharper than surrounding text and/or graphics	
❑ Align it with related elements	
❑ Arrange elements in a left to right progression	

Identify the Function of Your Graphics

Which of the following functions will you emphasize in your design of graphics? Use the table below to help you plan.

Function	Type of Graphic to Support the Function		
Showing use of tools	❏ Screens ❏ Drawings ❏ Icons	❏ Reports ❏ Charts ❏ Matrices	❏ Photographs ❏ Diagrams ❏ Tables
Showing results of actions	❏ Screens ❏ Drawings ❏ Icons	❏ Reports ❏ Charts ❏ Matrices	❏ Photographs ❏ Diagrams ❏ Tables
Presenting overviews	❏ Screens ❏ Drawings ❏ Icons	❏ Reports ❏ Charts ❏ Matrices	❏ Photographs ❏ Diagrams ❏ Tables
Suggesting functions and uses	❏ Screens ❏ Drawings ❏ Icons	❏ Reports ❏ Charts ❏ Matrices	❏ Photographs ❏ Diagrams ❏ Tables
Explaining processes	❏ Screens ❏ Drawings ❏ Icons	❏ Reports ❏ Charts ❏ Matrices	❏ Photographs ❏ Diagrams ❏ Tables
Supporting text explanations	❏ Screens ❏ Drawings ❏ Icons	❏ Reports ❏ Charts ❏ Matrices	❏ Photographs ❏ Diagrams ❏ Tables

Elements of Graphics

Use the list below to evaluate your designs for consistency and effectiveness.

Titles

❏ Titles numbered sequentially
❏ Numbers and titles listed in the front of the manual
❏ Titles cued by boldface or enlarging in the text

Labels

❏ Components of screens labeled to correspond with user tasks
❏ Captions kept brief
❏ Caption terminology consistent with the text

Placement

❏ Graphic frames obey the margins of the page grid
❏ Graphics assigned to a specific column or page region
❏ Graphics placed close to and following the text they relate to

Rules and Lines

❏ Rules and lines kept straight and neat
❏ Rules conform to the style of headers and other cues

❑ Rules clearly indicate hierarchies of information
❑ Greyscale rules used to save disk space and ink
❑ Rules used sparingly in help screens

Size

❑ Illustrations kept one page
❑ Oversized illustrations turned 90 degrees
❑ Pictures cropped for focus
❑ Illustrations organized around a clear hierarchy

Colors

❑ Colors related to patterns of information
❑ Colors kept in the same families of intensity: pastels, primaries, earth tones, etc.
❑ Single colors used for bleeders or cuing
❑ Reserved colors avoided

Practice/Problem Solving

1. Practice Using Different Kinds of Graphics

Find an example of a procedure in a manual or help system for a program you use. Rewrite the procedure using graphics. For example, you could choose the task of putting a disk in a computer and illustrate it in the following ways:

• With process boxes
• Using icons to represent the task
• With drawings
• With a scanned photograph

Follow the guidelines in the chapter for clear, focused graphics.

2. Evaluate Graphics in a Manual

Use the checklist in this chapter to evaluate the graphics in the computer manual or help system of your choice. Analyze the appropriateness of the graphics in terms of meeting the needs of users. What recommendations would you make to the authors of the document for revising their graphics styles?

3. Practice Describing Graphics Styles

Find a software manual that you think has an effective graphics scheme. Using Guideline 2, describe the styles that you find so that another writer could re-create the styles. Then exchange your descriptions with another person and see if they can recreate the graphics following your descriptions. (Hint: You may have to use some stand-in screen shots for this exercise.)

CHAPTER 14

Designing Indexes

Of all the features of software documentation that you can offer your user, the index ranks among one of the most valuable and most popular. It allows you to guide the user to just the right information in your manual or help system. This chapter helps you decide what index methodology to use to create your index: manual or electronic. But the main focus of the chapter falls on deciding what to index and setting levels of detail, phrasing, and techniques for building and proofreading.

How to Read This Chapter

This chapter follows a straightforward approach to indexing and building keyword searches. Necessarily limited in scope, it nevertheless provides you with the basic ideas behind these valuable task-oriented tools.

- The project-oriented reader can begin with the Guidelines and then read the rest of the chapter.

- The reader seeking greater understanding of indexes should also start with the Guidelines because they proceed through a demonstration-type sequence that can help those uninitiated into the mysteries of indexing compare one element to another.

Example

The index in Figure 14.1 exhibits many of the features of an index designed to support easy processing of information and to accommodate a variety of workplace uses. In particular, it demonstrates how you can orchestrate various kinds of information in the index.

Guidelines

Figure 14.2 lists guidelines for designing indexes, all of which will be discussed later in this chapter.

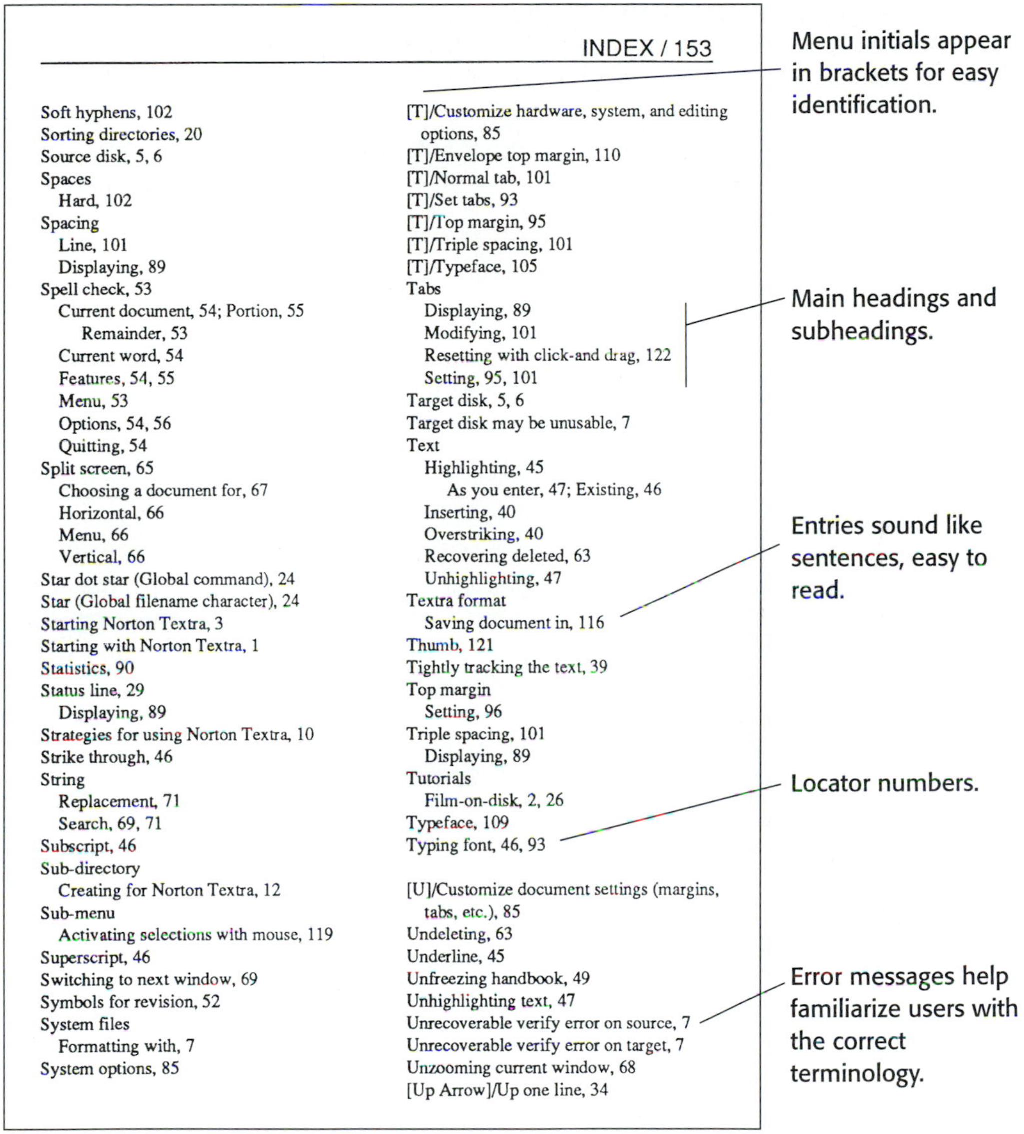

INDEX / 153

Menu initials appear in brackets for easy identification.

Main headings and subheadings.

Entries sound like sentences, easy to read.

Locator numbers.

Error messages help familiarize users with the correct terminology.

Soft hyphens, 102
Sorting directories, 20
Source disk, 5, 6
Spaces
    Hard, 102
Spacing
    Line, 101
    Displaying, 89
Spell check, 53
    Current document, 54; Portion, 55
        Remainder, 53
    Current word, 54
    Features, 54, 55
    Menu, 53
    Options, 54, 56
    Quitting, 54
Split screen, 65
    Choosing a document for, 67
    Horizontal, 66
    Menu, 66
    Vertical, 66
Star dot star (Global command), 24
Star (Global filename character), 24
Starting Norton Textra, 3
Starting with Norton Textra, 1
Statistics, 90
Status line, 29
    Displaying, 89
Strategies for using Norton Textra, 10
Strike through, 46
String
    Replacement, 71
    Search, 69, 71
Subscript, 46
Sub-directory
    Creating for Norton Textra, 12
Sub-menu
    Activating selections with mouse, 119
Superscript, 46
Switching to next window, 69
Symbols for revision, 52
System files
    Formatting with, 7
System options, 85

[T]/Customize hardware, system, and editing
    options, 85
[T]/Envelope top margin, 110
[T]/Normal tab, 101
[T]/Set tabs, 93
[T]/Top margin, 95
[T]/Triple spacing, 101
[T]/Typeface, 105
Tabs
    Displaying, 89
    Modifying, 101
    Resetting with click-and drag, 122
    Setting, 95, 101
Target disk, 5, 6
Target disk may be unusable, 7
Text
    Highlighting, 45
        As you enter, 47; Existing, 46
    Inserting, 40
    Overstriking, 40
    Recovering deleted, 63
    Unhighlighting, 47
Textra format
    Saving document in, 116
Thumb, 121
Tightly tracking the text, 39
Top margin
    Setting, 96
Triple spacing, 101
    Displaying, 89
Tutorials
    Film-on-disk, 2, 26
Typeface, 109
Typing font, 46, 93

[U]/Customize document settings (margins,
    tabs, etc.), 85
Undeleting, 63
Underline, 45
Unfreezing handbook, 49
Unhighlighting text, 47
Unrecoverable verify error on source, 7
Unrecoverable verify error on target, 7
Unzooming current window, 68
[Up Arrow]/Up one line, 34

FIGURE 14.1   A User-Oriented Index
This index page from the Norton TextraWriter word processing program illustrates standard features of a two-level outline.

1 Plan Your Indexing Strategy

Basically, you have two methods for making an index: manual or electronic. A manual index requires that you go through the document and write index entries on cards or in a document file. An electronic index uses the indexing feature of your software

FIGURE 14.2   **Guidelines for Designing Indexes**

1. Plan your indexing strategy.
2. Decide what to index.
3. Identify the level of detail.
4. Decide on phrasing and format.
5. Edit and proofread.

program to accomplish the task. Below you will find a discussion of each of these methods to help you decide which best suits your project.

Manual Indexes

Most manual indexes require that you have essentially finished the document because you need to write down the page numbers on which the terms you index occur. After doing this, you organize the entries alphabetically. You would use this method for indexing projects that others have done, where you don't have access to the automatic features of a word processor. This situation occurs when you don't have access to the software that is used to make the manual itself. In this case you have the actual pages in page proof version (showing the actual page numbers) and have to go through the pages one at a time reading and identifying the important terms and concepts and typing them into a document file.

While there is no one acknowledged process for creating an index, the basic procedure (which is covered in greater detail below) is as follows.

1. Review the user analysis. This step refreshes your memory about what the user's main activities and actions are with the software and can be used as a reference when you're deciding on what terms are important to the reader.

2. Read or scan the page for entries. Pick out the terms or phrases that you want to index. Typically with a manual or help system you would look for the following:

   - Tables, figures, captions
   - Examples and figures
   - Definitions of terms
   - Acronyms or abbreviations

   - Main topics
   - Important concepts
   - Main tasks

   - Tool buttons
   - Keyboard shortcuts
   - Menu names

3. Record the locations. Record the items in an open word processing file, along with the page number on which the item occurred.

4. Alphabetize and edit the index. This stage requires both revision, formatting, and proofreading of the completed index, making sure all the entries are consistent and useful.

Electronic Indexes

Word processing software programs usually contain functions for indexing. This type of indexing is called *"embedded indexing."* These require that you identify a term on a specific page that you want to index, and then highlight it, along with the category it pertains to. This, of course, implies that you have the entire document in one file or a series of files in which you specify the page numbers. You can use this method from the very beginning of working on a document, and then compile the index at the end of the project. The advantages of this method include the following:

- Automatic alphabetizing. You don't have to rely on your abcs.
- Automatic formatting. You don't have to bother with indenting by hand.
- Ease of revision. You can change the index at the last minute.

Figure 14.3 shows an example of how electronic indexing works in a word processing program. The process for electronic indexing is resembles that of manual editing, but is in some ways very different.

1. Review the user analysis. This step refreshes your memory about what the user's main activities and actions are with the software and can be used as a reference when you're deciding on what terms are important to the reader. This step is essentially the same as with manual editing however since you are indexing all the way through the document as you write it or after your alpha draft, you have to maintain this audience awareness all along.

2. Mark the index entries. As you read or scan a page you identify the entries it contains and use the electronic marking feature of the software. The advantage of marking entries in this way is that you can make decisions as to what terms have importance for the users as you're writing or editing.

3. Build the index. When you have marked all the entries in the document you select the location of the index and use the software to physically create the index. The program will read the text and collect the marks you have inserted. Then it will create index entries for each instance of market text, recording the page number where it occurs. The entries will appear in columns at the point you created it.

4. Edit the index. After the program has created your index you must edit for inconsistencies: doubled entries, missing entries, and so on (see below). However, instead of editing the actual index itself, you have to go back into the marked text, make your changes there, and then recreate the index. This process of revising and rebuilding the index continues until you get it right. At the end of the process you might perform some formatting work on the finished product to make the columns line up and so on, but most of the editing and revision work is done by manipulating the entries themselves.

2   Decide What to Index

As you will see below, you can index a number of different types of information. You should try to match your choice of what to index with your user analysis. The goal is to make the index support the activities and actions that the user will undertake in

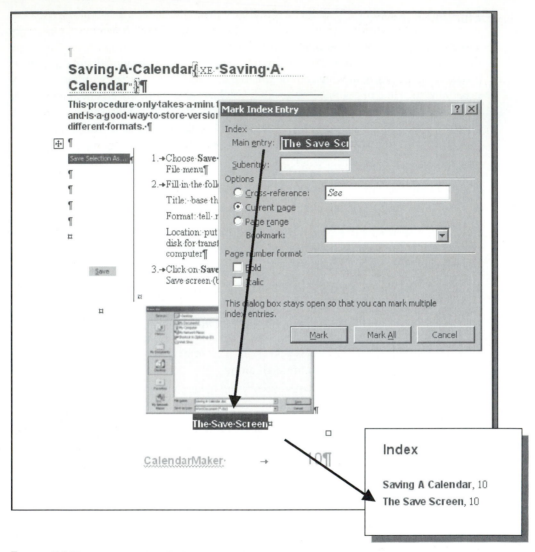

FIGURE 14.3   An Example of Electronic Indexing
Using Microsoft Word you can mark terms in the text that can later be assembled and edited to make an index.

applying the program to workplace needs. Consult your documentation plan for the specifications set up earlier in the project for the kind of index that you need for your users and your project. If you don't have an index specified in your documentation plan, make up some index specifications now and have the developer, users, and other writers review them.

Consider the following elements you could index, and determine the ones you need. Keep the list in front of you as you go through the document making your index or marking your index items. If you do your index electronically, enter these items into your style guide early in the project and refine the list as you go along. Also, if you have others working on the project, make sure they know the elements to index and the categories they belong in.

Commands and Functions

Commands and command sets and functions consist of all the terms that you find on menus. It also includes Control or Alt key sequences. Usually you would format the commands differently in the index because they don't sound like parts of sentences. This helps the user identify them as commands and spot them while scanning the index. In the example below, the commands under one menu were included in a special section under the name of the menu. You would also cite the occurrence of the command or function at another place under the name of the function.

COMMANDS AND FUNCTIONS

Occurrence in the Document	Result in the Index
Use the Make Report command (Report menu) to arrange your data into one of the many reports available in MarketMaster. (p.48)	Make Report (Report Menu), 48, x, 4 Report Menu Make Report, 48, x, 4

Concepts

Concepts refer to the ideas related to the subject matter of the program. All users have to understand these basic concepts in order to understand how the program works. Concepts make up an important element of your index because when the user encounters the term in various parts of the book, he or she will look the term up in the index.

CONCEPTS

Occurrence in the Document	Result in the Index
AccountMaster uses the idea of client levels as a way of organizing your client database. Basically, a client level identifies the category in which the client's record information resides, Client levels. . . . (p.15)	Client levels, xi, 15, 24, 33–38, 111 Levels, client, xi, 15, 24, 33–38, 111

User Terms and Questions

The index functions to connect words the user may know to synonyms or related words used in the program. For example the user may say "quit a program" but the program's function uses the term Exit. The index should list *quit* with an entry saying See *Exit, 23.* You can think of many examples of this type: *Blank line above a paragraph,* see *Line spacing; Start a program,* see *Load; Search,* see *Find.* Many of these terms come from users' experience with other software that uses parallel terms.

Users also have terms they use in their work that they need to correlate to the program. For example, the term *sku* (pronounced "skew") in retailing stands for *stock keeping unit* and represents an actual product (such as Sparkling Dawn Dish Detergent, 10oz., lemon scent #5167x443704). In a program that analyzes business data, let's say, the term *item* refers to the lowest level of a hierarchy of categories of data. A user would come to the index thinking "How do I display information at the sku level," and look under *s* for *sku.* A good index, in this case, would include the entry *Sku,* see *Item,* **45.** Such an entry would direct the user to the correct term and the correct page.

When you do your user analysis, you should watch out for these kinds of terms. Begin making a list while you have access to your users so you can then incorporate these terms into your index. You need this list, because in the case of user terms you will index words that don't actually occur in your manual. So if you don't have a list of them, you won't find them while scanning the pages.

USER TERMS AND QUESTIONS

Occurrence in the Document	Result in the Index
You can easily delete files from your directory by dragging the filename into the Recycling Bin. The recycling bin stores files you've deleted until you empty it using the Shred command. (p.19)	The recycling bin is almost full, see Shred, 19 Trash can, see Recycling Bin, 19

Glossary Terms

Indexing glossary terms simply means that you include a reference in the index for all the terms in your glossary.

GLOSSARY TERMS

Occurrence in the Document	Result in the Index

Scratch sheet. A scratch sheet refers to a quick print of account details. It has a set format, and you can use it to quickly print out data about a client to use in an office visit. (p.112)	Scratch sheet
	  printing, 67–68
	  deleting, 69
	  defined, 112

Proper Names of Products and Companies

With some programs you will find yourself documenting names of other programs or companies. Often your program will write files in formats usable by other programs, or you will refer to other companies that do business with your users. A manual that supports information-oriented work should do its best to acknowledge connections to the larger business environment surrounding the user's workplace. It should also acknowledge that users use other programs in their work. Users will consult the index with terms relating to this context and need to know where to get information about them.

PROPER NAMES OF PRODUCTS AND COMPANIES

Occurrence in the Document	Result in the Index
If you run AccountMaster on a computer with an Pentium 6 chip, you need to install a 3Com Ethernet Card. (p.2)	3Com Ethernet card, 2

Tasks and Procedures

You may want to have a special index for tasks and procedures, or you may want to include them in the main index. Often you will see tasks highlighted in some way to indicate to the user that this entry will lead them to guidance or step-by-step information.

You may want to include section names as part of the scheme for indexing tasks. Section names usually correspond to main areas or categories of tasks that the user would most likely look up in the manual or help system. You usually include them in an index to direct the user to the section containing these kinds of tasks.

TASKS AND PROCEDURES

Occurrence in the Document	Result in the Index
Deleting an Account	Accounts
You should delete accounts when you find that you have inadvertently created more than one account for a client, or a client goes out of business. It's a good practice to maintain your account lists in these ways. Follow the steps below to delete an account. 1. Choose Delete . . . from the Utilities menu. (p.37)	creating, 30, 22–25 deleting, 37 formatting records in, 20, 25–27 Delete a report, 60 an account, 37 a list record, 20, 22 a report field, 60, 61

3   ## Identify the Level of Detail

You can define the level of detail in two ways, one by the number of levels in the index (see Table 14.1), and the other by the number of items you index per page.

Levels of Detail

A very simple index will contain only one level and will probably only show the main headings within the text. Such an index, usually for a shorter document, basically

TABLE 14.1   Levels of Indexes

One Level	Two Levels	Three Levels
Getting Help, **3**	Getting Help, **3**	Getting Help, 3
Graphics, **30,** 32–35	Graphics	Graphics
Greeting, 12, 33, **51**	importing, 30, 32	importing
	in reports, 30, 32–35	bitmaps, 30
	Greeting	converting, 32
	deleting, 33	in reports
	example, 12	example, 30
	setting, 51	inserting 32–35
		Greeting
		deleting, 33
		example, 12
		setting, 51

puts the table of contents into alphabetical order and allows the user to access information in that way. A two-level index organizes terms by categories, where appropriate, and represents a more sophisticated organization of the material. The most extensive index usually goes to three or more levels and indicates a very complex way of organizing material.

Another way to describe the amount of detail in an index focuses on the number of indexable items per page. Of course, the number of items per page will vary from one part of a document to another, depending on the density of your pages, the number of figures, and the kinds of things you need to index, but basically you can rely on the following averages.

- Light index   two to three items per page
- Medium index   five to seven items per page
- Heavy index   eight to nine items per page

4   Decide on Phrasing and Format

You should give some time to the planning of the phrasing and format of your index entries. You can consult your documentation plan and coordinate the cuing of commands and other items in the index with your overall notational conventions. The following list discusses the kinds of issues you will face in this regard.

- Cue the primary locator numbers. In some indexes you will find the primary locator number cued in some way, usually by bolding. Example: Greeting, 12, 33, 51, where the main information for the entry occurs on page 51. This requires some decision making on your part as to the main entry. Usually you would use the primary locator to refer the user to the procedure related to the task, or to its definition.

- Capitalize terms consistently. In some indexes you capitalize all terms as a matter of editorial principle, and in others you don't. Consider these examples:

CAPITALIZATION

Uncapitalized terms	Capitalized terms
getting help, 3	Getting Help, 3
graphics	Graphics
importing, 30, 32	Importing, 30, 32
in reports, 30, 32–35	In reports, 30, 32–35
greeting	Greeting
deleting, 33	Deleting, 33
example, 12	Example, 12
setting, 51	Setting, 51

- Make entries sound like sentences. You want to make your entries sound like sentences when you put them together with their heading. Consider the following examples of how to do this.

ENTRIES THAT SOUND LIKE SENTENCES

Short Phrases	Complete Phrases
Getting Help, 3	Getting Help, 3
Graphics	Graphics
Format, 36	Format of, 36
Import, 30, 32	Importing from other programs, 30, 32
Reports, 30, 32–35	Using in reports, 30, 32–35
Greeting	Greeting
Delete, 33	Deleting, 33
Example, 12	Example, 12
Set, 51	Setting, 51

- Cue special terms. Identify a special format for commands, tasks, function keys, and other kinds of things beyond just words. In the example below, the toolbars appear in brackets and the command appears in bold face, followed by the menu it appears on.

CUING PATTERNS FOR COMMANDS AND OTHER SPECIAL ENTRIES

Toolbars
   [Accounts], 98
   [Format], 100
   [Utilities], 102
Tracing Activities
   Defined, 28
   Reasons for, 29
   Tracing Activities . . . Utilities menu, 102

5   Edit and Proofread

Once you have created your index, you should edit and proofread it carefully for format mistakes. Check specifically for indentation problems if you have two or more levels. You should also spot-check the page references to make sure they take the user to the right pages. If you have a user available, you should test the index with the user. You can do this simply by having the user review it, or by asking specific questions about terms you included. Finally, check the index for inconsistencies of reference, things like:

Accounts, opening

Accounts, open

Accounts, to open

There is more about proofreading indexes in Chapter 9, "Editing and Fine Tuning."

Editing an Index

Creating the index is one thing but editing and proofreading it is an important next step. According to one writer, editing an index can take up to 1/3 of the total time devoted to index creation.[1] Collins also points out that editing indexes takes its start in an awareness of the user's characteristics, as the indexer has to know what content to make more accessible, to decide which entries represent significant knowledge (as opposed to passing mention), and how to group and direct users to important information (that might otherwise just be scattered throughout the document).[2] Table 14.2 shows some of the tasks that an editor can perform on an index.[3]

Discussion

Indexing is the kind of task that is often left to the end of a project, and for good reason. You can't fill in the page numbers until you have the final document in the form in which it will be published. This is often the worst time for the creation of an index as the printers, binders, and sales persons can often impatiently pressure the indexer

TABLE 14.2   Editing Tasks for Indexes

Index Editing Task	Before	After
correct inconsistencies	Open the file, 18 Launch the file, 42	Open the file, 18, 42
adding missing references	Backup, 12	Backup, 12, 24–25, 40
combining entries	Open, 12 opening, 15 open, 42	Opening, 12, 15, 42
creating double postings	Backing up disks, 14	Disks, backing up, 14 Backing up disks, 14
eliminating subentries	Installing installing the program, 5	Installing, 5

to finish quickly.[4] Indexing also requires specialized knowledge and objectivity that often writers don't possess. According to the *Chicago Manual of Style* a writer, because of his or her familiarity with the material, creates a concordance of important terms rather than an index that arranges word frequency information in ways readers can best use. For these reasons, the index is often seen as a needless chore or redundant to the manual ("Why not just use the table of contents?")

Why an Index?

Seeing an index as a chore or as redundant shows that you don't have a correct understanding of the importance of this element of software documentation. In fact, you would do well to see the index as an integral part of the access system to your information. Consider the following justifications for the index.

- **Performs a unique function.** The index does not just repeat the information in the table of contents. The table of contents expresses the organization of the document, but the index does something completely different. It accesses the document from the user's point of view, in the user's language. It directs the user straight to the material in the manual and, for this reason, forms one of the most important elements of the manual. In fact, some manuals even print the index in the front of the manual before the table of contents.
- **Meeting place of multiple users.** Almost all programs have users in various business areas or professions. These users bring a multifaceted vocabulary to their work with the program. Also, you may have both beginners and advanced users who know other kinds of programs relating to your application and who use different terms to refer to different functions. The index serves as the meeting place of the vocabulary used by all these kinds of users. If you design and build it right, all users will find the terms they use in it and it will direct them to the information they need.

Online Index versus Print Index versus Keywords

When you design your indexing system, you need to consider the three main forms of index material in a situation where you have manuals and online help. Your manual will contain an index, possibly at the front but usually at the back, and your help system will probably contain an online index and a capability to do a keyword search. Consider the following differences among these three forms of the index information.

- **Print index.** Contains the terms printed in two- or three-column format usually at the back of the manual, divided into sections by alphabet. This is often referred to as the "back-of-the-book" index.
- **Online index.** Contains terms printed in a long, scrolling list with the alphabet displayed at the top of the page or in a frame at the left. The user can click on an alphabet entry to display the terms listed under that letter. This is often the way you can index a web site.

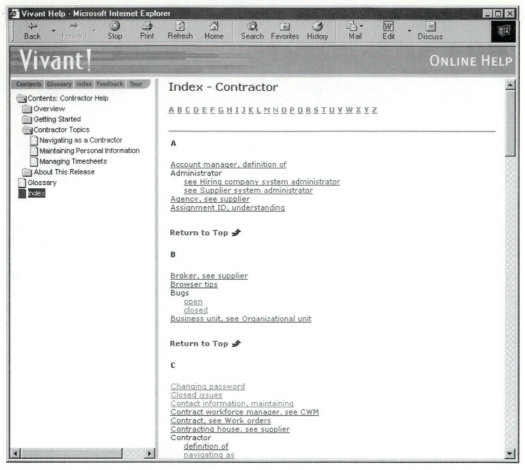

FIGURE 14.4   An Example of a Web Index
This example shows an easy-to-use index for a web-based manual.

In Figure 14.4 the user can select a letter of the alphabet and the index links will display that portion of the index. Entries in this index go to pages or to bookmarked sections of pages.

- **Keyword searches.** Consists of a list of terms and phrases that the user can scroll through or enter manually, after which the program, or search engine, will scan the list and find the closest related topic. The designer (you) has created electronic links to certain topics, a list of which appears for the user to choose from. The keyword search list differs from the index information because it limits terms to those relating to topics in the help system. The keyword search list usually doesn't contain user questions, user synonyms, or other material, and focuses strictly on the help topics.

The essential element here lies in the interchangeability of the information in all three of these lists. The words that trigger the keyword search among the topics should also appear in your print and online indexes. Careful planning on your part can help you achieve a high degree of consistency among these three forms of index information.

Automatic Indexing Software Programs

A number of programs exist to create indexes, or drafts of indexes, automatically. Microsoft Word, Adobe Framemaker and other programs that create documents have indexing capabilities built in. Some programs also work with web and HTML files (Figure 14.5) to create automatic indexes. Help programs such as e-Help, Sevensteps, or ForeHelp create index lists automatically for compiled WinHelp or HTML Help programs.

Among the programs available for creating indexes for manuals on web sites, the basic operating principle is the same. Such indexes do the work of examining the pages and creating a central repository of links to them, the indexer has control over the finished product and the links almost always produce relevant information in the form of web pages or sections of web pages. Some of them are limited to an operating system (run on Windows 2000 or Windows NT) and some also create search capabilities along with the alphabetical index. The more sophisticated of these programs, such as HTML Indexer embed indexing tags in your html documents so that if you add or alter an existing document the index automatically includes or excludes the relevant pages. The result of such embedding is a much more flexible product for both the index preparer and the index user.

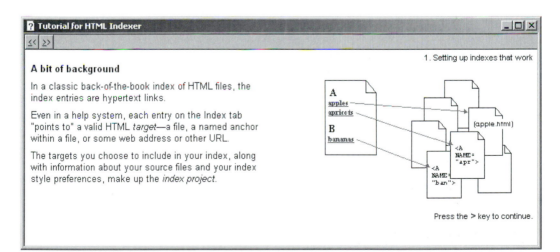

FIGURE 14.5   Automatic Indexing of Web Pages
This example from the tutorial for HTML Indexer shows how the program creates links to help files and then compiles the links automatically in an index page.

Indexing with Search Engines

Where automatic indexing begins to lose value is in search engines that examine the entire text of web pages and produce often voluminous results with very little relevancy to the user. If a user wanted to know how to open a database file, for example, and did a search on the term "file" using such an index, then the results would be many, but few of them would point to the procedure for opening a file. To remedy this problem many search sites try to train users to insert more logical terms or to use complicated boolean operators ("and," "or," etc.). Short of that solution, some search engines themselves filter out non-relevant content by examining "meta" tags (inserted keywords in HTML files) to extract hits that were intended by the page designer. Thus, if the user searched for "file" a more sophisticated search engine would know only to extract the page in which "file" was inserted as a keyword ("open a file") rather than some other page where the word "file" simply appeared in a sentence (". . . **file** on a hard drive . . ." or ". . . stored in a pro**file** of the user . . .") Because of misspellings, partial word retrievals, and confusingly formatted results, some researchers have found that sometimes using a search engine to find information on a web site can actually take more time than using conventional contents or navigation tools.[5] For the documenter the lesson is to examine the search engine included with your indexing software or HTML help generating program to see assess its capabilities.

Tools for Indexers

Being the brokers of specialized vocabularies indexers can rely on resources (usually reference books and web sites) to help them decide on terminology for indexes. The American Society of Indexers (http://www.asindexing.org), Figure 14.6, lists a number of these reference tools. Among them you will find the following:

- Dictionaries, Thesauri, and Other Language Tools
- Encyclopedias, Collections of Information, and Fact Books
- Phone Directories, Organizational Listings, and Geographical Maps

Beyond these sources, the ASI lists more specialized reference works in medicine, business, law, computer technology, science, and agriculture. These kinds of resources can help the indexer make the crucial matches between the vocabularies of a software program (represented by the words that make up the interface) and the terminology used by and familiar to the user.

Glossary

electronic links: software functions within hypertext programs, such as WinHelp or HTML Help, that allow the user to jump to another designated position within a text. The position or target to which the user moves is highlighted in the text, usually with blue as on Web pages, or green as in Windows Help documents.

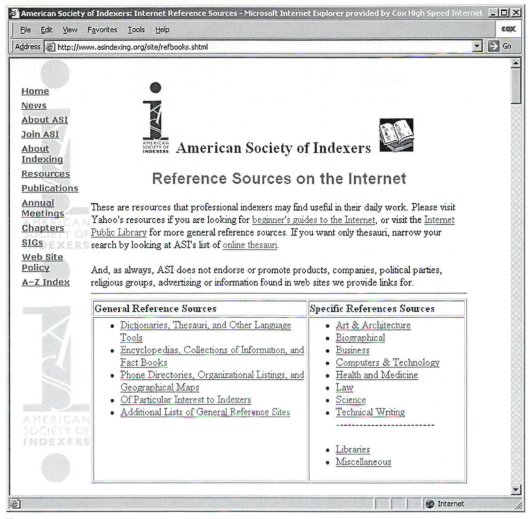

FIGURE 14.6   The American Society of Indexers
The ASI provides up-to-date links to a number of language reference sources for preparers of indexes.

keyword searches: electronic and automatic searches of the topics in a help system to find preidentified words relating to certain topics. Example: the keyword "File" might call up the following topics: Opening files, Saving files, and Deleting files.

notational conventions: conventions relating to how terms, commands, menus, and other interface elements appear in a manual. For example, often manuals will use italics (as in dir, copy) as the notational convention for commands.

primary locator number: the number in an index entry that tells the user which of the selection of numbers contains the main information about the entry.

search engine: a kind of computer program that searches an online document for words, parts of words, or search keywords provided by the writer.

☑ Checklist

Use the following checklist to evaluate your index.

Index Checklist
Decide on the Indexing Methodology

For your index, which methodology will you use?

❏ Manual: working by hand using   ❏ Electronic: using tools provided by
   page proofs   the word processing or publishing
                                         software

Decide What to Index

Which of the following elements will you include in your index?

❏ Commands and functions   ❏ Glossary terms
❏ Concepts relevant to user actions   ❏ Proper names of products and companies
❏ User synonyms for program terms   ❏ Tasks and procedures

Identify the Level of Detail

How many levels do you plan to include in your index?

❏ One level   ❏ Two levels   ❏ Three levels
How many indexable items do you plan to do for each page (on the average)?

❏ Light index   2 to 3 items per page
❏ Medium index   5 to 7 items per page
❏ Heavy index   8 to 9 items per page

Decide on Phrasing and Format

Which of the following format options do you plan to use?

❏ Cues for primary locator numbers
❏ Capitalization of terms
❏ "Sentence-like" entries
❏ Cues for special terms

Build and Proofread

Proofread and check the index for the following items:

❏ Page references (spot check)
❏ Cross references (spot check)
❏ Review terms with users
❏ Correct inconsistencies
❏ Add missing references
❏ Combine entries
❏ Create double postings
❏ Eliminate needless subentries

Practice/Problem Solving

1. Practice Indexing a Page

Choose a page of a manual and write a set of index items for it. Make sure to include terms on other pages that would make references to your page. Use your judgment about what level of detail the user would need to guide your indexing. After you have written the index, check with the existing index and consider the differences between your index for that page and the existing one. How would your index differ?

2. Analyze an Index

Find an entry in an index to a computer manual with three or more locator numbers. Analyze each locator number (entry) for task orientation, rank, cross referencing, and accurateness. Copy the sentences and evaluate the index for overall usability.

ENDNOTES

Chapter 1

1. Hackos, JoAnn T. 1998. "Choosing a Minimalist Approach for Expert Users," in *Minimalism Beyond the Nurnberg Funnel,* John M. Carroll, ed., Cambridge, MA: The MIT Press.
2. Mirel, Barbara. 1998 "Minimalism for Complex Tasks," in *Minimalism Beyond the Nurnberg Funnel,* John M. Carroll, ed., Cambridge, MA: The MIT Press.
3. Redish, Janice C. 2000. "What is Information Design," *Technical Communication,* 47:2, May, p. 163.
4. Nua Internet Surveys. Available on the WWW.
   http://www.nua.ie/surveys/how_many_online/ n_america.html
5. Mirel, Barbara. 1998. "'Applied Constructivism' for User Documentation: Alternatives to Conventional Task Orientation," *Journal of Business and Technical Communication,* 12:1, p. 7.
6. Anson, Patricia A. H. 1998. "Exploring Minimalism Today," in *Minimalism Beyond the Nurnberg Funnel,* John M. Carroll, ed., Cambridge, MA: The MIT Press, p. 92.
7. Yaverbaum, Gayle. 1988. "Critical factors in the user environment: An experimental study of users, organizations and tasks." *MIS Quarterly* (March): 75–88.
8. Zuboff, Shoshana. 1988. *In the age of the smart machine: The Future of work and power.* New York: Basic Books.
9. Sume, David, Loring Leifer, and Richard Saul Wurman, 2001. *Information Anxiety 2.*
10. Mirel, Barbara. 1998. op. cit. pp. 10–11.
11. Bannon, Liam J. and Susanne Bødker. 1991. "Beyond the Interface: Encountering Artifacts in Use," John M. Carroll (ed.) *Designing Interaction: Psychology at the human-computer interface.* (New York: Cambridge University Press).

Chapter 2

1. Susan Wiedenbeck, Patti L. Zila, and Daniel S. McConnell, End-user training: an empirical study comparing on-line practice methods, ACM, CHI 1995 Proceedings. Available on the WWW: http://www.acm.org/sigchi/ chi95/Electronic/documnts/papers/ sw_bdy.htm
2. Steven Lieb, "Principles of Adult Learning" Available on the WWW:
   http://www.hcc.hawaii.edu/ intranet/committees/FacDevCom/guidebk/teachtip/ adults-2.htm
3. Charney, Davida, Lynne Reder, and Gail Wells. 1988. "Studies of elaboration in instructional texts." In *Effective documentation: What we have learned from the research,* ed. - Stephen Doheny-Farina. Cambridge, MA: The MIT Press: 47–72.
4. van der Meij, Hans and John M. Carroll, "Principles and Heuristics for Designing Minimalist Instruction," in *Minimalism beyond the Nurnberg Funnel,* ed. John M. Carroll, Cambridge, MA: The MIT Press, 1998, pp. 19–46.
5. Ibid, p. 31.

Chapter 3

1. Marion, Craig. 2000. "Make Way for Interactive Assistance." Available on the WWW: http:// www.chesco.com/~cmarion/PCD/MakeWayforInteractiveAsst.html

Chapter 4

1. Rossett, Allison and Jeannette Gautier-Downes, 1991. *A Handbook of Job Aids.* San Francisco, CA: Pfeiffer and Company, p. 4.
2. Carroll, John M., and hans van der Meij. 1998. "Ten misconceptions about Minimalism." in *Minimalism Beyond the Nurnberg Funnel,* John M. Carroll, ed. Cambridge, MA: The MIT Press.
3. Bottka, Hary, Patricia McDaniels, Karla McMaster, Don Rasky, and Richard Wrye, 2001. "STC 2000–2001 International Competitions: Best of Show and Distinguished Technical Communication Winners," *Technical communication,*" 48:4, p. 467.

Chapter 5

1. Henderson, Kathryn. 1995. "The Visual Culture of Engineers," in *The Cultures of Computing,* New York: Blackwell Press, pp. 196–218.
2. Mirel, Barbara. 1998. "Applied Constructivism for User Documentation: Alternatives to Conventional Task Orientation," *Journal of Business and Technical Communication,* 12:1, p. 31.
3. Adapted from Ting, Toomey, Stella. 1999. *Communicating Across Cultures.* New York: The Guilford Press, pp. 101–106.
4. Gudykunst, William and Young Yun Kim. 1992. *Communicating with Strangers: An approach to International Communication.* New York: McGraw-Hill, Inc., pp. 202–205.
5. Gudykunst, op. cit. p. 75 ff.
6. Rosenbaum, Stephanie, L. 1992. "Collecting usability data: Alternatives to testing." In *Conference Record, IPPC,* pp. 248–253. Santa Fe, NM. These steps and the tips that follow are adapted from pp. 251–253.
7. Dourish, Paul. 2001. *Where the Action Is: The Foundations of Embodied Interaction,* Cambridge, MA: The MIT Press.
8. Honeyman, David S. and Warren J. White. 1987. "Computer anxiety in educators learning to use the computer: A preliminary report.*" Journal of Research on Computing in Education* 20:2. 129–138.
9. Bracey, Gerald W 1988. "Computers and anxiety in education: Round two." *Electronic Learning* 8:3. 26–28.
10. Foss, Donald J., Penny L. Smith-Kerker, and Mary Beth Rosson. 1987. "On comprehending a computer manual: Analysis of variables affecting performance." *International Journal of Man—Machine Studies* 26. 277–300.
11. Norman, Donald. 1988. *The psychology of everyday things.* New York: Harper Collins Press: 11.
12. interBiz web site. Available on the WWW at http://interbiz.ca.com/UserGroups/default.asp
13. Mumford, Alan. 1997. *Action Learning at Work,* Brookfield, VT: Gower Press.
14. Santhanam, Radhika and Susan Weidenbeck. 1990. "Modeling the intermittent user of word-processing software." *Journal of the American Society of Information Science* 42:3. 185–196.

Chapter 6

1. Hackos, JoAnn T. 1994. *Managing Your Documentation Projects.* New York, John Wiley & Sons, Inc.

2. For more information on cross functional teams I would suggest the special issue on "Communication in Cross-functional Teams," *Technical communication*, 43:1, March, 2000.
3. Hackos, p. 123.

Chapter 7

1. Sellin, Robert, and Elaine Winters. 2000. *Cultural Issues in Business Communication,* Berkeley, CA: Program Facilitating and Consulting, p. 45 ff.
2. Sellin and Winters, op. cit. p. 51.
3. Chatfield, Carl S. 1994. "Improving the Documentation Process through Structured Walkthroughs." *STC Proceedings:* 90–92.

Chapter 8

1. Masuda, Tadashi. 1994. "Using customer inquiries as a basis for revising and editing user manuals." In *Proceedings, 4lst Annual Conference of the STC,* p. 83
2. Gould, Emile and Stephen Doheny-Farina. 1988. "Studying usability in the field: Qualitative research techniques for technical communicators." In *Effective documentation: What we have learned from research,* Stephen Doheny-Farina, ed. Cambridge, MA: The MIT Press: 329–343.
3. Mirel, Barbara. 2000. "Product, Process, and Profit: The Politics of Usability in a Software Venture," ACM Journal of Computer Documentation, 24: 4, November 2000, pp. 185–203.
4. Some of these advantages are pointed out in Gould, p. 333.
5. The following description of Q-sorts adapts information in Claudia M. Hunter, "Pretesting the usability and task orientation of computer documentation," in *Conference Record,* IPCC, Santa Fe, NM, 1992: 80–85.
6. Donald A. Norman. 1994. *Things that make us smart.* New York: Addison-Wesley Publishing Company, p. 229.
7. Lauren Baker. 1988. "The relationship of product design to document design." In *Effective documentation: What we have learned from research,* Stephen Doheny-Farina, ed. pp. 317–328. Cambridge, MA: The MIT Press.
8. Baker, op. cit. pp. 317–328.

Chapter 9

1. Hall, Annette, and Jay Schram 2000. "Editing Windows 2000," Proceedings of the WinWriters Online Help Conference, San Diego, CA, March.
2. Weber, Jean Hollis. 2000. *Editing Online Help.* Airlie Beach: Australia. http://www.wrevenge.com.au.
3. Collins, William, and Karen J. Hamilton. 2001. "Editing and Index," *intercom,* 48:2, p. 11.
4. Hackos, JoAnn. 1994. *Managing Your Documentation Projects.* New York: John Wiley & Sons, p. 555–557.
5. Rude, Carolyn. 2001. *Technical Editing.* 3rd ed. New York: Allyn & Bacon, p. 436.
6. Rude, op. cit. p. 436. Rude cites Joyce Lasecke (1996) in *intercom,* 43: 9, p. 6.
7. Dayton, David. 2001. "Electronic Editing," in Carolyn Rude, *Technical Editing,* 3rd op. cit.
8. Weber, op. cit. "How much time is required for editing online help?"
9. Tarutz, Judith A. 1992. *Technical Editing: A practical guide for editors and writers.* New York: Addison-Wesley Publishing Company, Inc.
10. McNiell, Angie. 2001. "Technical Editing 101," *intercom,* 48:10, p. 10–11.
11. Sullivan, Bill. 1996. "How Do I Develop a Style Guide?" *intercom,* 43: 8, p. 25.

12. Hansen, James B. 1997. "Editing Your Own Writing," *intercom,* 44: 2, p. 15.

13. Tarutz, Judith. 1992. *Technical Editing.* New York: Addison-Wesley.

14. Rude, op. cit. p. 114.

15. Rude, op. cit. p. 115.

16. Restive, Katherine, and Phillip R. Shelton. 1994. "From editing to writing: Learning the write stuff." In *STC Proceedings,* pp. 70–72.

17. Boeing Simplified English Checker. 2002. "Simplified English." Available on the WWW: http://www.boeing.com/assocproducts/sechecker/se.html (site accessed February 2, 2002).

18. Andrews, Deborah, C. 2001. *Technical Communication in the Global Community.* NJ: Prentice Hall, p. 271.

19. Berry, Robert, Falpy Earle, and Michelle Corbin Nichols. 1994. "Good online indexing: It doesn't happen automatically." In *STC Proceedings,* pp. 110–112.

20. Masse, Roger E. 1985. "Theory and practice of editing processes in technical communication." *IEEE Transactions in Professional Communication.* 28: 1. 34–42.

Chapter 10

1. Zubak, Cheryl. 2000. "Rethinking Help as Software," Proceedings of the WinWriters Online Help Conference, San Diego, CA, March, 2000: H 48.

2. Sullivan, Patricia, and Linda Flower. 1989. "How do users read computer manuals? Some protocol contributions to writers' knowledge." In B. T. Petersen, ed. *Convergence's: Transactions in reading and writing.* Urbana, IL: National Council of Teachers of English: 163–178.

Chapter 11

1. Weiss, Edmund. 1991. *How to write usable user documentation.* Phoenix, AZ: Oryx Press.

Chapter 14

1. Collins, William. 2001. "Editing an Index," *intercom,* 48:2, February, pp. 10–13.

2. Collins, op. cit.

3. This information is based largely on Collins, op. cit., p. 12.

4. The Chicago Manual of Style, Chicago, IL, The University of Chicago Press, p. 519.

5. User Interface Engineering, "Why On-Site Searching Stinks," Available on the WWW: http://world.std.com/~uieweb/searchar.htm (last accessed 02/10/02).

CREDITS

Figures 1.1 and 10.4 used with permission of Visual Numerics, Inc.

Figure 1.2 reprinted with permission of Strohl Systems, Inc.

Figures 1.4, 2.2, 3.10, 3.11, 3.17, 4.11, 5.3, 5.17, 6.6, 10.2, 10.14, 10.21, 11.5, 11.6, 11.11, 13.1, 13.6, 13.7, 13.9, 13.14, and 13.15 reprinted by permission from Microsoft Corporation.

Figure 1.5 © Ed Stein, reprinted by permission of Newspaper Enterprise Association, Inc.

Figure 1.7 reprinted with permission of North Texas Linux User's Group, Inc.

Figures 2.1, 2.9, and 13.5 reprinted with permission of Inscriber Technology Corporation.

Figure 2.4 © 2000 David Deckert (dgd-filt@visar.com)

Figures 2.5, 6.4, 6.13, 10.5, 10.6, and 10.22 reprinted with permission of Sevensteps, Inc.

Figure 2.6 reprinted with permission of ICICI Group.

Figure 2.7 reprinted with permission of Gateway, Inc.

Figures 2.8 and 2.13 from Authorware Models for Instruction Design by Allen Interactions. Reprinted by permission of Pearson Education, Inc., Upper Saddle River, NJ.

Figure 2.11 reprinted with permission of Ballard Spahr Andrews & Ingersoll, LLP.

Figure 2.12 used with permission from Isys Information Architects, Inc.

Figures 3.1 and 13.2 © Corel Corporation. Used with permission. Corel is a registered trademark and Corel-DRAW is a trademark of Corel Corporation.

Figure 3.2 reprinted with permission of Marconi Australia Pty Ltd.

Figure 3.5 reprinted with permission of Cognitive Science Laboratory, Princeton University.

Figures 3.6, 13.17, and 13.20 © 1987–1999 Adobe Systems Incorporated. Used with express permission. All rights reserved. Adobe, Acrobat, and PageMaker is/are either [a] registered trademark[s] of Adobe Systems Incorporated in the United States and/or other countries.

Figure 3.13 reprinted with permission of Opto 22.

Figures 3.14 and 3.15 © 2002 Lawson Software. All rights reserved. Reprinted by permission of Lawson Software.

Figures 4.1 and 4.9 reprinted with permission of Borland International, Inc.

Figures 4.2 and 13.13 reprinted with permission of Ceridian Corporation.

INDEX